T0186137

SystemVerilog for Verification

Chris Spear · Greg Tumbush

SystemVerilog for Verification

A Guide to Learning the Testbench Language Features

Third Edition

 Springer

Chris Spear
Synopsys, Inc.
Marlborough, MA, USA

Greg Tumbush
University of Colorado, Colorado Springs
Colorado Springs, CO, USA

ISBN 978-1-4899-9500-1   ISBN 978-1-4614-0715-7 (eBook)
DOI 10.1007/978-1-4614-0715-7
Springer New York Dordrecht Heidelberg London

© Springer Science+Business Media, LLC 2012
Softcover reprint of the hardcover 3rd edition 2012
All rights reserved. This work may not be translated or copied in whole or in part without the written permission of the publisher (Springer Science+Business Media, LLC, 233 Spring Street, New York, NY 10013, USA), except for brief excerpts in connection with reviews or scholarly analysis. Use in connection with any form of information storage and retrieval, electronic adaptation, computer software, or by similar or dissimilar methodology now known or hereafter developed is forbidden.
The use in this publication of trade names, trademarks, service marks, and similar terms, even if they are not identified as such, is not to be taken as an expression of opinion as to whether or not they are subject to proprietary rights.

Printed on acid-free paper

Springer is part of Springer Science+Business Media (www.springer.com)

This book is dedicated to my wife Laura, who takes care of everything, my daughter Allie, long may you travel, my son Tyler, welcome back, and all the mice.

– Chris Spear

This book is dedicated to my wife Carolye, who shrugged off my "I need to work on the book" requests with a patient smile, and to my toddler son Lucca who was always available for play time.

– Greg Tumbush

Preface

What is this Book About?

This book should be the first one you read to learn the SystemVerilog verification language constructs. It describes how the language works and includes many examples on how to build a basic coverage-driven, constrained-random, layered testbench using Object-Oriented Programming (OOP). The book has many guidelines on building testbenches, to help you understand how and why to use classes, randomization, and functional coverage. Once you have learned the language, pick up some of the methodology books listed in the References section for more information on building a testbench.

Who Should Read this Book?

If you create testbenches, you need this book. If you have only written tests using Verilog or VHDL and want to learn SystemVerilog, this book shows you how to move up to the new language features. Vera and Specman users can learn how one language can be used for both design and verification. You may have tried to read the SystemVerilog Language Reference Manual but found it loaded with syntax but no guidelines on which construct to choose.

Chris originally wrote this book because, like many of his customers, he spent much of his career using procedural languages such as C and Verilog to write tests, and had to relearn everything when OOP verification languages came along. He made all the typical mistakes, and wrote this book so you won't have to repeat them.

Before reading this book, you should be comfortable with Verilog-1995. You do not need to know about Verilog-2001 or SystemVerilog design constructs, or SystemVerilog Assertions in order to understand the concepts in this book.

What is New in the Third Edition?

This new edition of SystemVerilog for Verification has many improvements over the first two editions, written in 2006 and 2008, respectively.

- Our universities need to train future engineers in the art of verification. This edition is suitable for the academic environment, with exercise questions at the end of each chapter to test your understanding.
- Qualified instructors should visit http://extras.springer.com for additional materials such as slides, tests, homework problems, solutions, and a sample syllabus suitable for a semester-long course.
- The 2009 version of the IEEE 1800 SystemVerilog Language Reference Manual (LRM) has many changes, both large and small. This book tries to include the latest relevant information.
- Accellera created UVM (Universal Verification Methodology) with ideas from VMM (Verification Methodology Manual), OVM (Open Verification Methodology), eRM (e Reuse Methodology), and other methodologies. Many of the examples in this book are based on VMM because its explicit calling of phases is easier to understand if you are new to verification. New examples are provided that show UVM concepts such as the test registry and configuration database.
- When looking for a specific topic, engineers read books backwards, starting with the index, so we boosted the number of entries.
- Lastly, a big thanks to all the readers who spotted mistakes in the previous editions, from poor grammar to code that was obviously written on the morning after an 18-hour flight from Asia to Boston, or, even worse, changing a diaper. This edition has been checked and reviewed many times over, but once again, all mistakes are ours.

Why was SystemVerilog Created?

In the late 1990s, the Verilog Hardware Description Language (HDL) became the most widely used language for describing hardware for simulation and synthesis. However, the first two versions standardized by the IEEE (1364-1995 and 1364-2001) had only simple constructs for creating tests. As design sizes outgrew the verification capabilities of the language, commercial Hardware Verification Languages (HVLs) such as OpenVera and e were created. Companies that did not want to pay for these tools instead spent hundreds of man-years creating their own custom tools.

This productivity crisis, along with a similar one on the design side, led to the creation of Accellera, a consortium of EDA companies and users who wanted to create the next generation of Verilog. The donation of the OpenVera language formed the basis for the HVL features of SystemVerilog. Accellera's goal was met in November 2005 with the adoption of the IEEE standard 1800-2005 for SystemVerilog, IEEE (2005). In December 2009, the latest Verilog LRM, 1364-2005, was merged with the aforementioned 2005 SystemVerilog standard to create the IEEE standard 1800-2009 for SystemVerilog. Merging these two standards into a single one means there is now one language, SystemVerilog, for both design and verification.

Importance of a Unified Language

Verification is generally viewed as a fundamentally different activity from design. This split has led to the development of narrowly focused languages for verification and to the bifurcation of engineers into two largely independent disciplines. This specialization has created substantial bottlenecks in terms of communication between the two groups. SystemVerilog addresses this issue with its capabilities for both camps. Neither team has to give up any capabilities it needs to be successful, but the unification of both syntax and semantics of design and verification tools improves communication. For example, while a design engineer may not be able to write an object-oriented testbench environment, it is fairly straightforward to read such a test and understand what is happening, enabling both the design and verification engineers to work together to identify and fix problems. Likewise, a designer understands the inner workings of his or her block, and is the best person to write assertions about it, but a verification engineer may have a broader view needed to create assertions between blocks.

Another advantage of including the design, testbench, and assertion constructs in a single language is that the testbench has easy access to all parts of the environment without requiring a specialized Application Programming Interface (API). The value of an HVL is its ability to create high-level, flexible tests, not its loop constructs or declaration style. SystemVerilog is based on the Verilog, VHDL, and C/C++ constructs that engineers have used for decades.

Importance of Methodology

There is a difference between learning the syntax of a language and learning how to use a tool. This book focuses on techniques for verification using constrained-random tests that use functional coverage to measure progress and direct the verification. As the chapters unfold, language and methodology features are shown side by side. For more on methodology, see Bergeron et al. (2006).

The most valuable benefit of SystemVerilog is that it allows the user to construct reliable, repeatable verification environments, in a consistent syntax, that can be used across multiple projects.

Overview of the Book

The SystemVerilog language includes features for design, verification, assertions, and more. This book focuses on the constructs used to verify a design. There are many ways to solve a problem using SystemVerilog. This book explains the trade-offs between alternative solutions.

Chapter 1, **Verification Guidelines**, presents verification techniques to serve as a foundation for learning and using the SystemVerilog language. These guidelines emphasize coverage-driven random testing in a layered testbench environment.

Chapter 2, **Data Types**, covers the new SystemVerilog data types such as arrays, structures, enumerated types, and packed arrays and structures.

Chapter 3, **Procedural Statements and Routines**, shows the new procedural statements and improvements for tasks and functions.

Chapter 4, **Connecting the Testbench and Design**, shows the new SystemVerilog verification constructs, such as program blocks, interfaces, and clocking blocks, and how they are used to build your testbench and connect it to the design under test.

Chapter 5, **Basic OOP**, is an introduction to Object-Oriented Programming, explaining how to build classes, construct objects, and use handles.

Chapter 6, **Randomization**, shows you how to use SystemVerilog's constrained-random stimulus generation, including many techniques and examples.

Chapter 7, **Threads and Interprocess Communication**, shows how to create multiple threads in your testbench, use interprocess communication to exchange data between these threads and synchronize them.

Chapter 8, **Advanced OOP and Testbench Guidelines**, shows how to build a layered testbench with OOP so that the components can be shared by all tests.

Chapter 9, **Functional Coverage**, explains the different types of coverage and how you can use functional coverage to measure your progress as you follow a verification plan.

Chapter 10, **Advanced Interfaces**, shows how to use virtual interfaces to simplify your testbench code, connect to multiple design configurations, and create interfaces with procedural code so your testbench and design can work at a higher level of abstraction.

Chapter 11, **A Complete SystemVerilog Testbench**, shows a constrained random testbench using the guidelines shown in Chapter 8. Several tests are shown to demonstrate how you can easily extend the behavior of a testbench without editing the original code, which always carries risk of introducing new bugs.

Chapter 12, **Interfacing with C / C++**, describes how to connect your C or C++ Code to SystemVerilog using the Direct Programming Interface.

Icons used in this book

Table i.1 Book icons

The compass shows verification methodology to guide your usage of SystemVerilog testbench features.

The bug shows common coding mistakes such as syntax errors, logic problems, or threading issues.

About the Authors

Chris Spear has been working in the ASIC design and verification field for 30 years. He started his career with Digital Equipment Corporation (DEC) as a CAD Engineer on DECsim, connecting the first Zycad box ever sold, and then a hardware Verification engineer for the VAX 8600, and a hardware behavioral simulation accelerator. He then moved on to Cadence where he was an Application Engineer for Verilog-XL, followed a a stint at Viewlogic. Chris is currently employed at Synopsys Inc. as a Verification Consultant, a title he created a dozen years ago. He has authored the first and second editions of SystemVerilog for Verification. Chris earned a BSEE from Cornell University in 1981. In his spare time, Chris enjoys road biking in the mountains and traveling with his wife.

Greg Tumbush has been designing and verifying ASICs and FPGAs for 13 years. After working as a researcher in the Air Force Research Labs (AFRL) he moved to beautiful Colorado to work with Astek Corp as a Lead ASIC Design Engineer. He then began a 6 year career with Starkey Labs, AMI Semiconductor, and ON Semiconductor where he was an early adopter of SystemC and SystemVerilog. In 2008, Greg left ON Semiconductor to form Tumbush Enterprises, where he has been consulting clients in the areas of design, verification, and backend to ensure first pass success. He is also a 1/2 time Instructor at the University of Colorado, Colorado Springs where he teaches senior and graduate level digital design and verification courses. He has numerous publications which can be viewed at www.tumbush.com. Greg earned a PhD from the University of Cincinnati in 1998.

Final comments

If you would like more information on SystemVerilog and Verification, you can find many resources at: `http://chris.spear.net/systemverilog`. This site has the source code for many of the examples in this book. Academics who want to use this book in their classes can access slides, tests, homework problems, solutions, and a sample syllabus on the book's webpage at `http://www.springer.com`.

Most of the code samples in the book were verified with Synopsys' Chronologic VCS, Mentor's QuestaSim, and Cadence Incisive. Any errors were caused by Chris' evil twin, Skippy. If you think you have found a mistake in this book, please check his web site for the Errata page. If you are the first to find a technical mistake in a chapter, we will send you a free, autographed book. Please include "SystemVerilog" in the subject line of your email.

Chris Spear
Greg Tumbush

Acknowledgments

We thank all the people who spent countless hours helping us learn SystemVerilog and reviewing the book that you now hold in your hands. We especially would like to thank all the people at Synopsys and Cadence for their help. Thanks to Mentor Graphics for supplying Questa licenses through the Questa Vanguard program, and to Tim Plyant at Cadence who checked hundreds of examples for us.

A big thanks to Mark Azadpour, Mark Barrett, Shalom Bresticker, James Chang, Benjamin Chin, Cliff Cummings, Al Czamara, Chris Felton, Greg Mann, Ronald Mehler, Holger Meiners, Don Mills, Mike Mintz, Brad Pierce, Tim Plyant, Stuart Sutherland, Thomas Tessier, and Jay Tyer, plus Professor Brent Nelson and his students who reviewed some very rough drafts and inspired many improvements. However, the mistakes are all ours!

Janick Bergeron provided inspiration, innumerable verification techniques, and top-quality reviews. Without his guidance, this book would not exist.

The following people pointed out mistakes in the second edition, and made valuable suggestions on areas where the book could be improved: Alok Agrawal, Ching-Chi Chang, Cliff Cummings, Ed D'Avignon, Xiaobin Chu, Jaikumar Devaraj, Cory Dearing, Tony Hsu, Dave Hamilton, Ken Imboden, Brian Jensen, Jim Kann, John Keen, Amirtha Kasturi, Devendra Kumar, John Mcandrew, Chet Nibby, Eric Ohana, Simon Peter, Duc Pham, Hani Poly, Robert Qi, Ranbir Rana, Dan Shupe, Alex Seibulescu, Neill Shepherd, Daniel Wei, Randy Wetzel, Jeff Yang, Dan Yingling and Hualong Zhao.

Lastly, a big thanks to Jay Mcinerney for his brash pronoun usage.

All trademarks and copyrights are the property of their respective owners. If you can't take a joke, don't sue us.

Contents

List of Figures

List of Tables

List of Samples

Chapter 1
Verification Guidelines

> *Some believed we lacked the programming language*
> *to describe your perfect world...*
>
> (The Matrix, 1999)

Imagine that you are given the job of building a house for someone. Where should you begin? Do you start by choosing doors and windows, picking out paint and carpet colors, or selecting bathroom fixtures? Of course not! First you must consider how the owners will use the space, and their budget, so you can decide what type of house to build. Questions you should consider are: Do they enjoy cooking and want a high-end kitchen, or will they prefer watching movies in their home theater room and eating takeout pizza? Do they want a home office or an extra bedroom? Or does their budget limit them to a more modest house?

Before you start to learn details of the SystemVerilog language, you need to understand how you plan to verify your particular design and how this influences the testbench structure. Just as all houses have kitchens, bedrooms, and bathrooms, all testbenches share some common structure of stimulus generation and response checking. This chapter introduces a set of guidelines and coding styles for designing and constructing a testbench that meets your particular needs. These techniques use some of the same concepts that are shown in the *Verification Methodology Manual for SystemVerilog* (VMM), Bergeron et al. (2006), but without the base classes. Other methodologies such as UVM and OVM share the same concepts.

The most important principle you can learn as a verification engineer is: "Bugs are good." Don't shy away from finding the next bug, do not hesitate to ring a bell each time you uncover one, and furthermore, always keep track of the details of each bug found. The entire project team assumes there are bugs in the design, so each bug found before tape-out is one fewer that ends up in the customer's hands. At each stage in the design cycle such as specification, coding, synthesis, manufacturing, the cost of fixing a bug goes up by a factor of 10, so find those bugs early and often. You need to be as devious as possible, twisting and torturing the design to

C. Spear and G. Tumbush, *SystemVerilog for Verification: A Guide to Learning the Testbench Language Features*, DOI 10.1007/978-1-4614-0715-7_1,
© Springer Science+Business Media, LLC 2012

extract all possible bugs now, while they are still easy to fix. Don't let the designers steal all the glory — without your craft and cunning, the design might never work!

This book assumes you already know the Verilog language and want to learn the System Verilog Hardware Verification Language (HVL). Some of the typical features of an HVL that distinguish it from a Hardware Description Language such as Verilog or VHDL are:

- Constrained-random stimulus generation
- Functional coverage
- Higher-level structures, especially Object-Oriented Programming, and transaction-level modeling
- Multi-threading and interprocess communication (IPC)
- Support for HDL types such as Verilog's 4-state values
- Tight integration with event-simulator for control of the design

There are many other useful features, but these allow you to create testbenches at a higher level of abstraction than you are able to achieve with an HDL or a programming language such as C.

1.1 The Verification Process

What is the goal of verification? If you answered, "Finding bugs," you are only partly correct. The goal of hardware design is to create a device that performs a particular task, such as a DVD player, network router, or radar signal processor, based on a design specification. Your purpose as a verification engineer is to make sure the device can accomplish that task successfully — that is, the design is an accurate representation of the specification. Bugs are what you get when there is a discrepancy. The behavior of the device when used outside of its original purpose is not your responsibility, although you want to know where those boundaries lie.

The process of verification parallels the design creation process. A designer reads the hardware specification for a block, interprets the human language description, and creates the corresponding logic in a machine-readable form, usually RTL code. To do this, he or she needs to understand the input format, the transformation function, and the format of the output. There is always ambiguity in this interpretation, perhaps because of ambiguities in the original document, missing details, or conflicting descriptions. As a verification engineer, you must also read the hardware specification, create the verification plan, and then follow it to build tests showing the RTL code correctly implements the features. Therefore, as a verification engineer, not only do you have to understand the design and its intent, but also, you have to consider all the corner test cases that the designer might not have thought about.

By having more than one person perform the same interpretation, you have added redundancy to the design process. As the verification engineer, your job is to read the same hardware specifications and make an independent assessment of what they mean. Your tests then exercise the RTL to show that it matches your interpretation.

1.1.1 Testing at Different Levels

What types of bugs are lurking in the design? The easiest ones to detect are at the block level, in modules created by a single person. Did the ALU correctly add two numbers? Did every bus transaction successfully complete? Did all the packets make it through a portion of a network switch? It is almost trivial to write directed tests to find these bugs, as they are contained entirely within one block of the design.

After the block level, the next place to look for discrepancies is at boundaries between blocks. This is known as the integration phase. Interesting problems arise when two or more designers read the same description yet have different interpretations. For a given protocol, what signals change and when? The first designer builds a bus driver with one view of the specification, while a second builds a receiver with a slightly different view. Your job is to find the disputed areas of logic and maybe even help reconcile these two different views.

To simulate a single design block, you need to create tests that generate stimuli from all the surrounding blocks — a difficult chore. The benefit is that these low-level simulations run very fast. However, you may find bugs in both the design and testbench, as the latter requires a great deal of code to provide stimuli from the missing blocks. As you start to integrate design blocks, they can stimulate each other, reducing your workload. These multiple block simulations may uncover more bugs, but they also run slower. Analyzing the behavior to determine the root cause of a bug is more time consuming at higher levels.

At the highest level of the Design Under Test (DUT), the entire system is tested, but the simulation performance is greatly reduced. Your tests should strive to have all blocks performing interesting activities concurrently. All I/O ports are active, processors are crunching data, and caches are being refilled. With all this action, data alignment and timing bugs are sure to occur.

At this level you are able to run sophisticated tests that have the DUT executing multiple operations concurrently so that as many blocks as possible are active. What happens if an MP3 player is playing music and the user tries to download new music from the host computer? Then, during the download, the user presses several of the buttons on the player? You know that when the real device is being used, someone is going to do all this, so why not try it out before it is built? This testing makes the difference between a product that is seen as easy to use and one that repeatedly locks up.

Once you have verified that the DUT performs its designated functions correctly, you need to see how it operates when there are errors. Can the design handle a partial transaction, or one with corrupted data or control fields? Just trying to enumerate all the possible problems is difficult, not to mention determining how the design should recover from them. Error injection and handling can be the most challenging part of verification.

As you move to system-level verification, the challenges also move to a higher level. At the block level, you can show that individual cells flow through the blocks of an ATM router correctly, but at the system level you might have to consider what happens if there are streams of different priority. Which cell should be chosen next is not always obvious at the highest level. You may have to analyze the statistics from thousands of cells to see if the aggregate behavior is correct.

One last point: you can never prove there are no bugs left, so you need to constantly come up with new verification tactics.

1.1.2   The Verification Plan

The verification plan is derived from the hardware specification and contains a description of what features need to be exercised and the techniques to be used. These steps may include directed or random testing, assertions, HW/SW co-verification, emulation, formal proofs, and use of verification IP. For a more complete discussion on verification see Bergeron (2006).

1.2   The Verification Methodology Manual

The book in your hands draws upon the VMM that has its roots in a methodology developed by Janick Bergeron and others at Qualis Design. They started with industry-standard practices and refined them based on their experience on many projects. VMM's techniques were originally developed for use with the OpenVera language and were extended in 2005 for SystemVerilog. VMM and its predecessor, the Reference Verification Methodology (RVM) for Vera, have been used successfully to verify a wide range of hardware designs, from networking devices to processors. Newer methodologies such as OVM and UVM use many similar ideas. This book is based on many of the same concepts as all these methodologies, though greatly simplified.

This book serves as a user guide for the SystemVerilog language. It describes many language constructs and provides guidelines for choosing the ones best suited to your needs. If you are new to verification, have little experience with Object-Oriented Programming (OOP), or are unfamiliar with constrained-random tests (CRT), this book can show you the right path to choose. Once you are familiar with them, you will find UVM and VMM to be an easy step up.

So why doesn't this book teach you UVM or VMM? Like any advanced tool, these methodologies were designed for use by an experienced user, and excel on difficult problems. Are you in charge of verifying a 100 million-gate design with many communication protocols, complex error handling, and a library of IP? If so, UVM or VMM are the right tools for the job. However, if you are working on smaller modules with a single protocol, you may not need such a robust methodology. Just remember that your block is part of a larger system; UVM- or VMM-compliant code is reusable both during a project and on later designs. The cost of verification goes beyond your immediate project.

The UVM and VMM have a set of base classes for data and environment, utilities for managing log files and interprocess communication, and much more. This book is an introduction to SystemVerilog and shows the techniques and tricks that go into these classes and utilities, giving you insight into their construction.

1.3   Basic Testbench Functionality

The purpose of a testbench is to determine the correctness of the DUT. This is accomplished by the following steps.

- Generate stimulus
- Apply stimulus to the DUT
- Capture the response
- Check for correctness
- Measure progress against the overall verification goals

Some steps are accomplished automatically by the testbench, while others are manually determined by you. The methodology you choose determines how the preceding steps are carried out.

1.4   Directed Testing

Traditionally, when faced with the task of verifying the correctness of a design, you probably used directed tests. Using this approach, you look at the hardware specification and write a verification plan with a list of tests, each of which concentrated on a set of related features. Armed with this plan, you write stimulus vectors that exercise these features in the DUT. You then simulate the DUT with these vectors and manually review the resulting log files and waveforms to make sure the design does what you expect. Once the test works correctly, you check it off in the verification plan and move to the next one.

This incremental approach makes steady progress, which is always popular with managers who want to see a project making headway. It also produces almost immediate results, since little infrastructure is needed when you are guiding the creation of every stimulus vector. Given ample time and staffing, directed testing is sufficient to verify many designs.

Figure 1.1 shows how directed tests incrementally cover the features in the verification plan. Each test is targeted at a very specific set of design elements. If you had enough time, you could write all the tests needed for 100% coverage of the entire verification plan.

What if you do not have the necessary time or resources to carry out the directed testing approach? As you can see, while you may always be making forward progress, the slope remains the same. When the design complexity doubles, it takes twice as long to complete or requires twice as many people to implement it.

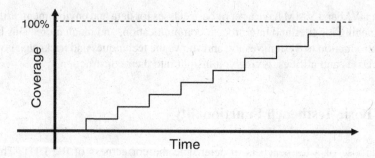

Fig. 1.1 Directed test progress over time

Neither of these situations is desirable. You need a methodology that finds bugs faster in order to reach the goal of 100% coverage. Brute force does not work; if you tried to verify every combination of inputs for a 32-bit adder, your simulations would still be running years after the project should have shipped.

Figure 1.2 shows the total design space and features that are covered by directed test cases. In this space there are many features, some of which have bugs. You need to write tests that cover all the features and find the bugs.

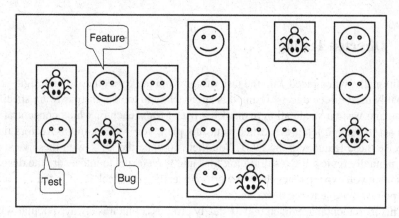

Fig. 1.2 Directed test coverage

1.5   Methodology Basics

This book uses the following principles.

- Constrained-random stimulus
- Functional coverage
- Layered testbench using transactors
- Common testbench for all tests
- Test case-specific code kept separate from testbench

All these principles are related. Random stimulus is crucial for exercising complex designs. A directed test finds the bugs you expect to be in the design, whereas a random test can find bugs you never anticipated. When using random stimuli, you need functional coverage to measure verification progress. Furthermore, once you start using automatically generated stimuli, you need an automated way to predict the results — generally a scoreboard or reference model. Building the testbench infrastructure, including self-prediction, takes a significant amount of work. A layered testbench helps you control the complexity by breaking the problem into manageable pieces. Transactors provide a useful pattern for building these pieces. With appropriate planning, you can build a testbench infrastructure that can be shared by all tests and does not have to be continually modified. You just need to leave "hooks" where the tests can perform certain actions such as shaping the stimulus and injecting disturbances. Conversely, code specific to a single test must be kept separate from the testbench to prevent it from complicating the infrastructure.

Building this style of testbench takes longer than a traditional directed testbench — especially the self-checking portions. As a result, there may be a significant delay before the first test can be run. This gap can cause a manager to panic, so make this effort part of your schedule. In Fig. 1.3, you can see the initial delay before the first random test runs.

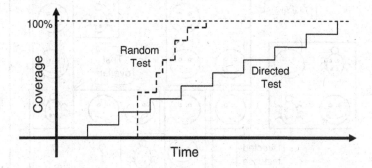

Fig. 1.3 Constrained-random test progress over time vs. directed testing

While this up-front work may seem daunting, the payback is high. Every random test you create shares this common testbench, as opposed to directed tests where each is written from scratch. Each random test contains a few dozen lines of code to constrain the stimulus in a certain direction and cause any desired exceptions, such as creating a protocol violation. The result is that your single constrained-random testbench is now finding bugs faster than the many directed ones.

As the rate of discovery begins to drop off, you can create new random constraints to explore new areas. The last few bugs may only be found with directed tests, but the vast majority of bugs will be found with random tests. If you create a random testbench, you can always constrain it to created directed tests, but a directed testbench can never be turned into a true random testbench.

1.6 Constrained-Random Stimulus

Although you want the simulator to generate the stimulus, you don't want totally random values. You use the SystemVerilog language to describe the format of the stimulus ("address is 32-bits; opcode is ADD, SUB or STORE; length < 32 bytes"), and the simulator picks values that meet the constraints. Constraining the random values to become relevant stimuli is covered in Chapter 6. These values are sent into the design, and are also sent into a high-level model that predicts what the result should be. The design's actual output is compared with the predicted output.

Figure 1.4 shows the coverage for constrained-random tests over the total design space. First, notice that a random test often covers a wider space than a directed one. This extra coverage may overlap other tests, or may explore new areas that you did not anticipate. If these new areas find a bug, you are in luck! If the new area is not legal, you need to write more constraints to keep random generation from creating illegal design functionality. Lastly, you may still have to write a few directed tests to find cases not covered by any other constrained-random tests.

Figure 1.5 shows the paths to achieve complete coverage. Start at the upper left with basic constrained-random tests. Run them with many different seeds.

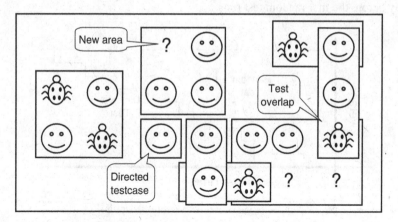

Fig. 1.4 Constrained-random test coverage

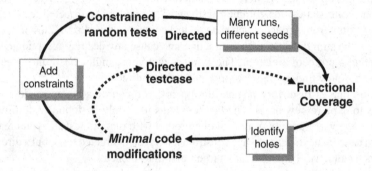

Fig. 1.5 Coverage convergence

When you look at the functional coverage reports, find the holes where there are gaps in the coverage. Then you make minimal code changes, perhaps by using new constraints, or by injecting errors or delays into the DUT. Spend most of your time in this outer loop, writing directed tests for only the few features that are very unlikely to be reached by random tests.

1.7   What Should You Randomize?

When you think of randomizing the stimulus to a design, you might first pick the data fields. These values are is the easiest to create — just call $random (). The problem is that this choice gives a very low payback in terms of bugs found. The primary types of bugs found with random data are data path errors, perhaps with bit-level mistakes. You need to find bugs in the control logic, source of the most devious problems.

Think broadly about all design inputs, such as the following.

- Device configuration
- Environment configuration
- Input data
- Protocol exceptions
- Errors and violations
- Delays

These are discussed in sections 1.7.1 through 1.7.4.

1.7.1   Device and Environment Configuration

What is the most common reason why bugs are missed during testing of the RTL design? Not enough different configurations are tried. Most tests just use the design as it comes out of reset, or apply a fixed set of initialization vectors to put it into a known state. This is like testing a PC's operating system right after it has been installed, but without any of the applications installed. Of course the performance is fine and there aren't any crashes.

In a real world environment, the DUT's configuration becomes more random the longer it is in use. For example, I helped a company verify a time-division multiplexor switch that had 2000 input channels and 12 output channels. The verification engineer said, "These channels could be mapped to various configurations on the other side. Each input could be used as a single channel, or further divided into multiple channels. The tricky part is that although a few standard ways of breaking it down are used most of the time, any combination of breakdowns is legal, leaving a huge set of possible customer configurations."

To test this device, the engineer had to write several dozen lines of directed test-bench code to configure each channel. As a result, she was never able to try configurations with more than a handful of channels. Together, we wrote a testbench that randomized the parameters for a single channel and then put this in a loop to configure all the switch's channels. Now she had confidence that her tests would uncover configuration-related bugs that would have been missed before.

In the real world, your device operates in an environment containing other components. When you are verifying the DUT, it is connected to a testbench that mimics this environment. You should randomize the entire environment configuration, including the length of the simulation, number of devices, and how they are configured. Of course you need to create constraints to make sure the configuration is legal.

In another Synopsys customer example, a company created an I/O switch chip that connected multiple PCI buses to an internal memory bus. At the start of simulation they randomly chose the number of PCI buses (1–4), the number of devices on each bus (1–8), and the parameters for each device (master or slave, CSR addresses, etc.). They kept track of the tested combinations using functional coverage so that they could be sure that they had covered almost every possible one.

Other environment parameters include test length, error injection rates, and delay modes. See Bergeron (2006) for more examples.

1.7.2   Input Data

When you read about random stimulus, you probably thought of taking a transaction such as a bus write or ATM cell and filling the data fields with random values. Actually, this approach is fairly straightforward as long as you carefully prepare your transaction classes as shown in Chapters 5 and 8. You need to anticipate any layered protocols and error injection, plus scoreboarding and functional coverage.

1.7.3   Protocol Exceptions, Errors, and Violations

There are few things more frustrating than when a device such as a PC or cell phone locks up. Many times, the only cure is to shut it down and restart. Chances are that deep inside the product there is a piece of logic that experienced some sort of error condition from which it could not recover and thus prevented the device from working correctly.

How can you prevent this from happening to the hardware you are building? If something can go wrong in the real hardware, you should try to simulate it. Look at all the errors that can occur. What happens if a bus transaction does not complete? If an invalid operation is encountered? Does the design specification state that two signals are mutually exclusive? Drive them both and make sure the device continues to operate properly.

Just as you are trying to provoke the hardware with ill-formed commands, you should also try to catch these occurrences. For example, recall those mutually exclusive signals. You should add checker code to look for these violations. Your code should at least print a warning message when this occurs, and preferably generate an error and wind down the test. It is frustrating to spend hours tracking back through code trying to find the root of a malfunction, especially when you could have caught it close to the source with a simple assertion. See Vijayaraghavan [2005] for more guidelines on writing assertions in your testbench and design code. Just make sure that you can disable the code that stops simulation on error so that you can easily test error handling.

1.7.4   Delays and Synchronization

How fast should your testbench send in stimulus? You should pick random delays to help catch protocol bugs. A test with the shortest delays is easy to write, but won't create all possible stimulus combinations. Subtle bugs around boundary conditions are often revealed when realistic delays are chosen.

A block may function correctly for all possible permutations of stimulus from a single interface, but subtle errors may occur when transactions are flowing into multiple inputs. Try to coordinate the various drivers so they can communicate at different timing rates. What if the inputs arrive at the fastest possible rate, but the output is being throttled back to a slower rate? What if stimulus arrives at multiple inputs concurrently? What if it is staggered with different delays? Use functional coverage, which will be discussed in Chapter 9, to measure what combinations have been randomly generated.

1.7.5   Parallel Random Testing

How should you run the tests? A directed test has a testbench that produces a unique set of stimulus and response vectors. To change the stimulus, you need to change the test. A random test consists of the testbench code plus a random seed. If you run the same test 50 times, each time with a unique seed, you will get 50 different sets of stimuli. Running with multiple seeds broadens the coverage of your test and leverages your work.

You need to choose a unique seed for each simulation. Some people use the time of day, but that can still cause duplicates. What if you are using a batch queuing system across a CPU farm and tell it to start 10 jobs at midnight? Multiple jobs could start at the same time but on different computers, and will thus get the same random seed and run the same stimulus. You should blend in the processor name to the seed. If your CPU farm includes multiprocessor machines, you could have two jobs start running at midnight with the same seed, so you should also throw in the process ID. Now all jobs get unique seeds.

You need to plan how to organize your files to handle multiple simulations. Each job creates a set of output files, such as log files and functional coverage data. You can run each job in a different directory, or you can try to give a unique name to each file. The easiest approach is to append the random seed value to the directory name.

1.8 Functional Coverage

Sections 1.6 and 1.7 showed how to create stimuli that can randomly walk through the entire space of possible inputs. With this approach, your testbench visits some areas often, but takes too long to reach all possible states. Unreachable states will never be visited, even given unlimited simulation time. You need to measure what has been verified in order to check off items in your verification plan.

The process of measuring and using functional coverage consists of several steps. First, you add code to the testbench to monitor the stimulus going into the device, and its reaction and response, to determine what functionality has been exercised. Run several simulations, each with a different seed. Next, merge the results from these simulations into a report. Lastly, you need to analyze the results and determine how to create new stimulus to reach untested conditions and logic. Chapter 9 describes functional coverage in SystemVerilog.

1.8.1 Feedback from Functional Coverage to Stimulus

A random test evolves using feedback. The initial test can be run with many different seeds, thus creating many unique input sequences. Eventually the test, even with a new seed, is less likely to generate stimulus that reaches areas of the design space. As the functional coverage asymptotically approaches its limit, you need to change the test to find new approaches to reach uncovered areas of the design. This is known as "coverage-driven verification" and is shown in Fig. 1.6.

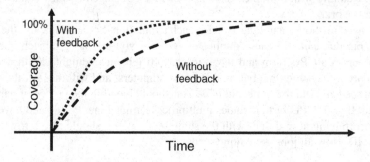

Fig. 1.6 Test progress with and without feedback

What if your testbench were smart enough to do this for you? In a previous job, I wrote a test that generated every bus transaction for a processor and additionally fired every bus terminator (Success, Parity Error, Retry) in every cycle. This was before HVLs, so I wrote a long set of directed tests and spent days lining up the terminator code to fire at just the right cycles. After much hand analysis I declared success — 100% coverage. Then the processor's timing changed slightly! Now I had to reanalyze the test and change the stimuli.

A more productive testing strategy uses random transactions and terminators. The longer you run it, the higher the coverage. As a bonus, the test can be made flexible enough to create valid stimuli even if the design's timing changed. You can accomplish this by adding a feedback loop that looks at the stimulus created so far (generated all write cycles yet?) and then change the constraint weights (drop write weight to zero). This improvement would greatly reduce the time needed to get to full coverage, with little manual intervention.

This is not a typical situation however, because of the trivial feedback from functional coverage to the stimulus. In a real design, how should you change the stimulus to reach a desired design state? This requires deep knowledge of the design and powerful formal techniques. There are no easy answers, so dynamic feedback is rarely used for constrained-random stimulus. Instead, you need to manually analyze the functional coverage reports and alter your random constraints.

Feedback is used in formal analysis tools such as Magellan (Synopsys 2003). It analyzes a design to find all the unique, reachable states. It then runs a short simulation to see how many states were visited. Lastly, it searches from the state machine to the design inputs to calculate the stimulus needed to reach any remaining states and then Magellan applies this to the DUT.

1.9 Testbench Components

In simulation, the testbench wraps around the DUT, just as a hardware tester connects to a physical chip, as shown in Fig. 1.7. Both the testbench and tester provide stimulus and capture responses. The difference between them is that your testbench needs to work over a wide range of levels of abstraction, creating transactions and sequences, which are eventually transformed into bit vectors. A tester just works at the bit level.

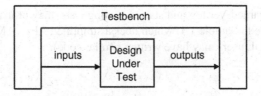

Fig. 1.7 The testbench — design environment

What goes into that testbench block? It is comprised of many Bus Functional Models (BFM), which you can think of as testbench components — to the DUT they look like real components, but they are part of the testbench, not the RTL design. If the real device connects to AMBA, USB, PCI, and SPI buses, you have to build equivalent components in your testbench that can generate stimulus and check the response, as shown in Fig. 1.8. These are not detailed, synthesizable models, but instead highlevel transactors that obey the protocol, and execute more quickly. On the other hand, if you are prototyping using FPGAs or emulation, the BFMs do need to be synthesizable.

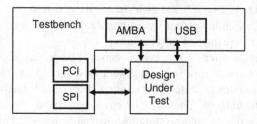

Fig. 1.8 Testbench components

1.10   Layered Testbench

A key concept for any modern verification methodology is the layered testbench. Although this process may seem to make the testbench more complex, it actually helps to make your task easier by dividing the code into smaller pieces that can be developed separately. Don't try to write a single routine that can randomly generate all types of stimuli, both legal and illegal, plus inject errors with a multi-layer protocol. The routine quickly becomes complex and unmaintainable. In addition, a layered approach allows reuse and encapsulation of Verification IP (VIP) which are OOP concepts.

1.10.1   A Flat Testbench

When you first learned Verilog and started writing tests, they probably looked like the low-level code in Sample 1.1, which does a simplified APB (AMBA Peripheral Bus) Write. (VHDL users may have written similar code).

Sample 1.1   Driving the APB pins

```
module test(PAddr, PWrite, PSel, PWData, PEnable, Rst, clk);
// Port declarations omitted...

  initial begin
    // Drive reset
    Rst <= 0;
    #100 Rst <= 1'b1;

    // Drive the control bus
    @(posedge clk)
    PAddr   <= 16'h50;
    PWData <= 32'h50;
    PWrite <= 1'b1;
    PSel    <= 1'b1;

    // Toggle PEnable
    @(posedge clk)
      PEnable <= 1'b1;
    @(posedge clk)
      PEnable <= 1'b0;

    // Check the result
    if (top.mem.memory[16'h50] == 32'h50)
      $display("Success");
    else
      $display("Error, wrong value in memory");
    $finish;
  end
endmodule
```

After a few days of writing code like this, you probably realized that it is very repetitive, so you created tasks for common operations such as a bus write, as shown in Sample 1.2.

Sample 1.2 A task to drive the APB pins

```
task write(reg [15:0] addr, reg [31:0] data);
  // Drive Control bus
  @(posedge clk)
  PAddr   <= addr;
  PWData <= data;
  PWrite <= 1'b1;
  PSel   <= 1'b1;

  // Toggle Penable
  @(posedge clk)
    PEnable <= 1'b1;
  @(posedge clk)
    PEnable <= 1'b0;
endtask
```

Now your testbench became simpler, as shown in Sample 1.3

Sample 1.3 Low-level Verilog test

```
module test(PAddr,PWrite,PSel,PWData,PEnable,Rst,clk);
  // Port declarations omitted...

  // Tasks as shown in Sample 1-2

  initial begin
    reset();                    // Reset the device
    write(16'h50, 32'h50);     // Write data into memory

    // Check the result
    if (top.mem.memory[16'h50] == 32'h50)
      $display("Success");
    else
      $display("Error, wrong value in memory");
    $finish;
  end
endmodule
```

By taking the common actions (such as reset, bus reads and writes) and putting them in a routine, you became more efficient and made fewer mistakes. This creation of the physical and command layers is the first step to a layered testbench.

1.10.2   The Signal and Command Layers

Figure 1.9 shows the lower layers of a testbench.

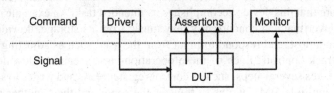

Fig. 1.9 Signal and command layers

At the bottom is the signal layer that contains the design under test and the signals that connect it to the testbench.

The next higher level is the command layer. The DUT's inputs are driven by the driver that runs single commands, such as bus read or write. The DUT's output drives the monitor that takes signal transitions and groups them together into commands. Assertions also cross the command/signal layer, as they look at individual signals and also changes across an entire command.

1.10.3   The Functional Layer

Figure 1.10 shows the testbench with the functional layer added, which feeds down into the command layer. The agent block (called the transactor in the VMM) receives higher-level transactions such as DMA read or write and breaks them into individual commands or transactions. These commands are also sent to the scoreboard that predicts the results of the transaction. The checker compares the commands from the monitor with those in the scoreboard.

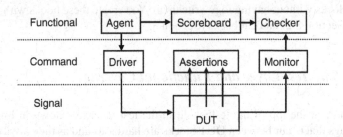

Fig. 1.10 Testbench with functional layer added

1.10.4 The Scenario Layer

The functional layer is driven by the generator in the scenario layer, as shown in
Fig. 1.11. What is a scenario? Remember that your job as a verification engineer is
to make sure that this device accomplishes its intended task. An example device is
an MP3 player that can concurrently play music from its storage, download new
music from a host, and respond to input from the user, such as adjusting the vol-
ume and track controls. Each of these operations is a scenario. Downloading a
music file takes several steps, such as control register reads and writes to set up the
operation, multiple DMA writes to transfer the song, and then another group of
reads and writes. The scenario layer of your testbench orchestrates all these steps
with constrained-random values for parameters such as track size and memory
location.

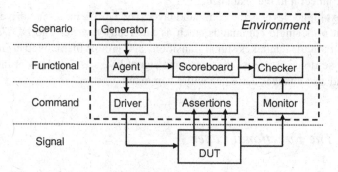

Fig. 1.11 Testbench with scenario layer added

   The blocks in the testbench environment (inside the dashed line of Fig. 1.11) are
written at the beginning of development. During the project they may evolve and
you may add functionality, but these blocks should not change for individual tests.
This is done by leaving "hooks" in the code so that a test can change the behavior
of these blocks without having to rewrite them. You create these hooks with factory
patterns (Section 8.2) and callbacks (Section 8.7).

1.10.5 The Test Layer and Functional Coverage

You are now at the top of the testbench, in the test layer, as shown in Fig. 1.12.
Design bugs that occur between DUT blocks are harder to find as they involve multiple people reading and interpreting multiple specifications.
   This top-level test is the conductor: he does not play any musical instrument, but
instead guides the efforts of others. The test contains the constraints to create the
stimulus.
   Functional coverage measures the progress of all tests in fulfilling the verification plan requirements. The functional coverage code changes through the project

as the various criteria complete. This code is constantly being modified and thus it is not part of the environment.

You can create a directed test in a constrained-random environment. Simply insert a section of directed test code into the middle of a random sequence, or put the two pieces of code in parallel. The directed code performs the work you want, but the random "background noise" may cause a bug to become visible, perhaps in a block that you never considered.

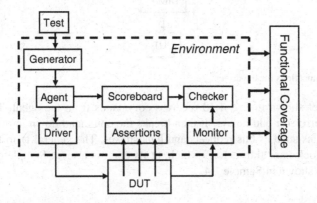

Fig. 1.12 Full testbench with all layers

Do you need all these layers in your testbench? The answer depends on what your DUT looks like. A complicated design requires a sophisticated testbench. You always need the test layer. For a simple design, the scenario layer may be so simple that you can merge it with the agent. When estimating the effort to test a design, don't count the number of gates; count the number of designers. Every time you add another person to the team, you increase the chance of different interpretations of the specifications. Typical hardware teams need more than two verification engineers for every designer.

You may need more layers. If your DUT has several protocol layers, each should get its own layer in the testbench environment. For example, if you have TCP traffic that is wrapped in IP and sent in Ethernet packets, consider using three separate layers for generation and checking. Better yet, use existing verification components.

One last note about Fig. 1.12. It shows some of the possible connections between blocks, but your testbench may have a different set. The test may need to reach down to the driver layer to force physical errors. What has been described here is just guidelines — let your needs guide what you create.

1.11 Building a Layered Testbench

Now it is time to take the preceding figures and learn how to map the components into SystemVerilog constructs.

1.11.1   Creating a Simple Driver

First, take a closer look at one of the blocks, the driver.

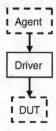

Fig. 1.13 Connections for the driver

The driver shown in Fig. 1.13 receives commands from the agent. The driver may inject errors or add delays. It then breaks the command down into individual signal changes such as bus requests and handshakes. The general term for such a testbench block is a "transactor," which, at its core, is a loop. Sample code for a transactor is shown in Sample 1.4.

Sample 1.4   Basic transactor code

```
task run();
  done = 0;
  while (!done) begin
    // Get the next transaction
    // Make transformations
    // Send out transactions
  end
endtask
```

Chapter 5 presents basic OOP and how to create an object that includes the routines and data for a transactor. Another example of a transactor is the agent. It might break apart a complex transaction such as a DMA read into multiple bus commands. Also in Chapter 5, you will see how to build an object that contains the data and routines that make up a command. These objects are sent between transactors using SystemVerilog mailboxes. In Chapter 7, you will learn about many ways to exchange data between the different layers and to synchronize the transactors.

1.12   Simulation Environment Phases

Up until now you have been learning what parts make up the environment. When do these parts execute? You want to clearly define the phases to coordinate the testbench so that all the code for a project works together. The three primary phases are Build, Run, and Wrap-up. Each is divided into smaller steps. These three are a subset of the many phases of the UVM and VMM.

The Build phase is divided into the following steps:
- *Generate configuration*: Randomize the configuration of the DUT and the surrounding environment.
- *Build environment*: Allocate and connect the testbench components based on the configuration. A testbench component is one that only exists in the testbench, as opposed to physical components in the design that are built with RTL code. For example, if the configuration chose three bus drivers, the testbench would allocate and initialize them in this step.
- *Reset the DUT*.
- *Configure the DUT*: Based on generated configuration from the first step, load the DUT command registers.

The Run phase is where the test actually runs. It has the following steps:
- *Start environment*: Run the testbench components such as BFMs and stimulus generators.
- *Run the test*: Start the test and then wait for it to complete. It is easy to tell when a directed test has completed, but doing so can be complex for a random test. You can use the testbench layers as a guide. Starting from the top, wait for a layer to drain all the inputs from the previous layer (if any), wait for the current layer to become idle, and then wait for the next lower layer. You should also use time-out checkers to ensure that the DUT or testbench does not lock up.

The Wrap-up phase has two steps:
- *Sweep*: After the lowest layer completes, you need to wait for the final transactions to drain out of the DUT.
- *Report*: Once the DUT is idle, sweep the testbench for lost data. Sometimes the scoreboard holds transactions that never came out, perhaps because they were dropped by the DUT. Armed with this information, you can create the final report on whether the test passed or failed. If it failed, be sure to delete any functional coverage results, as they may not be correct.

As shown in Fig. 1.12, the test starts the environment, which, in turn, runs each of the steps. More details can be found in Chapter 8.

1.13   Maximum Code Reuse

To verify a complex device with hundreds of features, you have to write hundreds of directed tests. If you use constrained-random stimulus, you would write fewer tests. Instead, the real work is put into constructing the testbench, which contains all the lower testbench layers: scenario, functional, command, and signal. This testbench code is used by all the tests, so it remains generic.

These guidelines appear to recommend an overly complicated testbench, but remember that every line that you put into a testbench can eliminate a line in every single test. If you know you will be creating a few dozen tests, there is a high pay-

back in making a more sophisticated testbench. Keep this in mind when you read
Chapter 8.

1.14   Testbench Performance

If this is the first time you have seen this methodology, you probably have some
qualms about how it works compared to directed testing. A common objection is
testbench performance. A directed test often simulates in a few seconds, whereas
constrained-random tests will wander around through the state space for minutes or
even hours. The problem with this argument is that it ignores a real verification
bottleneck: the time required by you to create a test. You may be able to hand-craft
a directed test in a day and debug it and manually verify the results by hand in
another day or two. The actual simulation run time is dwarfed by the amount of time
that you personally invested.

There are several steps to creating a constrained-random test. The first and most
significant step is building the layered testbench, including the self-checking portion. The benefit of this work is shared by all tests, so it is well worth the effort. The
second step is creating the stimulus specific to a goal in the verification plan. You
may be crafting random constraints, or devious ways of injecting errors or protocol
violations. Building one of these may take more time than making several directed
tests, but the payoff will be much higher. A constrained-random test that tries thousands of different protocol variations is worth more than the handful of directed
tests that could have been created in the same amount of time.

The third step in constrained-random testing is functional coverage. This task
starts with the creation of a strong verification plan with clear goals that can be easily
measured. Next you need to create the SystemVerilog code that adds instrumentation
to the environment and gathers the data. Finally, it is essential that you analyze the
results to determine if you have met the goals, and, if not, how you should modify
the tests.

1.15   Conclusion

The continuous growth in complexity of electronic designs requires a modern, sys-
tematic, and automated approach to creating testbenches. The cost of fixing a bug
grows by tenfold as a project moves from each step of specification to RTL coding,
gate synthesis, fabrication, and finally into the user's hands. Directed tests only test
one feature at a time and cannot create the complex stimulus and configurations that
the device would be subjected to in the real world. To produce robust designs, you
must use constrained-random stimulus combined with functional coverage to create
the widest possible range of stimuli.

1.16   Exercises

1. Write a verification plan for an Arithmetic Logic Unit (ALU) with:

   • Asynchronous active high input reset
   • Input clock
   • 4-bit signed inputs, A and B
   • 5-bit signed output C that is registered on the positive edge of input clock.
   • 4 opcodes

     – Add: A + B
     – Sub: A – B
     – Bit-wise invert: A
     – Reduction Or: B

2. What are the advantages and disadvantages to testing at the block level? Why?
3. What are the advantages and disadvantages to testing at the system level? Why?
4. What are the advantages and disadvantages to directed testing? Why?
5. What are the advantages and disadvantages to constrained random testing? Why?

Chapter 2
Data Types

SystemVerilog offers many improved data structures compared with Verilog. Some of these were created for designers but are also useful for testbenches. In this chapter you will learn about the data structures most useful for verification.

System Verilog introduces new data types with the following benefits.

- Two-state: better performance, reduced memory usage
- Queues, dynamic and associative arrays: reduced memory usage, built-in support for searching and sorting
- Classes and structures: support for abstract data structures
- Unions and packed structures: allow multiple views of the same data
- Strings: built-in string support
- Enumerated types: code is easier to write and understand

2.1 Built-In Data Types

Verilog-1995 has two basic data types: variables and nets, both which hold 4-state values: 0, 1, Z, and X. RTL code uses variables to store combinational and sequential values. Variables can be unsigned single or multi-bit (`reg [7:0] m`), signed 32-bit variables (`integer`), unsigned 64-bit variables (`time`), and floating point numbers (`real`). Variables can be grouped together into arrays that have a fixed size. A net is used to connect parts of a design such as gate primitives and module instances. Nets come in many flavors, but most designers use scalar and vector wires to connect together the ports of design blocks. Lastly, all storage is static, meaning that all variables are alive for the entire simulation and routines cannot use a stack to hold arguments and local values. Verilog-2001 allows you to switch between static and dynamic storage, such as stacks.

System Verilog adds many new data types to help both hardware designers and verification engineers.

C. Spear and G. Tumbush, *SystemVerilog for Verification: A Guide to Learning the Testbench Language Features*, DOI 10.1007/978-1-4614-0715-7_2,
© Springer Science+Business Media, LLC 2012

2.1.1   The Logic Type

The one thing in Verilog that always leaves new users scratching their heads is the difference between a `reg` and a `wire`. When driving a port, which should you use? How about when you are connecting blocks? SystemVerilog improves the classic `reg` data type so that it can be driven by continuous assignments, gates, and modules, in addition to being a variable. It is given the synonym `logic` as some people new to Verilog thought that `reg` declared a digital register, and not a signal. A `logic` signal can be used anywhere a net is used, except that a `logic` variable cannot be driven by multiple structural drivers, such as when you are modeling a bidirectional bus. In this case, the variable needs to be a net type such as `wire` so that SystemVerilog can resolve the multiple values to determine the final value.

Sample 2.1 shows the SystemVerilog `logic` type.

Sample 2.1   Using the logic type

```
module logic_data_type(input logic rst_h);
  parameter CYCLE = 20;
  logic q, q_l, d, clk, rst_l;
  initial begin
    clk = 0;                       // Procedural assignment
    forever #(CYCLE/2) clk = ~clk;
  end

  assign rst_l = ~rst_h;       // Continuous assignment
  not n1(q_l, q);              // q_l is driven by gate
  my_dff d1(q, d, clk, rst_l); // q is driven by module

endmodule
```

You can use the `logic` type to find netlist bugs as this type can only have a single driver. Rather than trying to choose between `reg` and `wire`, declare all your signals as `logic`, and you'll get a compilation error if it has multiple drivers. Of course, any signal that you do want to have multiple drivers, such as a bidirectional bus, should be declared with a net type such as `wire` or `tri`.

2.1.2   2-State Data Types

SystemVerilog introduces several 2-state data types to improve simulator performance and reduce memory usage, compared to variables declared as 4-state types. The simplest type is the `bit`, which is always unsigned. There are four signed 2-state types: `byte`, `shortint`, `int`, and `longint`. as shown in Sample 2.2.

Sample 2.2 Signed data types

```
bit b;              // 2-state, single-bit
bit [31:0] b32;     // 2-state, 32-bit unsigned integer
int unsigned ui;    // 2-state, 32-bit unsigned integer
int i;              // 2-state, 32-bit signed integer
byte b8;            // 2-state, 8-bit signed integer
shortint s;         // 2-state, 16-bit signed integer
longint l;          // 2-state, 64-bit signed integer
integer i4;         // 4-state, 32-bit signed integer
time t;             // 4-state, 64-bit unsigned integer
real r;             // 2-state, double precision floating point
```

 You might be tempted to use types such as byte to replace more verbose declarations such as logic [7:0]. Hardware designers should be careful as these new types are signed variables, so a byte variable can only count up to 127, not the 255 you may expect. (It has the range −128 to +127.) You could use byte unsigned, but that is more verbose than just bit [7:0]. Signed variables can also cause unexpected results with randomization, as discussed in Chapter 6.

 Be careful connecting 2-state variables to the design under test, especially its outputs. If the hardware tries to drive an X or Z, these values are converted to a 2-state value, and your testbench code may never know. Don't try to remember if they are converted to 0 or 1; instead, always check for propagation of unknown values. Use the $isunknown() operator that returns 1 if any bit of the expression is X or Z, as shown in Sample 2.3.

Sample 2.3 Checking for 4-state values

```
if ($isunknown(iport) == 1)
  $display("@%0t: 4-state value detected on iport %b",
           $time, iport);
```

The format %0t and the argument $time print the current simulation time, formatted as specified with the $timeformat() routine. Time values are explored in more detail in Section 3.7.

2.2   Fixed-Size Arrays

SystemVerilog offers several flavors of arrays beyond the single-dimension, fixed-size Verilog-1995 arrays. Additionally, many new features have been added to support these data types.

2.2.1 Declaring and Initializing Fixed-Size Arrays

Verilog requires that the low and high array limits must be given in the declaration. Since almost all arrays use a low index of 0, SystemVerilog lets you use the shortcut of just giving the array size, which is similar to C's style, as shown in Sample 2.4.

Sample 2.4 Declaring fixed-size arrays

```
int lo_hi[0:15];          // 16 ints [0]..[15]
int c_style[16];          // 16 ints [0]..[15]
```

How can you compute the number of bits needed to address a given array size? SystemVerilog has the `$clog2()` function that calculates the ceiling of log base 2, as shown in Sample 2.5.

Sample 2.5 Calculating the address width for a memory

```
parameter int MEM_SIZE = 256;
parameter int ADDR_WIDTH = $clog2(MEM_SIZE); // $clog2(256) = 8
bit [15:0] mem[MEM_SIZE];
bit [ADDR_WIDTH-1:0] addr;                    // [7:0]
```

You can create multi-dimensional fixed-size arrays by specifying the dimensions after the variable name. Sample 2.6 creates several two-dimensional arrays of integers, 8 entries by 4, and sets the last entry to 1. Multi-dimensional arrays were introduced in Verilog-2001, but the compact declaration style is new.

Sample 2.6 Declaring and using multi-dimensional arrays

```
int array2 [0:7][0:3];   // Verbose declaration
int array3 [8][4];       // Compact declaration
array2[7][3] = 1;        // Set last array element
```

If your code accidently tries to read from an out-of-bounds address, System-Verilog will return the default value for the array element type. That just means that an array of 4-state types, such as `logic`, will return X's, whereas an array of 2-state types, such as `int` or `bit`, will return 0. This applies for all array types – fixed, dynamic, associative, or queue, and also if your address has an X or Z. An undriven net is Z.

Many SystemVerilog simulators store each element on a 32-bit word boundary. So a `byte`, `shortint`, and `int` are all stored in a single word, whereas a `longint` is stored in two words.

An unpacked array, such as the one shown in Sample 2.7, stores the values in the lower portion of the word, whereas the upper bits are unused. The array of bytes, `b_unpack`, is stored in three words, as shown in Fig. 2.1.

Sample 2.7 Unpacked array declarations

```
bit [7:0] b_unpack[3];    // Unpacked
```

Fig. 2.1 Unpacked array storage

Packed arrays are explained in Section 2.2.6.

Simulators generally store 4-state types such as `logic` and `integer` in two or more consecutive words, using twice the storage as 2-state variables.

2.2.2 The Array Literal

Sample 2.8 shows how to initialize an array using an array literal, which is an apostrophe followed by the values in curly braces. (This is not the accent grave used for compiler directives and macros.) You can set some or all elements at once. You can replicate values by putting a count before the curly braces.

Sample 2.8 Initializing an array

```
initial begin
  static int ascend[4] = '{0,1,2,3}; // Initialize 4 elements
  int descend[5];

  descend = '{4,3,2,1,0};          // Set 5 elements
  descend[0:2] = '{7,6,5};         // Set just first 3 elements
  ascend = '{4{8}};                // Four values of 8
  ascend = '{default:42};          // All elements are set to 42
end
```

Notice that in Sample 2.8, the declaration of the array ascend includes an initial value. The 2009 LRM states that these variables must be declared either in a static block, or have the `static` keyword. Since this book recommends always declaring your test modules and programs as `automatic`, you need to add the `static` keyword to a declaration plus initialization when it is inside an `initial` block.

A great new feature in the 2009 LRM is printing with the `%p` format specifier. This prints an assignment pattern that is equivalent to the data object's value. You can print any data type in SystemVerilog including arrays, structures, classes, and more. Sample 2.9 shows how to print an array with the %p format specifier.

Sample 2.9 Printing with %p print specifier

```
initial begin
  ascend = '{0,1,2,3};
  $display("%p", ascend);        // '{0, 1, 2, 3}
  ascend = '{4{8}};
  $display("%p", ascend);        // '{8, 8, 8, 8}
end
```

2.2.3   Basic Array Operations — for and Foreach

The most common way to manipulate an array is with a for or foreach loop. In Sample 2.10, the variable i is declared local to the for loop. The SystemVerilog function $size returns the size of the array. In the foreach loop, you specify the array name and an index in square brackets, and SystemVerilog automatically steps through all the elements of the array. The index variable is automatically declared for you and is local to the loop.

Sample 2.10 Using arrays with for- and foreach loops

```
initial begin
  bit [31:0] src[5], dst[5];
  for (int i=0; i<$size(src); i++)
    src[i] = i;                 // Initialize src array
  foreach (dst[j])
    dst[j] = src[j] * 2;   // Set dst array to 2 * src
end
```

Note that in Sample 2.11, the syntax of the foreach loop for multi-dimensional arrays may not be what you expected. Instead of listing each subscript in separate square brackets, [i][j], they are combined with a comma: [i,j].

Sample 2.11 Initialize and step through a multi-dimensional array

```
int md[2][3] = '{'{0,1,2}, '{3,4,5}};
initial begin
  $display("Initial value:");
  foreach (md[i,j])    // Yes, this is the right syntax
    $display("md[%0d][%0d] = %0d", i, j, md[i][j]);

  $display("New value:");
  // Replicate last 3 values of 5
  md = '{'{9, 8, 7}, '{3{5}}};
  foreach (md[i,j])     // Yes, this is the right syntax
    $display("md[%0d][%0d] = %0d", i, j, md[i][j]);
end
```

The output from Sample 2.11 is shown in Sample 2.12.

Sample 2.12 Output from printing multi-dimensional array values

```
Initial value:
md[0][0] = 0
md[0][1] = 1
md[0][2] = 2
md[1][0] = 3
md[1][1] = 4
md[1][2] = 5
New value:
md[0][0] = 9
md[0][1] = 8
md[0][2] = 7
md[1][0] = 5
md[1][1] = 5
md[1][2] = 5
```

You can omit some dimensions in the `foreach` loop if you don't need to step through all of them. Sample 2.13 prints a two-dimensional array in a rectangle. It steps through the first dimension in the outer loop, and then through the second dimension in the inner loop.

Sample 2.13 Printing a multi-dimensional array

```
initial begin
  byte twoD[4][6];
  foreach(twoD[i,j])
    twoD[i][j] = i*10+j;

  foreach (twoD[i]) begin      // Step through first dim.
    $write("%2d:", i);
    foreach(twoD[,j])             // Step through second
      $write("%3d", twoD[i][j]);
    $display;
  end
end
```

Sample 2.13 produces the output shown in Sample 2.14.

Sample 2.14 Output from printing multi-dimensional array values

```
0:  0  1  2  3  4  5
1: 10 11 12 13 14 15
2: 20 21 22 23 24 25
3: 30 31 32 33 34 35
```

Lastly, a `foreach` loop iterates using the ranges in the original declaration. The array f[5] is equivalent to f[0:4], and a foreach (f[i]) is equivalent to for (int i=0;i<=4; i++). With the array rev[6:2], the statement foreach(rev[i]) is equivalent to for(int i=6; i>=2; i--).

2.2.4   Basic Array Operations – Copy and Compare

You can perform aggregate compare and copy of arrays without loops. (An aggregate operation works on the entire array as opposed to working on just an individual element.) Comparisons are limited to just equality and inequality. Sample 2.15 shows several examples of compares. The ? : conditional operator is a mini if-else statement. In Sample 2.15, it is used to choose between two strings. The final compare uses an array slice, src[1:4], which creates a temporary array with 4 elements.

Sample 2.15   Array copy and compare operations

```
initial begin
  bit [31:0] src[5] = '{0,1,2,3,4},
             dst[5] = '{5,4,3,2,1};

  // Aggregate compare the two arrays
  if (src==dst)
    $display("src == dst");
  else
    $display("src != dst");

  // Aggregate copy all src values to dst
  dst = src;

  // Change just one element
  src[0] = 5;

  // Are all values equal (no!)
  $display("src %s dst", (src == dst) ? "==" : "!=");

  // Use array slice to compare elements 1-4 (they are equal)
  $display("src[1:4] %s dst[1:4]",
           (src[1:4] == dst[1:4]) ? "==" : "!=");
end
```

A copy between fixed arrays of different sizes causes a compile error. You can not perform aggregate arithmetic such as addition or subtraction on arrays, for example, a = b + c. Instead, use foreach loops. For logical operations such as xor, you have to either use a loop or use packed arrays as described in Section 2.2.6.

2.2.5   Bit and Array Subscripts, Together at Last

A common annoyance in Verilog-1995 is that you cannot use array and bit subscripts together. Verilog-2001 removes this restriction for fixed-size arrays. Sample 2.16 prints the first array element (binary 101), its lowest bit (1), and the next two higher bits (binary 10).

Sample 2.16 Using word and bit subscripts together

```
initial begin
  bit [31:0] src[5] = '{5{5}};
  $displayb(src[0],,           // 'b101 or 'd5
           src[0][0],,         // 'b1
           src[0][2:1]);       // 'b10
end
```

Although this change is not new to SystemVerilog, many users may not know about this useful improvement in Verilog-2001. FYI - a double comma in a $display statement inserts a space.

2.2.6 Packed Arrays

For some data types, you may want both to access the entire value and also to divide it into smaller elements. For example, you may have a 32-bit register that sometimes you want to treat as four 8-bit values and at other times as a single, unsigned value. A SystemVerilog packed array is treated as both an array and a single value. It is stored as a contiguous set of bits with no unused space, unlike an unpacked array.

2.2.7 Packed Array Examples

The packed bit and array dimensions are specified as part of the type, before the variable name. These dimensions must be specified in the [msb:lsb] format, not [size]. Sample 2.17 shows the variable bytes, a packed array of four bytes that are stored in a single 32-bit word as shown in Fig. 2.2.

Sample 2.17 Packed array declaration and usage

```
bit [3:0] [7:0] bytes;    // 4 bytes packed into 32-bits
bytes = 32'hCAFE_DADA;
$displayh(bytes,,           // Show all 32-bits
          bytes[3],,        // Most significant byte "CA"
          bytes[3][7]);     // Most significant bit "1" of "CA"
```

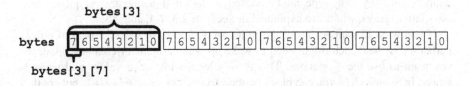

Fig. 2.2 Packed array layout

You can mix packed and unpacked dimensions. You may want to make an array that represents a memory that can be accessed as bits, bytes, or longwords. Sample 2.18 shows `barray`, an unpacked array of five packed elements, each four bytes wide, which are stored in memory as shown in Fig. 2.3.

Sample 2.18 Declaration for a mixed packed/unpacked array

```
bit [3:0] [7:0] barray [5];     // 5 elements: packed 4-bytes
bit [31:0] lw = 32'h0123_4567;  // Word
bit [7:0] [3:0] nibbles;        // Packed array of nibbles
barray[0] = lw;
barray[0][3] = 8'h01;
barray[0][1][6] = 1'b1;
nibbles = barray[2];            // Copy packed values
```

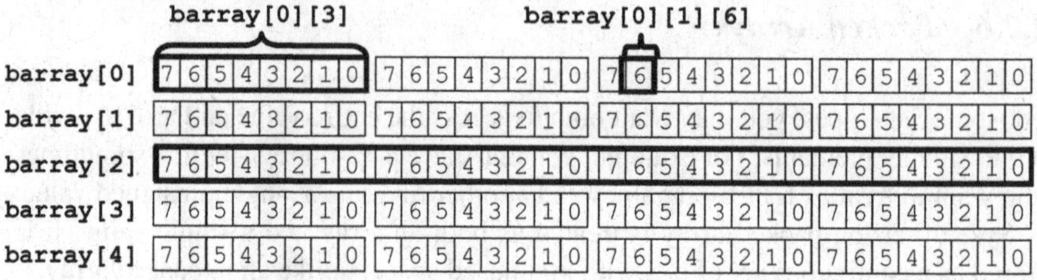

Fig. 2.3 Packed array bit layout

With a single subscript, you get a word of data, `barray[0]`. With two subscripts, you get a byte of data, `barray[0][3]`. With three subscripts, you can access a single bit, `barray[0][1][6]`. Because one dimension is specified after the name, `barray[5]`, that dimension is unpacked, so you must always give at least one subscript.

The last line of Sample 2.18 copies between two packed arrays. Since the underlying values are just bits, you can copy even if the arrays have different dimensions.

2.2.8 Choosing Between Packed and Unpacked Arrays

Which should you choose — a packed or an unpacked array? A packed array is handy if you need to convert to and from scalars. For example, you might need to reference a memory as a byte or as a word. The `barray` in Fig. 2.3 can handle this requirement. Any array type can be packed, including dynamic arrays, queues and associative arrays, which are explained in Sections 2.3, 2.4, and 2.5.

If you need to wait for a change in an array, you have to use a packed array. Perhaps your testbench might need to wake up when a memory changes value, so you want to use the @ operator. This is only legal with scalar values and packed arrays. In Sample 2.18 you can block on the variables `lw` or `barray[0]`, but not the entire array `barray` unless you expand it: @(barray[0] or barray[1] or barray[2] or barray[3] or barray[4]).

2.3   Dynamic Arrays

The basic Verilog array type shown so far is known as a fixed-size array, as its size is set at compile time. What if you do not know the size of the array until run time? For example, you may want generate a random number of transactions at the start of simulation. If you stored the transactions in a fixed-size array, it would have to be large enough to hold the maximum number of transactions, but would typically hold far fewer, thus wasting memory. SystemVerilog provides a dynamic array that can be allocated and resized during simulation so your simulation consumes a minimal amount of memory.

A dynamic array is declared with empty word subscripts [ ]. This means that you do not specify the array size at compile time; instead, give it at run time. The array is initially empty, so you must call the new[] constructor to allocate space, passing in the number of entries in the square brackets. If you pass an array name to the new[] constructor, the values are copied into the new elements, as shown in Sample 2.19.

Sample 2.19   Using dynamic arrays

```
int dyn[], d2[];                 // Declare dynamic arrays

initial begin
  dyn = new[5];                  // A: Allocate 5 elements
  foreach (dyn[j]) dyn[j] = j;   // B: Initialize the elements
  d2 = dyn;                      // C: Copy a dynamic array
  d2[0] = 5;                     // D: Modify the copy
  $display(dyn[0],d2[0]);        // E: See both values (0 & 5)
  dyn = new[20](dyn);            // F: Allocate 20 ints & copy
  dyn = new[100];                // G: Allocate 100 new ints
                                 //    Old values are lost
  dyn.delete();                  // H: Delete all elements
end
```

In Sample 2.19, Line A calls new[5] to allocate 5 array elements. The dynamic array dyn now holds 5 int's. Line B sets the value of each element of the array to its index value. Line C allocates another array and copies the contents of dyn into it. Lines D and E show that the arrays dyn and d2 are separate. Line F allocates 20 new elements, and copies the existing 5 elements of dyn to the beginning of the array. Then the old 5-element dyn array is deallocated. The result is that dyn points to a 20-element array. The last call to new[] allocates 100 elements, but the existing values are not copied. The old 20-element array is deallocated. Finally, line H deletes the dyn array.

The $size function returns the size of a fixed or dynamic array. Dynamic arrays have several built-in routines, such as delete and size.

If you want to declare a constant array of values but do not want to bother counting the number of elements, use a dynamic array with an array literal. In Sample 2.20 there are 9 mask elements of 8-bits each. You should let SystemVerilog count them, rather than making a fixed-size array and accidently choosing the wrong array size.

Sample 2.20 Using a dynamic array for an uncounted list

```
bit [7:0] mask[] = '{8'b0000_0000, 8'b0000_0001,
                     8'b0000_0011, 8'b0000_0111,
                     8'b0000_1111, 8'b0001_1111,
                     8'b0011_1111, 8'b0111_1111,
                     8'b1111_1111};
```

You can make assignments between fixed-size and dynamic arrays as long as they have the same base type such as `int`. You can assign a dynamic array to a fixed array as long as they have the same number of elements.

When you copy a fixed-size array to a dynamic array, SystemVerilog calls the `new[]` constructor to allocate space, and then copies the values.

You can have multi-dimensional dynamic arrays, so long as you are careful when constructing the sub-arrays. Remember, a multi-dimensional array in SystemVerilog can be thought of as an array of other arrays. First you need to construct the left-most dimension. Then construct the sub-arrays. In Sample 2.21, each sub-array has a different size.

Sample 2.21 Multi-dimensional dynamic array

```
// A dynamic array of dynamic arrays
  int d[][];

initial begin
  // Construct the first or left-most dimension
  d = new[4];

  // Construct the 2nd dimension, each array a different size
  foreach(d[i])
    d[i] = new[i+1];

  // Initialize the elements.  d[4][2] = 42;
  foreach(d[i,j])
    d[i][j] = i*10 + j;
end
```

2.4   Queues

SystemVerilog introduces a new data type, the queue, which combines the best of a linked list and array. Like a linked list, you can add or remove elements anywhere in a queue, without the performance hit of a dynamic array that has to allocate a new array and copy the entire contents. Like an array, you can directly access any element with an index, without linked list's overhead of stepping through the preceding elements.

A queue is declared with word subscripts containing a dollar sign: [$]. The elements of a queue are numbered from 0 to $. Sample 2.22 shows how you can add and remove values from a queue using methods. Note that queue literals only have curly braces, and are missing the initial apostrophe of array literals.

The SystemVerilog queue is similar to the Standard Template Library's deque data type. You create a queue by adding elements. SystemVerilog typically allocates extra space so you can quickly insert additional elements. If you add enough elements that the queue runs out of that extra space, SystemVerilog automatically allocates more. As a result, you can grow and shrink a queue without the performance penalty of a dynamic array, and SystemVerilog keeps track of the free space for you. Note that you never call the new[] constructor for a queue.

Sample 2.22 Queue methods

```
int j = 1,
    q2[$] = {3,4},             // Queue literals do not use '
    q[$] = {0,2,3};            // {0,2,3}

initial begin
  q.insert(1, j);              // {0,1,2,3}  Insert j before ele #1
  q.delete(1);                 // {0,2,3}    Delete element #1

  // These operations are fast
  q.push_front(6);             // {6,0,2,3}  Insert at front
  j = q.pop_back;              // {6,0,2}    j = 3
  q.push_back(8);              // {6,0,2,8}  Insert at back
  j = q.pop_front;             // {0,2,8}    j = 6
  foreach (q[i])
    $display(q[i]);            //            Print entire queue
  q.delete();                  // {}         Delete queue
end
```

The LRM does not allow inserting a queue in another queue using the above methods, though some simulators permit this.

You can use word subscripts and concatenation instead of methods. As a shortcut, if you put a $ on the left side of a range, such as [$:2], the $ stands for the minimum value, [0:2]. A $ on the right side, as in [1:$], stands for the maximum value, [1:2], in first line of the initial block of Sample 2.23.

Sample 2.23 Queue operations

```
int j = 1,
    q2[$] = {3,4},          // Queue literals do not use '
    q[$] = {0,2,5};         // {0,2,5}

initial begin               // Result
  q = {q[0], j, q[1:$]};    // {0,1,2,5}    Insert 1 before 2
  q = {q[0:2], q2, q[3:$]}; // {0,1,2,3,4,5} Insert queue in q
  q = {q[0], q[2:$]};       // {0,2,3,4,5}  Delete elem. #1

  // These operations are fast
  q = {6, q};               // {6,0,2,3,4,5} Insert at front

  j = q[$];                 // j = 5        pop_back
  q = q[0:$-1];             // {6,0,2,3,4}     equivalent

  q = {q, 8};               // {6,0,2,3,4,8} Insert at back

  j = q[0];                 // j = 6        pop_front
  q = q[1:$];               // {0,2,3,4,8}     equivalent

  q = {};                   // {}           Delete contents
end
```

The queue elements are stored in contiguous locations, so it is efficient to push and pop elements from the front and back. This takes a fixed amount of time no matter how large the queue. Adding and deleting elements in the middle of a queue requires shifting the existing data to make room. The time to do this grows linearly with the size of the queue.

You can copy the contents of a fixed or dynamic array into a queue.

2.5   Associative Arrays

Dynamic arrays are good if you want to occasionally create a big array, but what if you want something really large? Perhaps you are modeling a processor that has a multi-gigabyte address range. During a typical test, the processor may only touch a few hundred or thousand memory locations containing executable code and data, so allocating and initializing gigabytes of storage is wasteful.

SystemVerilog offers associative arrays that store entries in a sparse matrix. This means that while you can address a very large address space, SystemVerilog only allocates memory for an element when you write to it. In the following picture, the associative array holds the values 0:3, 42, 1000, 4521, and 200,000. The memory used to store these is far less than would be needed to store a fixed or dynamic array with 200,000 entries, as shown in Figure 2.4.

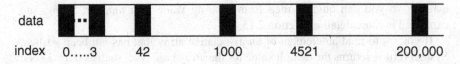

Fig. 2.4 Associative array

An associative array can be stored by the simulator as a tree or hash table. This additional overhead is acceptable when you need to store arrays with widely separated index values, such as packets indexed with 32-bit addresses or 64-bit data values. An associative array is declared with a data type in square brackets, such as [int]. or [Packet]. Sample 2.24 shows declaring, initializing, printing, and stepping through an associative array.

Sample 2.24 Declaring, initializing, and using associative arrays

```
byte assoc[byte], idx = 1;
initial begin
  // Initialize widely scattered values
  do begin
    assoc[idx] = idx;
    idx = idx << 1;
  end while (idx != 0);

  // Step through all index values with foreach
  foreach (assoc[i])
    $display("assoc[%h] = %h", i, assoc[i]);

  // Step through all index values with functions
  if (assoc.first(idx))    // Get first index
    do
      $display("assoc[%h]=%h", idx, assoc[idx]);
    while (assoc.next(idx));  // Get next index

  // Find and delete the first element
  void'(assoc.first(idx));
  void'(assoc.delete(idx));
  $display("The array now has %0d elements", assoc.num());
end
```

Sample 2.24 has the associative array, assoc, with very scattered elements: 1, 2, 4, 8, 16, etc. A simple for loop cannot step through them; you need to use a foreach loop. If you want finer control, you can use the first and next functions in a do...while loop. These functions modify the index argument, and return 0 or 1 depending on whether any elements are left in the array. You can find the number of elements in an associative array with the num or size functions.

Associative arrays can also be addressed with a string index, similar to Perl's hash arrays. Sample 2.25 reads a file with strings and builds the associative array switch so you can quickly map from a string value to a number. Strings are explained in more detail in Section 2.15.

If you try to read an element of an associative array that has not been written, SystemVerilog returns the default value for the array base type, such as 0 for 2-state types such as bit or int, or X for 4-state types such as logic. The simulator may also give a warning message. You can use the function exists() to check if an element has been allocated, as shown in Sample 2.25.

Sample 2.25 Using an associative array with a string index

```
/* Input file contains:
   42   min_address
   1492 max_address
*/

int switch[string], min_address, max_address, i, file;
initial begin
  string s;
  file = $fopen("switch.txt", "r");
  while (! $feof(file)) begin
    $fscanf(file, "%d %s", i, s);
    switch[s] = i;
  end
  $fclose(file);

  // Get the min address
  // If string not found, use default value of 0 for int array
  min_address = switch["min_address"];

  // Get the max address.
  // Use 1000 if max_address does not exist
  if (switch.exists("max_address"))
    max_address = switch["max_address"];
  else
    max_address = 1000;

  // Print all switches
  foreach (switch[s])
    $display("switch['%s']=%0d", s, switch[s]);
end
```

You can initialize an associative array with the array literal with index:element pairs as shown in Sample 2.26. When you print the array with %p, the elements are displayed in the same format.

Sample 2.26 Initializing and printing associative arrays

```
int power_of_2[int] = '{0:1, 1:2, 2:4};
initial begin
  for (int i=3; i<5; i++)
    power_of_2[i] = 1 << i;
  $display("%p", power_of_2);  // '{0:1, 1:2, 2:4, 3:8, 4:16}
end
```

You can also declare an associative array with wildcard subscripts, as in wild[*]. However, this style is not recommended as you are allowing an index of almost any data type. One of the many resulting problems is with foreach–loops: what type is the variable j in foreach(wild[j])? Integer, string, bit, or logic?

2.6 Array Methods

There are many array methods that you can use on any unpacked array types: fixed, dynamic, queue, and associative. These routines can be as simple as giving the current array size or as complex as sorting the elements. The parentheses are optional if there are no arguments.

2.6.1 Array Reduction Methods

A basic array reduction method takes an array and reduces it to a single value, as shown in Sample 2.27. You can calculate the sum, product, or perform a logical operation on all the elements.

Sample 2.27 Array reduction operations

```
byte b[$] = {2, 3, 4, 5};
int w;
w = b.sum();      //  14 = 2 + 3 + 4 + 5
w = b.product();  // 120 = 2 * 3 * 4 * 5
w = b.and();      // 0000_0000 = 2 & 3 & 4 & 5
```

Other array reduction methods are or, and xor.

SystemVerilog does not have a method specifically for choosing a random element from an array, so use the index $urandom_range(array.size()-1) for queues and dynamic arrays, and $urandom_range($size(array)-1) for fixed arrays, queues, dynamic and associative arrays. See Section 6.10 for more information on $urandom_range.

If you need to choose a random element from an associative array, you need to step through the elements one by one as there is no one-line way to access the Nth element. Sample 2.28 shows how to choose a random element from an associative array indexed by integers by first picking a random number, then stepping through the array. If the array was indexed by a string, just change the type of idx to string.

Sample 2.28 Picking a random element from an associative array

```
// Declare and initialize associative array with 7 elements
int aa[int] = '{0:1, 5:2, 10:4, 15:8, 20:16, 25:32, 30:64};
int idx, element, count;

element = $urandom_range(aa.size()-1);
foreach(aa[i])
  if (count++ == element) begin
    idx = i;            // Save the associative array index
    break;              //      and quit
  end

$display("element#%0d aa[%0d] = %0d",
         element, idx, aa[idx]);
```

2.6.2 Array Locator Methods

What is the largest value in an array? Does an array contain a certain value? The array locator methods find data in an unpacked array. At first you may wonder why these return a queue of values. After all, there is only one maximum value in an array. However, SystemVerilog needs a queue for the case when you ask for a value from an empty queue or dynamic array.

Sample 2.29 shows the array locator methods: min and max functions find the smallest and largest elements in an array. These methods also work for associative arrays. The unique method returns a queue of the unique values from the array — duplicate values are not included.

Sample 2.29 Array locator methods: min, max, unique

```
int f[6] = '{1,6,2,6,8,6};    // Fixed-size array
int d[]  = '{2,4,6,8,10};     // Dynamic array
int q[$] = {1,3,5,7},         // Queue
    tq[$];                    // Temporary queue for result

tq = q.min();                 // {1}
tq = d.max();                 // {10}
tq = f.unique();              // {1,6,2,8}
```

You could search through an array using a foreach loop, but SystemVerilog can do this in one operation with a locator method. The with expression tells SystemVerilog how to perform the search, as shown in Sample 2.30. These methods return an empty queue if the value you are searching for does not exist in the array.

Sample 2.30 Array locator methods: find

```
int d[] = '{9,1,8,3,4,4}, tq[$];

// Find all elements greater than 3
tq = d.find with (item > 3);                  // {9,8,4,4}
// Equivalent code
tq.delete();
foreach (d[i])
  if (d[i] > 3)
    tq.push_back(d[i]);


tq = d.find_index with (item > 3);        // {0,2,4,5}
tq = d.find_first with (item > 99);       // {} - none found
tq = d.find_first_index with (item==8);   // {2} d[2]=8
tq = d.find_last with (item==4);          // {4}
tq = d.find_last_index with (item==4);    // {5} d[5]=4
```

In a with clause, the name item is called the iterator argument and represents a single element of the array. You can specify your own name by putting it in the argument list of the array method as shown in Sample 2.31.

Sample 2.31 Declaring the iterator argument

```
tq = d.find_first with (item==4);        // These
tq = d.find_first() with (item==4);      // are
tq = d.find_first(item) with (item==4);  // all
tq = d.find_first(x) with (x==4);        // equivalent
```

Sample 2.32 shows various ways to total up a subset of the values in the array. The first line compares the item with 7. This relational returns a 1 (true) or 0 (false) so the calculation is a sum of the array {1,0,1,0,0,0}. The second multiplies the boolean result with the array element being tested. So the total is the sum of {9,0,8,0,0,0}, which is 17. The third calculates the total of elements less than 8. The fourth total is computed using the ? : conditional operator. The last counts the number of 4's.

Sample 2.32 Array locator methods

```
int count, total, d[] = '{9,1,8,3,4,4};

count = d.sum(x) with (x >  7);         //  2=sum{1,0,1,0,0,0}
total = d.sum(x) with ((x > 7) * x);    // 17=sum{9,0,8,0,0,0}
count = d.sum(x) with (x <  8);         //  4=sum{0,1,0,1,1,1}
total = d.sum(x) with (x < 8 ? x : 0);  // 12=sum{0,1,0,3,4,4}
count = d.sum(x) with (x == 4);         //  2=sum{0,0,0,0,1,1}
```

When you combine an array reduction such as sum using the with clause, the results may surprise you. In Sample 2.32, the sum operator totals the number of times that the expression is true. For the first statement in Sample 2.32, there are two array elements that are greater than 7 (9 and 8) so count is set to 2.

The array locator methods that return an index, such as **find_index**, return a queue of type **int**, not **integer**. Your code may not compile if you use the wrong queue type with these statements.

Be careful of SystemVerilog's rules for the width of operations. Normally, if you were to add a set of single bit values, SystemVerilog would make the calculations with enough precision not to lose any bits. But the sum method uses the width of the array. So, if you add the values of a single-bit array, the result is a single bit, which is probably not what you expected. The solution is to use a with expression as shown in Sample 2.33.

Sample 2.33 Creating the sum of an array of single bits

```
bit one[6];  // Array of single bits
int total;

initial begin
  foreach (one[i])
    one[i] = i;                                 // one[i] gets 0 or 1

  // Compute the single-bit sum
  total = one.sum();                 // total = 1 = (0+1+0+1+0+1) & 1

  // Compute with 32-bit signed arithimetic
  total = one.sum() with (int'(item));     // total = 3
end
```

2.6.3 Array Sorting and Ordering

SystemVerilog has several methods for changing the order of elements in an array. You can sort the elements, reverse their order, or shuffle the order as shown in Sample 2.34. Notice that these change the original array, unlike the array locator methods in Section 2.6.2, which create a queue to hold the results.

Sample 2.34 Sorting an array

```
int d[] =        '{9,1,8,3,4,4};
d.reverse();  // '{4,4,3,8,1,9}
d.sort();     // '{1,3,4,4,8,9}
d.rsort();    // '{9,8,4,4,3,1}
d.shuffle();  // '{9,4,3,8,1,4}
```

The `reverse` and `shuffle` methods have no `with`-clause, so they work on the entire array. Sample 2.35 shows how to sort a structure by sub-fields. Structures and packed structures are explained in Section 2.9.

Sample 2.35   Sorting an array of structures

```
struct packed { bit [7:0] r, g, b; } c[];
c = '{'{r:7, g:4, b:9}, '{r:3, g:2, b:9}, '{r:5, g:2, b:1}};

c.sort with (item.r); // sort using r only
//   '{'{r:3, g:2, b:9}, '{r:5, g:2, b:1}, '{r:7, g:4, b:9}}

c.sort(x) with ({x.g, x.b}); // Sort g first, then b
//   '{'{r:5, g:2, b:1}, '{r:3, g:2, b:9}, '{r:7, g:4, b:9}}
```

Only fixed and dynamic arrays, plus queues can be sorted, reversed, or shuffled. Associative arrays can not be reordered.

2.6.4   Building a Scoreboard with Array Locator Methods

The array locator methods can be used to build a scoreboard. Sample 2.36 defines the `Packet` structure, then creates a scoreboard made from a queue of these structures. Section 2.8 describes how to create structures with `typedef`.

Sample 2.36   A scoreboard with array methods

```
typedef struct packed
  {bit [7:0] addr;
   bit [7:0] pr;
   bit [15:0] data; } Packet;

Packet scb[$];

function void check_addr(bit [7:0] addr);
  int intq[$];

  intq = scb.find_index() with (item.addr == addr);
  case (intq.size())
  0: $display("Addr %h not found in scoreboard", addr);
  1: scb.delete(intq[0]);
  default:
    $display("ERROR: Multiple hits for addr %h", addr);
  endcase
endfunction : check_addr
```

The `check_addr()` function in Sample 2.36 looks up an address in the score-board. The `find_index()` method returns an `int` queue. If the queue is empty (size==0), no match was found. If the queue has one member (size==1), a single match was found, which the `check_addr()` function deletes. If the queue has multiple members (size > 1), there are multiple packets in the scoreboard whose address matching the requested one.

A better choice for storing packet information is a class, which is described in Chapter 5. You can read more about structures in Section 2.9.

2.7 Choosing a Storage Type

Here are some guidelines for choosing the right storage type based on flexibility, memory usage, speed, and sorting. These are just rules of thumb, and results may vary between simulators.

2.7.1 Flexibility

Use a fixed-size or dynamic array if it is accessed with consecutive positive integer indices: 0, 1, 2, 3… Choose a fixed-size array if the array size is known at compile time, or choose a dynamic array if the size is not known until run time. For example, variable-size packets can easily be stored in a dynamic array. If you are writing routines to manipulate arrays, consider using just dynamic arrays, as one routine can work with any size dynamic array as long as the element types match: `int`, `string`, etc. Likewise, you can pass a queue of any size into a routine as long as the element type matches the queue argument. Associative arrays can also be passed regardless of size. However, a routine with a fixed-size array argument only accepts arrays of the specified length.

Choose associative arrays for nonstandard indices such as widely separated values because of random values or addresses. Associative arrays can also be used to model content-addressable memories.

Queues are a good way to store values when the number of elements grows and shrinks a lot during simulation, such as a scoreboard that holds expected values.

2.7.2 Memory Usage

If you want to reduce the simulation memory usage, use 2-state elements. You should choose data sizes that are multiples of 32 bits to avoid wasted space. Simulators usually store anything smaller in a 32-bit word. For example, an array of 1024 bytes wastes ¾ of the memory if the simulator puts each element in a 32-bit word. Packed arrays can also help conserve memory.

For arrays that hold up to a thousand elements, the type of array that you choose does not make a big difference in memory usage (unless there are many instances of these arrays). For arrays with a thousand to a million active elements, fixed-size and dynamic arrays are the most memory efficient. You may want to reconsider your algorithms if you need arrays with more than a million active elements.

Queues are slightly less efficient to access than fixed-size or dynamic arrays because of additional pointers. However, if your data set grows and shrinks often, and you store it in a dynamic memory, you will have to manually call `new[]` to allocate memory and copy. This is an expensive operation and would wipe out any gains from using a dynamic memory.

Modeling memories larger than a few megabytes should be done with an associative array. Note that each element in an associative array can take several times more memory than a fixed-size or dynamic memory because of pointer overhead.

2.7.3 Speed

Choose your array type based on how many times it is accessed per clock cycle. For only a few reads and writes, you could use any type, as the overhead is minor compared with the DUT. As you use an array more often, its size and type matters.

Fixed-size and dynamic arrays are stored in contiguous memory, so any element can be found in the same amount of time, regardless of array size.

Queues have almost the same access time as a fixed-size or dynamic array for reads and writes. The first and last elements can be pushed and popped with almost no overhead. Inserting or removing elements in the middle requires many elements to be shifted up or down to make room. If you need to insert new elements into a large queue, your testbench may slow down, so consider changing how you store new elements.

When reading and writing associative arrays, the simulator must search for the element in memory. The LRM does not specify how this is done, but popular ways are hash tables and trees. These require more computation than other arrays, and therefore associative arrays are the slowest.

2.7.4 Data Access

Since SystemVerilog can sort any single-dimension array (fixed-size, dynamic, and associative arrays plus queues), you should pick the array type based on how often the values are added to it. If the values are received all at once, choose a fixed-size or dynamic array so that you only have to allocate the array once. If the data slowly dribbles in, choose a queue, as adding new elements to the head or tail is very efficient.

If you have unique and noncontiguous values, such as `'{1, 10, 11, 50}`, you can store them in an associative array by using them as an index. Using the routines

first, next, and prev, you can search an associative array for a value and find
successive values. Lists are doubly linked, so you can find values both larger and
smaller than the current value. Both of these support removing a value. However,
the associative array is much faster in accessing any given element given an index.

For example, you can use an associative array of bits to hold expected 32-bit
values. When the value is created, write to that location. When you need to see if a
given value has been written, use the exists function. When done with an element,
use delete to remove it from the associative array.

2.7.5   Choosing the Best Data Structure

Here are some suggestions on choosing a data structure.
*   *Network packets*. Properties: fixed size, accessed sequentially. Use a fixed-size or
    dynamic array for fixed- or variable-size packets.
*   *Scoreboard of expected values*. Properties: array size not known until run time,
    accessed by value, and a constantly changing size. In general, use a queue, as you
    are continually adding and deleting elements during simulation. If you can give
    every transaction a fixed ID, such as 1, 2, 3, ..., you could use this as an index
    into the queue. If your transaction is filled with random values, you can just push
    them into a queue and search for unique values. If the scoreboard may have hun-
    dreds of elements and you are often inserting and deleting them from the middle,
    an associative array may be faster. If you model your transactions as objects, the
    scoreboard can be a queue of handles. See Chapter 5 for more information of
    classes.
*   *Sorted structures*. Use a queue if the data comes out in a predictable order or an
    associative array if the order is unspecified. If the scoreboard never needs to be
    searched, just store the expected values in a mailbox as shown in Section 7.6.
*   *Modeling very large memories, greater than a million entries*. If you do not need
    every location, use an associative array as a sparse memory. If you do plan on
    accessing every location, try a different approach where you do not need so much
    live data. Be sure to use 2-state values packed into 32-bits to conserve simulation
    memory.
*   *Command names or opcodes from a file*. Property: translate a string to a fixed
    value. Read string from a file, and then look up the commands or opcodes in an
    associative array using the command as a string index.

2.8   Creating New Types with typedef

You can create new types using the typedef statement. For example, you may have
an ALU that can be configured at compile time to use 8, 16, 24, or 32-bit operands.
In Verilog you would define a macro for the operand width and another for the type
as shown in Sample 2.37.

Sample 2.37 User-defined type-macro in Verilog

```
// Old Verilog style
`define OPSIZE 8
`define OPREG reg [`OPSIZE-1:0]

`OPREG op_a, op_b;
```

You are not really creating a new type; you are just performing text substitution. In SystemVerilog you create a new type as shown in Sample 2.38. This book uses the convention that user-defined types use the suffix "_t" except for the basic `uint`.

Sample 2.38 User-defined type in SystemVerilog

```
// New SystemVerilog style
parameter OPSIZE = 8;
typedef logic [OPSIZE-1:0] opreg_t;

opreg_t op_a, op_b;
```

In general, SystemVerilog lets you copy between these basic types with no warning, either extending or truncating values if there is a width mismatch.

Note that `parameter` and `typedef` statements can be put in a package so they can be shared across the design and testbench, as shown in Section 2.10.

 One of the most useful types you can create is an unsigned, 2-state, 32-bit integer as shown in Sample 2.39. Most values in a testbench are positive integers such as field length or number of transactions received, and so having a signed integer can cause problems. Put the definition of `uint` in a package of common definitions so it can be used anywhere.

Sample 2.39 Definition of uint

```
typedef bit [31:0] uint;     // 32-bit unsigned 2-state
typedef int unsigned uint;   // Equivalent definition
```

The syntax for defining a new array type is not obvious. You need to put the array subscripts on the new name. Sample 2.40 creates a new type, `fixed_array5_t`, a fixed array with 5 elements. It then declares an array of this type and initializes it.

Sample 2.40   User-defined array type

```
typedef int fixed_array5_t[5];
fixed_array5_t f5;      // Equivalent to "int f5[5]"

initial begin
  foreach (f5[i])
    f5[i] = i;
end
```

A good use for a user defined type is an associative array, which must be declared with an index that is a simple type. You could change Sample 2.24 to use 64 bit values by changing the first line as shown in Sample 2.41.

Sample 2.41   User-defined associative array index

```
typedef bit[63:0] bit64_t;
bit64_t assoc[bit64_t], idx = 1;
```

2.9   Creating User-Defined Structures

One of the biggest limitations of Verilog is the lack of data structures. In SystemVerilog you can create a structure using the `struct` statement, similar to what is available in C. However, a `struct` has just a subset of the functionality of a class, so use a class instead for your testbenches, as shown in Chapter 5. Just as a Verilog module combines both data (signals) and code (always/initial blocks plus routines), a class combines data and routines to make an entity that can be easily debugged and reused. A `struct` just groups data fields together. Without the code that manipulates the data, you are only creating half of the solution.

Since a `struct` is just a collection of data, it can be synthesized. If you want to model a complex data type, such as a pixel, in your design code, put it in a `struct`. This can also be passed through module ports. Eventually, when you want to generate constrained random data, look to classes.

2.9.1   Creating a struct and a New Type

You can combine several variables into a structure. Sample 2.42 creates a structure called `pixel` that has three unsigned bytes for red, green, and blue.

Sample 2.42   Creating a single pixel type

```
struct {bit [7:0] r, g, b;} pixel;
```

The problem with the preceding declaration is that it creates a single pixel of this type. To be able to share pixels using ports and routines, you should create a new type instead, as shown in Sample 2.43.

Sample 2.43 The pixel `struct`

```
typedef struct {bit [7:0] r, g, b;} pixel_s;
pixel_s my_pixel;
```

Use the suffix "_s" when declaring a `struct`. This makes it easier to spot user-defined types, simplifying the process of sharing and reusing code.

2.9.2 Initializing a Structure

You can assign multiple values to a struct just like an array, either in the declaration or in a procedural assignment. Just surround the values with an apostrophe and braces, as shown in Sample 2.44.

Sample 2.44 Initializing a struct

```
initial begin
  typedef struct {int a;
                  byte b;
                  shortint c;
                  int d;} my_struct_s;
  my_struct_s st = '{32'haaaa_aaaa,
                     8'hbb,
                     16'hcccc,
                     32'hdddd_dddd};

  $display("str = %x %x %x %x ", st.a, st.b, st.c, st.d);
end
```

2.9.3 Making a Union of Several Types

In hardware, the interpretation of a set of bits in a register may depend on the value of other bits. For example, a processor instruction may have many layouts based on the opcode. Immediate-mode operands might store a literal value in the operand field. This value may be decoded differently for integer instructions than for floating point instructions. Sample 2.45 stores both the unsigned bit vector b and the integer i in the same location.

Sample 2.45 Using typedef to create a union

```
typedef union { bit [31:0] b; int i; } num_u;
num_u un;
un.i = -1; // set value using signed integer
```

Use the suffix "_u" when declaring a union.

Unions are useful when you frequently need to read and write a register in several different formats. However, don't go overboard, especially just to save memory. Unions may help squeeze a few bytes out of a structure, but at the expense of having to create and maintain a more complicated data structure. Instead, make a class with a discriminant variable, as shown in Section 8.4.4. This "kind" variable indicates which type of transaction you have, and thus which fields to read, write, and randomize. If you just need an array of values, plus all the bits, use a packed array as described Section in 2.2.6

2.9.4 Packed Structures

SystemVerilog allows you more control in how bits are laid out in memory by using packed structures. A packed structure is stored as a contiguous set of bits with no unused space. The `struct` for a pixel in Sample 2.43 has three values, so it is stored in three longwords, even though it only needs three bytes. You can specify that it should be packed into the smallest possible space with the `packed` keyword, as shown in Sample 2.46.

Sample 2.46 Packed structure

```
typedef struct packed {bit [7:0] r, g, b;} pixel_p_s;
pixel_p_s my_pixel;
```

Packed structures are used when the underlying bits represent a numerical value or when you are trying to reduce memory usage. For example, you could pack together several bit-fields to make a single register. Or you might pack together the opcode and operand fields to make a value that contains an entire processor instruction.

2.9.5 Choosing Between Packed and Unpacked Structures

When you are trying to choose between packed and unpacked structures, consider how the structure is most commonly used and the alignment of the elements. If you plan on making aggregate operations on the structure, such as copying the entire structure, a packed structure is more efficient. However, if your code accesses the individual members more than the entire structure, use an unpacked structure. The difference in performance is greater if the elements are not aligned on byte boundaries, have sizes that don't match the typical byte, or have word instructions used by processors. Reading and writing elements with odd sizes in a packed structure requires expensive shift and mask operations.

2.10 Packages

At the start of a project, you need to create new types and parameters. For example, if your processor communicates with your company's ABC bus, your testbench needs to define ABC data types, and parameters to specify the bus width and timing. Another project may want to use these types, plus those for the XYZ bus.

You could create separate files for each bus and use the `'include` statement to bring in the files during compilation. But then every name associated with each bus must be unique, even those that are internal variables, never intended to be visible. How can you organize these types to avoid name conflicts?

The SystemVerilog package allows you to share declarations among modules, packages, plus programs and interface, which are described in Chapter 4. Sample 2.47 shows the package for the ABC bus.

Sample 2.47 Package for ABC bus

```
package ABC;
  parameter int abc_data_width = 32;
  typedef logic [abc_data_width-1:0] abc_data_t;
  parameter time timeout = 100ns;
  string message = "ABC done";
endpackage // ABC
```

You import symbols from a package with the `import` statement. The compiler only looks in imported packages when a symbol is not defined in the usual search path. In Sample 2.48, the first `import` statement makes the symbols `abc_data_width`, `abc_data_t`, and `timeout` visible if there is no local variable with the same name. The variable `message` in ABC is hidden by the one in the module.

Sample 2.48 Importing packages

```
module test;
  import ABC::*;                          // Search ABC for symbols

  abc_data_t data;                        // From package ABC
  string message = "Test timed out"; // Hides message in ABC

  initial begin
    #(timeout);                           // From package ABC
    $display("Timeout - %s", message);
    $finish;
  end
endmodule
```

If you really want to see the `message` variable in ABC, use `ABC::message`.

You can import specific symbols from a package with the scope operator, `::`. Sample 2.49 imports all the symbols from ABC, plus just the `timeout` variable from XYZ.

Sample 2.49 Importing selected symbols from a package

```
module test;
  import ABC::*;                     // Search ABC for symbols
  import XYZ::timeout;               // Just import timeout
  string message = "Test timed out"; // Hides message in ABC

  initial begin
    #(timeout);                       // From package XYZ
    $display("Timeout - %s"", message);
    $finish;
  end
endmodule
```

Packages can only see symbols defined inside themselves, or packages that they import. You can not have hierarchical references to symbols such as signals, routines, or modules from outside the package. Think of a package as being completely standalone, able to plug in where needed, with no outside dependencies.

A package can contain routines, plus classes, as shown in Section 5.4.

2.11   Type Conversion

SystemVerilog has several rules to ensure that expressions are evaluated with little or no loss of accuracy. For example, if you add two 8-bit values, the addition is done with 9-bit precision to avoid overflow. Multiply two 8-bit values, and SystemVerilog calculates a 16-bit result.

The proliferation of data types in SystemVerilog means that you may need to convert between them. If the layout of the bits between the source and destination variables are the same, such as an integer and enumerated type, cast between the two values. If the bit layouts differ, such as an array of bytes and words, use the streaming operators to rearrange the bits as described in Section 2.12.

2.11.1   The Static Cast

The static cast operation converts between two types with no checking of values. You specify the destination type, an apostrophe, and the expression to be converted as shown in Sample 2.50. Note that Verilog has always implicitly converted between types such as integer and real, and also between different width vectors.

Sample 2.50   Converting between int and real with static cast

```
int i;
real r;
i = int '(10.0 - 0.1);  // cast is optional
r = real'(42);          // cast is optional
```

2.11.2   The Dynamic Cast

The dynamic cast, $cast, allows you to check for out-of-bounds values. See Section 2.13.3 for an explanation and example with enumerated types.

Use a static cast when you want SystemVerilog to use a type with more precision, like when using the sum method for a single bit array. Use the dynamic cast when converting from a type with a larger number of values than the destination, such as int to an enumerated variable.

2.12   Streaming Operators

When used on the right side of an assignment, the streaming operators << and >> take an expression, structure, or array, and packs it into a stream of bits. The >> operator streams data from left to right while << streams from right to left, as shown in Sample 2.51. You can also give a slice size, used to break up the source before being streamed. You can not assign the bit stream result directly to an unpacked array. Instead, use the streaming operators on the left side of an assignment to unpack the bit stream into an unpacked array.

Sample 2.51   Basic streaming operator

```
initial begin
  int h;
  bit [7:0] b, g[4], j[4] = '{8'ha, 8'hb, 8'hc, 8'hd};
  bit [7:0] q, r, s, t;

  h = { >> {j}};            // 0a0b0c0d pack array into int
  h = { << {j}};            // b030d050 reverse bits
  h = { << byte {j}};       // 0d0c0b0a reverse bytes
  {>>{g}} = { << byte {j}}; // 0d,0c,0b,0a unpack into array
  b = { << {8'b0011_0101}}; // 1010_1100 reverse bits
  b = { << 4 {8'b0011_0101}}; // 0101_0011 reverse nibble
  {>> {q, r, s, t}} = j;    // Scatter j into bytes
  h = {>>{t, s, r, q}};     // Gather bytes into h
end
```

You could do the same operations with many concatenation operators, { }, but the streaming operators are more compact and easier to read.

If you need to pack or unpack arrays, use the streaming operator to convert between arrays of different element sizes. For instance, you can convert an array of bytes to an array of words. You can use fixed size arrays, dynamic arrays, and queues. Sample 2.52 converts between queues, but would also work with dynamic arrays. Array elements are automatically allocated as needed.

Sample 2.52 Converting between queues with streaming operator

```
initial begin
  bit [15:0] wq[$] = {16'h1234, 16'h5678};
  bit [7:0]  bq[$];

  // Convert word array to byte
  bq = { >> {wq}};  // 12 34 56 78

  // Convert byte array to words
  bq = {8'h98, 8'h76, 8'h54, 8'h32};
  wq = { >> {bq}};  // 9876 5432
end
```

 A common mistake when streaming between arrays is mismatched array subscripts. The word subscript [256] in an array declaration is equivalent to [0:255], not [255:0]. Since many arrays are declared with the word subscripts [high:low], streaming them to an array with the subscript [size] would result in the elements ending up in reverse order. Likewise, streaming an unpacked array declared as bit [7:0] src[255:0] to the packed array declared as bit [7:0] [255:0] dst will scramble the order of values. The correct declaration for a packed array of bytes is bit [255:0] [7:0] dst.

You can also use the streaming operator to pack and unpack structures, such as an ATM cell, into an array of bytes. In Sample 2.53 a structure is streamed into a dynamic array of bytes, then the byte array is streamed back into the structure.

Sample 2.53 Converting between a structure and an array with streaming operators

```
initial begin
  typedef struct {int a;
                  byte b;
                  shortint c;
                  int d;} my_struct_s;
  my_struct_s st = '{32'haaaa_aaaa,
                     8'hbb,
                     16'hcccc,
                     32'hdddd_dddd};
  byte b[];

  // Covert from struct to byte array
  b = { >> {st}};        // {aa aa aa aa bb cc cc dd dd dd dd}

  // Convert from byte array to a struct
  b = '{8'h11, 8'h22, 8'h33, 8'h44, 8'h55, 8'h66, 8'h77,
        8'h88, 8'h99, 8'haa, 8'hbb};
  st = { >> {b}};        // st = 11223344, 55, 6677, 8899aabb
end
```

2.13 Enumerated Types

An enumerated type allows you to create a set of related but unique constants such as states in a state machine or opcodes. In classic Verilog, you had to use text macros. Their global scope is too broad, and their value might not be visible in the debugger. An enumeration creates a strongly typed variable that is limited to a set of specified names. For example, the names ADD, MOVE, or ROTW make your code easier to write and maintain than if you had used literals such as 8'h01 or macros. A weaker alternative for defining constants is a parameter. These are fine for individual values, but an enumerated type automatically gives a unique value to every name in the list.

The simplest enumerated type declaration contains a list of constant names and one or more variables as shown in Sample 2.54. This creates an anonymous enumerated type, but it cannot be used for any other variables than the ones in this declaration.

Sample 2.54 A simple enumerated type, not recommended

```
enum {RED, BLUE, GREEN} color;
```

It is recommended to create a named enumerated type so you can declare multiple variables of the same type, especially if these are used as routine arguments or module ports. You first create the enumerated type, and then the variables of this type, as shown in Sample 2.55. You can get the string representation of an enumerated variable with the built-in function name().

Sample 2.55 Enumerated types, recommended style

```
// Create data type for values 0, 1, 2
typedef enum {INIT, DECODE, IDLE} fsmstate_e;
fsmstate_e pstate, nstate;    // declare typed variables

initial begin
  case (pstate)
    IDLE:    nstate = INIT;      // data assignment
    INIT:    nstate = DECODE;
    default: nstate = IDLE;
  endcase
  $display("Next state is %s",
           nstate.name());       // Display symbolic state name
end
```

Use the suffix "_e" when declaring an enumerated type name.

2.13.1   Defining Enumerated Values

The actual values default to int starting at 0 and then increase. You can choose your own enumerated values. The code in Sample 2.56 uses the default value of 0 for INIT, then 2 for DECODE, and 3 for IDLE.

Sample 2.56 Specifying enumerated values

```
typedef enum {INIT, DECODE=2, IDLE} fsmtype_e;
```

Enumerated constants, such as INIT in Sample 2.56, follow the same scoping rules as variables. Consequently, if you use the same name in several enumerated types (such as INIT in different state machines), they have to be declared in different scopes such as modules, program blocks, packages, routines, or classes.

An enumerated type is stored as int unless you specify otherwise. Be careful when assigning values to enumerated constants, as the default value of an int is 0. In Sample 2.57, position is initialized to 0, which is not a legal ordinal_e variable. This behavior is <u>not</u> a tool bug – it is how the language is specified. So always specify an enumerated constant with the value of 0, as shown in Sample 2.58, just to catch the testbench error.

Sample 2.57 Incorrectly specifying enumerated values

```
typedef enum {FIRST=1, SECOND, THIRD} ordinal_e;
ordinal_e position;
```

Sample 2.58 Correctly specifying enumerated values

```
typedef enum {BAD_0=0, FIRST=1, SECOND, THIRD} ordinal_e;
ordinal_e position;
```

2.13.2 Routines for Enumerated Types

SystemVerilog provides several functions for stepping through enumerated types.

- first () returns the first member of the enumeration.
- last() returns the last member of the enumeration.
- next() returns the next element of the enumeration.
- next (N) returns the Nth next element.
- prev () returns the previous element of the enumeration.
- prev(N) returns the Nth previous element.

The functions next and prev wrap around when they reach the beginning or end of the enumeration.

Note that there is no clean way to write a for loop that steps through all members of an enumerated type if you use an enumerated loop variable. You get the starting member with first function and the next member with next. A for loop ends when the loop variable is outside the defined bounds, but the next function always returns a value inside the enumeration. If you use the test current != current.last(), the loop ends before using the last value. If you use current<=current.last(), you get an infinite loop, as next never gives you a value that is greater than the final value. This is similar to trying to make a for loop that steps through the values 0..3 with an index declared as bit [1:0]. The loop never exits! You can get around this limitation by either using an integer variable in the loop, or incrementing the enumerated variable, but both of these solutions can give illegal values if your enumerated values are not contigious, such as 1, 2, 3, 5, 8.

You can use a do...while loop to step through all the values, checking when the value wraps around, as shown in Sample 2.59.

Sample 2.59 Stepping through all enumerated members

```
typedef enum {RED, BLUE, GREEN} color_e;
color_e color;
color = color.first;
do
  begin
  $display("Color = %0d/%s", color, color.name());
  color = color.next;
  end
while (color != color.first);  // Done at wrap-around
```

2.13.3   Converting to and from Enumerated Types

The default type for an enumerated type is `int` (2-state). You can take the value of an enumerated variable and assign it to a non-enumerated variable such as an `int` with a simple assignment. SystemVerilog does not, however, let you store an integer value in an `enum` without explicitly changing the type. Instead, it requires you to explicitly cast the value to make you realize that you could be writing an out-of-bounds value.

Sample 2.60 Assignments between integers and enumerated types

```
typedef enum {RED, BLUE, GREEN} color_e;
color_e color, c2;
int c;

initial begin
  color = BLUE;            // Set to known good value
  c = color;              // Convert from enum to int (1)
  c++;                    // Increment int (2)
  if (!$cast(color, c))   // Cast int back to enum
    $display("Cast failed for c=%0d", c);
  $display("Color is %0d / %s", color, color.name());
  c++;                    // 3 is out-of-bounds for enum
  c2 = COLOR_E'(c);       // No type checking
  $display("c2 is %0d / '%s'", c2, c2.name());
end
```

When called as a function as shown in Sample 2.60, `$cast()` tried to assign the right value to the left variable. If the assignment succeeds, `$cast()` returns 1. If the assignment fails because of an out-of-bounds value, no assignment is made and the function returns 0. If you use `$cast()` as a task and the operation fails, SystemVerilog prints an error.

You can also cast the value using the `type'(val)` as shown in the example, but this does not do any type checking, so the result may be out-of-bounds. For example, after the static cast in Sample 2.60, c2 has an out-of-bounds value. You should avoid this style of casting with enumerated types.

2.14 Constants

There are several types of constants in SystemVerilog. The classic Verilog way to create a constant is with a text macro. On the plus side, macros have global scope and can be used for bit field definitions and type definitions. On the negative side, macros are global, so that they can cause conflicts if you just need a local constant. Lastly, a macro requires the ` character so that it is recognized and expanded by the compiler.

A Verilog parameter was loosely typed and was limited in scope to a single module. Verilog-2001 added typed parameters, but their limited scope kept parameters from being widely used. In SystemVerilog, parameters can be declared in a package so they can be used across multiple modules. This approach can replace most Verilog macros that were just being used as constants.

SystemVerilog also supports the const modifier that allows you to make a variable that can be initialized in the declaration but not written by procedural code.

Sample 2.61 Declaring a const variable

```
initial begin
  const byte colon = ":";
  ...
end
```

In Sample 2.61, the value of colon is initialized at run time, when the initial block is entered. In the next chapter, Sample 3.11 shows a const routine argument.

2.15 Strings

If you have ever tried to use a Verilog reg variable to hold a string of characters, your suffering is over. The SystemVerilog string type holds variable-length strings. An individual character is of type byte. The elements of a string of length N are numbered 0 to N-1. Note that, unlike C, there is no null character at the end of a string, and any attempt to use the character "\0" is ignored. Memory for strings is dynamically allocated, so you do not have to worry about running out of space to store the string.

Sample 2.62 shows various string operations. The function getc(N) returns the byte at location N, while toupper returns an upper-case copy of the string and tolower returns a lowercase copy. The curly braces {} are used for concatenation. The task putc(M, C) writes a byte C into a string at location M, that must be between 0 and the length as given by `len`. The `substr(start,end)` function extracts characters from location `start` to `end`.

Sample 2.62 String methods

```
string s;

initial begin
  s = "IEEE ";
  $display(s.getc(0));      // Display: 73, ASCII value of 'I'
  $display(s.tolower());    // Display: 'ieee '

  s.putc(s.len()-1, "-");   // change ' '-> '-'
  s = {s, "1800"};          // "IEEE-1800"

  $display(s.substr(2, 5)); // Display: EE-1

  // Create temporary string, note format
  my_log($sformatf("%s %5d", s, 42));
end

function void my_log(string message);
  // Print a message to a log
  $display("@%0t: %s", $time, message);
endfunction
```

Note how useful dynamic strings can be. In other languages such as C, you have to keep making temporary strings to hold the result from a function. In Sample 2.62, the `$sformatf` function is used instead of `$sformat`, from Verilog-2001. This new function returns a formatted temporary string that, as shown above, can be passed directly to another routine. This saves you from having to declare a temporary string and passing it between the formatting statement and the routine call. The undocumented function `$psprintf` has the same functionality as `$sformatf`, but is not in the LRM, even though most vendors support this non-standard system function.

 There are two ways to compare strings, but they behave differently. The equality operator, `s1==s2`, returns 1 if the strings are identical, and 0 if they are not. The string comparison function, `s1.compare(s2)`, returns 1 if s1 is greater than s2, 0 if they are equal, and −1 if s1 is less than s2. While this matches the ANSI C `strcmp()` behavior, it may not be what you expect.

2.16   Expression Width

A prime source for unexpected behavior in Verilog has been the width of expressions. Sample 2.63 adds 1+1 using four different styles. Addition A uses two 1-bit variables, so with this precision 1+1=0. Addition B uses 8-bit precision because there is an 8-bit variable on the left side of the assignment. In this case, 1+1=2. Addition c uses a dummy constant to force SystemVerilog to use 2-bit precision. Lastly, in addition D, the first value is cast to be a 2-bit value with the cast operator, so 1+1=2.

Sample 2.63   Expression width depends on context

```
bit [7:0] b8;
bit one = 1'b1;                 // Single bit
$displayb(one + one);           // A: 1+1 = 0

b8 = one + one;                 // B: 1+1 = 2
$displayb(b8);

$displayb(one + one + 2'b0);    // C: 1+1 = 2 with constant

$displayb(2'(one) + one);       // D: 1+1 = 2 with cast
```

There are several tricks you can use to avoid this problem. First, avoid situations where the overflow is lost, as in addition A. Use a temporary, such as b8, with the desired width. Or, you can add another value to force the minimum precision, such as 2'b0. Lastly, in SystemVerilog, you can cast one of the variables to the desired precision.

2.17   Conclusion

SystemVerilog provides many new data types and structures so that you can create high-level testbenches without having to worry about the bit-level representation. Queues work well for creating scoreboards for which you constantly need to add and remove data. Dynamic arrays allow you to choose the array size at run time for maximum testbench flexibility. Associative arrays are used for sparse memories and some scoreboards with a single index. Enumerated types make your code easier to read and write by creating groups of named constants.

Don't go off and create a procedural testbench with just these constructs. Explore the OOP capabilities of SystemVerilog in Chapter 5 to learn how to design code at an even higher level of abstraction, thus creating robust and reusable code.

2.18   Exercises

1. Given the following code sample:

```
byte my_byte;
integer my_integer;
int my_int;
bit [15:0] my_bit;
shortint my_short_int1;
shortint my_short_int2;

my_integer = 32'b000_1111_xxxx_zzzz;
my_int = my_integer;
my_bit = 16'h8000;
my_short_int1 = my_bit;
my_short_int2 = my_short_int1-1;
```

   a. What is the range of values `my_byte` can take?
   b. What is the value of `my_int` in hex?
   c. What is the value of `my_bit` in decimal?
   d. What is the value of `my_short_int1` in decimal?
   e. What is the value of `my_short_int2` in decimal?

2. Given the following code sample:

```
bit [7:0] my_mem[3];
logic [3:0] my_logicmem[4];
logic [3:0] my_logic;

my_mem = '{default:8'hA5};
my_logicmem = '{0,1,2,3};
my_logic = 4'hF;
```

   Evaluate the following statements in the given order and give the result for each assignment

   a. `my_mem[2] = my_logicmem[4];`
   b. `my_logic = my_logicmem[4];`
   c. `my_logicmem[3] = my_mem[3];`
   d. `my_mem[3] = my_logic;`
   e. `my_logic = my_logicmem[1];`
   f. `my_logic = my_mem[1];`
   g. `my_logic = my_logicmem[my_logicmem[41];`

3. Write the SystemVerilog code to:

  a. Declare a 2-state array, `my_array`, that holds four 12-bit values

  b. Initialize `my_array` so that:

```
* my_array[0] = 12'h012
* my_array[1] = 12'h345
* my_array[2] = 12'h678
* my_array[3] = 12'h9AB
```

  c. Traverse `my_array` and print out bits [5:4] of each 12-bit element

    * With a `for` loop
    * With a `foreach` loop

4. Declare a 5 by 31 multi-dimensional unpacked array, `my_array1`. Each element of the unpacked array holds a 4-state value.

  a. Which of the following assignment statements are legal and not out of bounds?

```
* my_array1[4][30] = 1'b1;
* my_array1[29][4] = 1'b1;
* my_array1[4]     = 32'b1;
```

  b. Draw `my_array1` after the legal assignments complete.

5. Declare a 5 by 31 multi-dimensional packed array, `my_array2`. Each element of the packed array holds a 2-state value.

  a. Which of the following assignment statements are legal and not out of bounds?

```
* my_array2[4][30] = 1'b1;
* my_array2[29][4] = 1'b1;
* my_array2[3]     = 32'b1;
```

  b. Draw `my_array2` after the assignment statements complete.

6. Given the following code, determine what will be displayed.

```
module test;
  string street[$];

  initial begin
    street = {"Tejon", "Bijou", "Boulder"};
    $display("Street[0] = %s", street[0]);
    street.insert(2, "Platte");
    $display("Street[2] = %s", street[2]);
    street.push_front("St. Vrain");
    $display("Street[2] = %s", street[2]);
    $display("pop_back = %s", street.pop_back);
    $display("street.size = %d", street.size);
  end
endmodule // test
```

7. Write code for the following problems.

   a. Create memory using an associative array for a processor with a word width of 24 bits and an address space of 2^{20} words. Assume the PC starts at address 0 at reset. Program space starts at 0×400. The ISR is at the maximum address.
   b. Fill the memory with the following instructions:

   ```
   * 24'hA50400; // Jump to location 0x400 for the main code
   * 24'h123456; // Instruction 1 located at location 0x400
   * 24'h789ABC; // Instruction 2 located at location 0x401
   * 24'h0F1E2D; // ISR = Return from interrupt
   ```

   c. Print out the elements and the number of elements in the array.

8. Create the SystemVerilog code for the following requirements

   a. Create a 3-byte queue and initialize it with 2, −1, and 127
   b. Print out the sum of the queue in the decimal radix
   c. Print out the min and max values in the queue
   d. Sort all values in the queue and print out the resulting queue
   e. Print out the index of any negative values in the queue
   f. Print out the positive values in the queue
   g. Reverse sort all values in the queue and print out the resulting queue

9. Define a user defined 7-bit type and encapsulate the fields of the following packet in a structure using your new type. Lastly, assign the header to 7'h5A.

10. Create the SystemVerilog code for the following requirements

    a. Create a user-defined type, nibble, of 4 bits
    b. Create a real variable, r, and initialize it to 4.33
    c. Create a short int variable, i_pack
    d. Create an unpacked array, k, containing 4 elements of your user defined type nibble and initialize it to 4'h0, 4'hF, 4'hE, and 4'hD
    e. Print out k
    f. Stream k into i_pack right to left on a bit basis and print it out
    g. Stream k into i_pack right to left on a nibble basis and print it out
    h. Type convert real r into a nibble, assign it to k[0], and print out k

11. An ALU has the opcodes shown in Table 2.1.

Table 2.1 ALU Opcodes

Opcode	Encoding
Add: A + B	2'b00
Sub: A − B	2'b01
Bit-wise invert: A	2'b10
Reduction Or: B	2'b11

Write a testbench that performs the following tasks.

a. Create an enumerated type of the opcodes: `opcode_e`
b. Create a variable, `opcode`, of type `opcode_e`
c. Loop through all the values of variable `opcode` every 10ns
d. Instantiate an ALU with one 2-bit input opcode

Chapter 3
Procedural Statements and Routines

As you verify your design, you need to write a great deal of code, most of which is in tasks and functions. SystemVerilog introduces many incremental improvements to make this easier by making the language look more like C, especially around argument passing. If you have a background in software engineering, these additions should be very familiar.

3.1 Procedural Statements

SystemVerilog adopts many operators and statements from C and C++. You can declare a loop variable inside a `for` loop that then restricts the scope of the loop variable and can prevent some coding bugs. The new auto-increment `++` and auto-decrement `--` operators are available in both pre- and post-forms. The compound assignments, `+=`, `-=`, `^=`, and many more make your code tighter. If you have a label on a `begin` or `fork` statement, you can put the same label on the matching `end` or `join` statement. This makes it easier to match the start and finish of a block. You can also put a label on other SystemVerilog end statements such as `endmodule`, `endtask`, `endfunction`, and others that you will learn in this book. Sample 3.1 demonstrates some of the new constructs.

C. Spear and G. Tumbush, *SystemVerilog for Verification: A Guide to Learning the Testbench Language Features*, DOI 10.1007/978-1-4614-0715-7_3, © Springer Science+Business Media, LLC 2012

Sample 3.1 New procedural statements and operators

```
initial
  begin : example
  integer array[10], sum, j;

  // Declare i in for statement
  for (int i=0; i<10; i++)          // Increment i
    array[i] = i;

  // Add up values in the array
  sum = array[9];
  j=8;
  do                                // do...while loop
    sum += array[j];                // Compound assignment
  while (j--);                      // Test if j=0
  $display("Sum=%4d", sum);         // %4d - specify width
end : example                       // End label
```

Two new statements help with loops. First, if you are in a loop, but want to skip over the rest of the statements and do the next iteration, use continue. If you want to leave the loop immediately, use break.

The compound assignment in Sample 3.1 is equivalent to sum = sum + array[j]; The loop in Sample 3.2 reads commands from a file using the file I/O system tasks that are part of Verilog-2001. If the command is just a blank line, the code does a continue, skipping any further processing of the command. If the command is "done," the code does a break to terminate the loop.

Sample 3.2 Using break and continue while reading a file

```
initial begin
  bit [127:0] cmd;
  int file, c;

  file = $fopen("commands.txt", "r");
  while (!$feof(file)) begin
    c = $fscanf(file, "%s", cmd);
    case (cmd)
      "":      continue;    // Blank line - skip to loop end
      "done": break;        // Done - leave loop
      ...                   // Process other commands here
    endcase // case(cmd)
    end
  $fclose(file);
end
```

SystemVerilog expands the `case` statement so that you no longer have to give every possible value, but can instead give a range values as shown in Sample 3.3. This is a version of the `inside` operator shown more in more detail in Section 6.4.5.

Sample 3.3 Case-inside statement with ranges

```
case (graduation_year) inside // <<< Note "inside" keyword
   [1950:1959]: $display("Do you like bobby sox?");
   [1960:1969]: $display("Did you go to Woodstock?");
   [1970:1979]: $display("Did you dance to disco?");
endcase
```

3.2   Tasks, Functions, and Void Functions

Verilog makes a very clear differentiation between tasks and functions. The most important difference is that a task can consume time whereas a function cannot. A function cannot have a delay, `#100`, a blocking statement such as `@(posedge clock)` or `wait(ready)`, or call a task. Additionally, a Verilog function must return a value and the value must be used, as in an assignment statement.

SystemVerilog relaxes this rule a little in that a function can call a task, but only in a thread spawned with the `fork...join_none` statement, which is described in Section 7.1.

If you have a SystemVerilog task that does not consume time, you should make it a `void function`, which is a function that does not return a value. Now it can be called from any task or function. For maximum flexibility, any debug routine should be a void function rather than a task so that it can be called from any task or function. Sample 3.4 prints values from a state machine.

Sample 3.4 Void function for debug

```
function void print_state();
   $display("@%0t: state = %s", $time, cur_state.name());
endfunction
```

In SystemVerilog, if you want to call a function and ignore its return value, cast the result to `void`, as shown in Sample 3.5. Some simulators, such as VCS, allow you to ignore the return value without using the `void` syntax. The LRM says this should be a warning.

Sample 3.5 Ignoring a function's return value

```
void'($fscanf(file, "%d", i));
```

3.3   Task and Function Overview

SystemVerilog makes several small improvements to tasks and functions to make them look more like C or C++ routines. In general, a routine definition or call with no arguments does not need the empty parentheses (). This book includes them for added clarity.

3.3.1   Routine `Begin`...`End` Removed

The first improvement you may notice in SystemVerilog routines is that `begin`...`end` blocks are optional, while Verilog-1995 required them on all but single-line routines. The `task` / `endtask` and `function` / `endfunction` keywords are enough to define the routine boundaries, as shown in Sample 3.6.

Sample 3.6   Simple task without `begin`...`end`

```
task multiple_lines();
  $display("First line");
  $display("Second line");
endtask : multiple_lines
```

3.4   Routine Arguments

Many of the SystemVerilog improvements for routines make it easier to declare arguments and expand the ways you can pass values to and from a routine.

3.4.1   C-style Routine Arguments

SystemVerilog and Verilog-2001 allow you to declare task and function arguments more cleanly and with less repetition. The following Verilog task requires you to declare some arguments twice: once for the direction, and once for the type, as shown in Sample 3.7.

Sample 3.7   Verilog-1995 routine arguments

```
task mytask1;
  output [31:0] x;
  reg    [31:0] x;
  input         y;
  ...
endtask
```

With SystemVerilog, you can use the less verbose C-style, shown in Sample 3.8. Note that you should use the universal input type of `logic`.

Sample 3.8 C-style routine arguments

```
task mytask2 (output logic [31:0] x,
                  input  logic y);
...
endtask
```

3.4.2   Argument Direction

You can take even more shortcuts with declaring routine arguments. The direction and type default to "input logic" and are sticky, so you don't have to repeat these for similar arguments. Sample 3.9 shows a routine header written using the Verilog-1995 style and SystemVerilog data types.

Sample 3.9 Verbose Verilog-style routine arguments

```
task t3;
  input a, b;
  logic a, b;
  output [15:0] u, v;
  bit [15:0] u, v;
  ...
endtask
```

You could rewrite this as shown in Sample 3.10.

Sample 3.10 Routine arguments with sticky types

```
task t3(a, b, output bit [15:0] u, v);  // Lazy declarations
  ...
endtask
```

The arguments a and b are input logic, 1-bit wide. The arguments u and v are 16-bit output bit types. Now that you know this, don't depend on the defaults, as your code will be infested with subtle and hard to find bugs, as explained in Section 3.4.6. Always declare the type and direction for every routine argument.

3.4.3   Advanced Argument Types

Verilog had a simple way to handle arguments: an `input` or `inout` was copied to a local variable at the start of the routine, whereas an `output` or `inout` was copied when the routine exited. No memories could be passed into a Verilog routine, only scalars.

In SystemVerilog, you can specify that an argument is passed by reference, rather than copying its value. This argument type, `ref`, has several benefits over `input`, `output`, and `inout`. First, you can now pass an array into a routine, here one that prints the checksum.

Sample 3.11 Passing arrays using `ref` and `const`

```
function automatic void print_csm11 (const ref bit [31:0] a[]);
  bit [31:0] checksum = 0;
  for (int i=0; i<a.size(); i++)
    checksum ^= a[i];
  $display("The array checksum is %h", checksum);
endfunction
```

The `^=` compound assignment in Sample 3.11 is a shorthand way of writing the statement: `checksum = checksum ^ a[i];`

SystemVerilog allows you to pass array arguments without the `ref` direction, but the array is copied onto the stack, an expensive operation for all but the smallest arrays.

The SystemVerilog LRM states that `ref` arguments can only be used in routines with automatic storage. If you specify the `automatic` attribute for programs and module, all the routines inside are automatic. See Section 3.6 for more details on storage.

Sample 3.11 also shows the `const` modifier. As a result, the array `a` points to the array in the routine call, but the contents of the array cannot be modified. If you try to change the contents, the compiler prints an error.

Always use `ref` when passing arrays to a routine for best performance. If you don't want the routine to change the array values, use the `const ref` type, which causes the compiler to check that your routine does not modify the array.

The second benefit of `ref` arguments is that a task can modify a variable and is instantly seen by the calling function. This is useful when you have several threads executing concurrently and want a simple way to pass information. See Chapter 7 for more details on using `fork-join`.

In Sample 3.12, the `thread2` block in the initial block can access the data from memory as soon as `enable` is asserted, even though the `bus_read` task does not return until the bus transaction completes, which could be several cycles later.

Sample 3.12   Using `ref` across threads

```
task automatic bus_read(input logic [31:0] addr,
                        ref   logic [31:0] data);

  // Request bus and drive address
  bus_request <= 1'b1;
  @(posedge bus_grant) bus_addr <= addr;

  // Wait for data from memory
  @(posedge bus_enable) data <= bus_data;

  // Release bus and wait for grant
  bus_request <= 1'b0;
  @(negedge bus_grant);
endtask

logic [31:0] addr, data;

initial
  fork
    bus_read(addr, data);
    begin : thread2
      @data;  // Trigger on data change
      $display("Read %h from bus", data);
    end
  join
```

The `data` argument is passed as `ref`, and as a result, the `@data` statement triggers as soon as `data` changes in the task. If you had declared `data` as `output`, the `@data` statement would not trigger until the end of the bus transaction.

3.4.4   *Default Value for an Argument*

As your testbench grows in sophistication, you may want to add additional controls to your code but not break existing code. For the function in Sample 3.11, you might want to print a checksum of just the middle values of the array. However, you don't want to go back and rewrite every call to add extra arguments. In SystemVerilog you can specify a default value that is used if you leave out an argument in the call. Sample 3.13 adds `low` and `high` arguments to the `print_csm` function so you can print a checksum of a range of values.

Sample 3.13 Function with default argument values

```
function automatic void print_csm (const ref bit [31:0] a[],
                                    input bit [31:0] low = 0,
                                    input int high = -1);
  bit [31:0] checksum = 0;

  if (high == -1 || high >= a.size())
    high = a.size()-1;

  for (int i=low; i<=high; i++)
    checksum ^= a[i];
  $display("The array checksum is %h", checksum);
endfunction
```

You can call this function in the following ways, as shown in Sample 3.14. Note that the first call is compatible with both versions of the print_csm routine.

Sample 3.14 Using default argument values

```
print_csm(a);         // Checksum a[0:size()-1] - default
print_csm(a, 2, 4);   // Checksum a[2:4]
print_csm(a, 1);      // Start at 1
print_csm(a,, 2);     // Checksum a[0:2]
print_csm();          // Compile error: a has no default
```

Using a default value of −1 (or any out-of-range value) is a good way to see if the call specified a value.

A Verilog for loop always executes the initialization (int i=low), and test (i<=high) before starting the loop. Thus, if you accidently passed a low value that was larger than high or the array size, the for loop would never execute the body.

3.4.5   Passing Arguments by Name

You may have noticed in the SystemVerilog LRM that the arguments to a task or function are sometimes called "ports," just like the connections for a module. If you have a task or function with many arguments, some with default values, and you only want to set a few of those arguments, you can specify a subset by specifying the name of the routine argument with a port-like syntax, as shown in Sample 3.15.

Sample 3.15   Binding arguments by name

```
task many (input int a=1, b=2, c=3, d=4);
  $display("%0d %0d %0d %0d", a, b, c, d);
endtask

initial begin         // a  b  c  d
  many(6, 7, 8, 9);   // 6  7  8  9   Specify all values
  many();             // 1  2  3  4   Use defaults
  many(.c(5));        // 1  2  5  4   Only specify c
  many(, 6, .d(8));   // 1  6  3  8   Mix styles
end
```

3.4.6   Common Coding Errors

The most common coding mistake that you are likely to make with a routine is forgetting that the argument type is sticky with respect to the previous argument, and that the default type for the first argument is a single-bit input. Start with the simple task header in Sample 3.16.

Sample 3.16   Original task header

```
task sticky(int a, b);
```

The two arguments are input integers. As you are writing the task, you realize that you need access to an array, so you add a new array argument, and use the ref type so it does not have to be copied. Your routine header now looks like Sample 3.17.

Sample 3.17   Task header with additional array argument

```
task automatic sticky(ref int array[50],
                      int a, b);  // What direction are these?
```

What argument types are a and b? They take the direction of the previous argument that is a ref. Using ref for a simple variable such as an int is not usually needed, but you would not get even a warning from the compiler, and thus would not realize that you were using the wrong direction.

If any argument to your routine is something other than the default input type, specify the direction for all arguments as shown in Sample 3.18.

Sample 3.18   Task header with additional array argument

```
task automatic sticky(ref    int array[50],
                      input int a, b);  // Be explicit
```

3.5   Returning from a Routine

Verilog had a primitive way to end a routine; after you executed the last statement in
a routine, it returned to the calling code. In addition, a function returned a value by
assigning that value to a variable with the same name as the function.

3.5.1   The Return Statement

SystemVerilog adds the `return` statement to make it easier for you to control the
flow in your routines. The task in Sample 3.19 needs to return early because of error
checking. Otherwise, it would have to put the rest of the task in an `else` clause,
which would cause more indentation and be more difficult to read.

Sample 3.19 Return in a task

```
task automatic load_array(input int len, ref int array[]);
  if (len <= 0) begin
    $display("Bad len");
    return;
  end

  // Code for the rest of the task
  ...
endtask
```

The `return` statement in Sample 3.20 can simplify your functions.

Sample 3.20 Return in a function

```
function bit transmit(input bit [31:0] data);
  // Send transaction
  ...
  return status; // Return status: 0=error
endfunction
```

3.5.2   Returning an Array from a Function

Verilog routines could only return a simple value such as a bit, integer, or vector. If
you wanted to compute and return an array, there was no simple way. In System
Verilog, a function can return an array, using several techniques.

The first way is to define a type for the array, and then use that in the function
declaration. Sample 3.21 uses the array type from Sample 2.40, and creates an function to initialize the array.

Sample 3.21 Returning an array from a function with a typedef

```
typedef int fixed_array5_t[5];
fixed_array5_t f5;

function fixed_array5_t init(input int start);
  foreach (init[i])
    init[i] = i + start;
endfunction

initial begin
  f5 = init(5);
  foreach (f5[i])
    $display("f5[%0d] = %0d", i, f5[i]);
end
```

One problem with the preceding code is that the function init creates an array, which is copied into the array f5. If the array was large, this could be a large performance problem.

The alternative is to pass the routine by reference. The easiest way is to pass the array into the function as a ref argument, as shown in Sample 3.22.

Sample 3.22 Passing an array to a function as a ref argument

```
function automatic void init(ref int f[5], input int start);
  foreach (f[i])
    f[i] = i + start;
endfunction

int fa[5];
initial begin
  init(fa, 5);
  foreach (fa[i])
    $display("fa[%0d] = %0d", i, fa[i]);
end
```

The last way for a function to return an array is to wrap the array inside of a class, and return a handle to an object. Chapter 5 describes classes, objects, and handles.

3.6   Local Data Storage

When Verilog was created in the 1980s, it was tightly tied to describing hardware. As a result, all objects in the language were statically allocated. In particular, routine arguments and local variables were stored in a fixed location, rather than pushing them on a stack like other programming languages. Why try to model dynamic code such as a recursive routine when there is no way to build this in silicon? However, software engineers verifying the designs, who were used to the behavior of stack-based languages such as C, were bitten by these subtle bugs, and were thus limited in their ability to create complex testbenches with libraries of routines.

3.6.1   *Automatic Storage*

In Verilog-1995, if you tried to call a task from multiple places in your testbench, the local variables shared common, static storage, and so the different threads stepped on each other's values. In Verilog-2001 you can specify that tasks, functions, and modules use automatic storage, which causes the simulator to use the stack for local variables.

 In SystemVerilog, routines still use static storage by default, for both modules and program blocks. You should always make program blocks (and their routines) use automatic storage by putting the `automatic` keyword in the program statement. In Chapter 4 you will learn about `program` blocks that hold the testbench code. Section 7.1.6 shows how automatic storage helps when you are creating multiple threads.

Sample 3.23 shows a task to monitor when data are written into memory.

Sample 3.23 Specifying automatic storage in program blocks

```
program automatic test();
  task wait_for_bus(input logic [31:0] addr, expect_data,
                    output logic success);
    while (bus_addr !== addr)
      @(bus_addr);
    success = (bus_data == expect_data);
  endtask

endprogram
```

You can call this task multiple times concurrently, as the `addr` and `expect_data` arguments are stored separately for each call. Without the `automatic` modifier, if you called `wait_for_bus` a second time while the first was still waiting, the second call would overwrite the two arguments.

3.6.2   *Variable Initialization*

 A similar problem occurs when you try to initialize a local variable in a declaration, as it is actually initialized before the start of simulation. The general solution is to avoid initializing a variable in a declaration to anything other than a constant. Use a separate assignment statement to give you better control over when initialization is done.

   The task in Sample 3.24 looks at the bus after five cycles and then creates a local
variable and attempts to initialize it to the current value of the address bus.

Sample 3.24   Static initialization bug

```
program initialization; // Buggy version

  task check_bus();
    repeat (5) @(posedge clock);
    if (bus_cmd === READ) begin
      // When is local_addr initialized?
      logic [7:0] local_addr = addr<<2;   // Bug
      $display("Local Addr = %h", local_addr);
    end
  endtask

endprogram
```

   The bug is that the variable local_addr is statically allocated, so it is actually initialized at the start of simulation, not when the begin...end block is entered. Once
again, the solution is to declare the program as automatic as shown in Sample 3.25.

Sample 3.25   Static initialization fix: use automatic

```
program automatic initialization; // Bug solved
...
endprogram
```

   Additionally, you can avoid this by never initializing a variable in the declaration,
but this is harder to remember, especially for C programmers. Sample 3.26 show the
recommended style of separating the declaration and initialization.

Sample 3.26   Static initialization fix: break apart declaration and initialization

```
logic [7:0] local_addr;
local_addr = addr << 2;   // Bug solved
```

3.7   Time Values

SystemVerilog has several new constructs to allow you to unambiguously specify
time values in your system.

3.7.1   Time Units and Precision

When you rely on the `timescale compiler directive, you must compile the files in
the proper order to be sure all the delays use the proper scale and precision. One way

avoiding this compilation ordering problem is to require that every file that starts with a `timescale compiler directive should end with one that resets it back to a company-specific default such as 1ns/1ns.

The timeunit and timeprecision declarations eliminate this ambiguity by precisely specifying the values for every module. Sample 3.27 shows these declarations. Note that if you use these instead of `timescale, you must put them in every module that has a delay. See the LRM for more on these declarations.

3.7.2   Time Literals

SystemVerilog allows you to unambiguously specify a time value plus units. Your code can use delays such as 0.1ns or 20ps. Just remember to use timeunit and timeprecision or `timescale. You can make your code even more time aware by using the classic Verilog $timeformat(), $time, and $realtime system tasks. The four arguments to $timeformat are the scaling factor (−9 for nanoseconds, −12 for picoseconds), the number of digits to the right of the decimal point, a string to print after the time value, and the minimum field width.

Sample 3.27 shows various delays and the result from printing the time when it is formatted by $timeformat() and the %t specifier.

Sample 3.27 Time literals and $timeformat

```
module timing;
  timeunit 1ns;
  timeprecision 1ps;
  initial begin
    $timeformat(-9, 3, "ns", 8);
    #1      $display("%t", $realtime); // 1.000ns
    #2ns    $display("%t", $realtime); // 3.000ns
    #0.1ns  $display("%t", $realtime); // 3.100ns
    #41ps   $display("%t", $realtime); // 3.141ns
  end
endmodule
```

3.7.3   Time and Variables

You can store time values in variables and use them in calculations and delays. The values are scaled and rounded according to the current time scale and precision. Variables of type time cannot hold fractional delays as they are just 64-bit integers, so delays will be rounded. You should use realtime variables if this is a problem.

Sample 3.28 shows how realtime variables are rounded when used as a delay.

Sample 3.28 Time variables and rounding

```
`timescale 1ns/100ps

module ps;

  initial begin
    realtime rtdelay = 800ps;     // Stored as 0.8 (800ps)
    time     tdelay  = 800ps;     // Rounded to 1

    $timeformat(-12, 0, "ps", 5);
    #rtdelay;                     // Delay of 800ps
    $display("%t", rtdelay);      // "800ps"
    #tdelay;                      // Delay another 1ns
    $display("%t", tdelay);       // "1000ps"
  end

endmodule
`timescale 1ns/1ns                // Reset to default
```

3.7.4 $time vs. $realtime

The system task $time returns an integer scaled to the time unit of the current module, but missing any fractional units, while $realtime returns a real number with the complete time value, including fractions. This book uses $time in the examples for brevity, but your testbenches may need to use $realtime.

3.8 Conclusion

The new SystemVerilog procedural constructs and task/function features make it easier for you to create testbenches by making the language look more like other programming languages such as C/C++. As a bonus, SystemVerilog has additional HDL constructs such as timing controls, simple thread control, and 4-state logic.

3.9 Exercises

1. Create the SystemVerilog code with the following requirements:

   a. Create a 512 element integer array
   b. Create a 9-bit address variable to index into the array
   c. Initialize the last location in the array to 5
   d. Call a task, my_task(), and pass the array and the address

e. Create `my_task()` that takes two inputs: a constant 512-element integer array passed by reference, and a 9-bit address. The task calls a function, `print_int()`, and passes the array element indexed by the address, pre-decrementing the address.

f. Create `print_int()` that prints out the simulation time and the value of the input. The function has no return value.

2. For the following SystemVerilog code, what is displayed if the task `my_task2()` is automatic?

```
int new_address1, new_address2;
bit clk;
initial begin
  fork
     my_task2(21, new_address1);
     my_task2(20, new_address2);
  join
  $display("new_address1 = %0d", new_address1);
  $display("new_address2 = %0d", new_address2);
end

initial
  forever #50 clk = !clk;

task my_task2(input int address, output int
new_address);
  @(clk);
  new_address = address;
endtask
```

3. For the same SystemVerilog code in Exercise 2, what is displayed if the task `my_task2()` is not automatic?

4. Create the SystemVerilog code to specify that the time should be printed in ps (picoseconds), display 2 digits to the right of the decimal point, and use as few characters as possible

5. Using the formatting system task from Exercise 4, what is displayed by the following code?

```
timeunit 1ns;
timeprecision 1ps;
parameter real t_real = 5.5;
parameter time t_time = 5ns;

initial begin
  #t_time $display("1 %t", $realtime);
  #t_real $display("1 %t", $realtime);
  #t_time $display("1 %t", $realtime);
  #t_real $display("1 %t", $realtime);
end

initial begin
  #t_time $display("2 %t", $time);
  #t_real $display("2 %t", $time);
  #t_time $display("2 %t", $time);
  #t_real $display("2 %t", $time);
end
```

Chapter 4
Connecting the Testbench and Design

There are several steps needed to verify a design: generate stimulus, capture responses, determine correctness, and measure progress. However, first you need the proper testbench, connected to the design, as shown in Fig. 4.1.

Your testbench wraps around the design, sending in stimulus and capturing the design's response. The testbench forms the "real world" around the design, mimicking the entire environment. For example, a processor model needs to connect to various buses and devices, which are modeled in the testbench as bus functional models. A networking device connects to multiple input and output data streams that are modeled based on standard protocols. A video chip connects to buses that send in commands, and then forms images that are written into memory models. The key concept is that the testbench simulates everything not in the design under test.

Your testbench needs a higher-level way to communicate with the design than Verilog's ports and the error-prone pages of connections. You need a robust way to describe the timing so that synchronous signals are always driven and sampled at the correct time and all interactions are free of the race conditions so common to Verilog models.

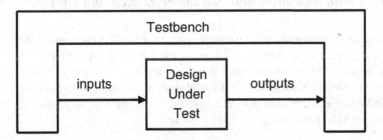

Fig. 4.1 The testbench – design environment

C. Spear and G. Tumbush, *SystemVerilog for Verification: A Guide to Learning the Testbench Language Features*, DOI 10.1007/978-1-4614-0715-7_4,
© Springer Science+Business Media, LLC 2012

4.1   Separating the Testbench and Design

In an ideal world, all projects have two separate groups: one to create the design and one to verify it. In the real world, limited budgets may require you to wear both hats. Each team has its own set of specialized skills, such as creating synthesizable RTL code, or figuring out new ways to find bugs in the design. These two groups each read the original design specification and make their own interpretations. The designer has to create code that meets that specification, whereas your job as the verification engineer is to create scenarios where the design does not match its description.

Likewise, your testbench code is in a separate block from design code. In classic Verilog, each goes in a separate module. However, using a module to hold the testbench often causes timing problems around driving and sampling, so SystemVerilog introduces the program block to separate the testbench, both logically and temporally. For more details, see Section 4.3.

As designs grow in complexity, the connections between the blocks increase. Two RTL blocks may share dozens of signals, which must be listed in the correct order for them to communicate properly. One mismatched or misplaced connection and the design will not work. You can reduce errors by using the connect-by-name syntax, but this more than doubles your typing burden. If it is a subtle error, such as swapping pins that only toggle occasionally, you may not notice the problem for some time. Worse yet is when you add a new signal between two blocks. You have to edit not only the blocks to add the new port but also the higher-level modules that wire up the devices. Again, one wrong connection at any level and the design stops working. Or worse, the system only fails intermittently!

The solution is the interface, the SystemVerilog construct that represents a bundle of wires. Additionally, you can specify timing, signal direction, and even add functional code. An interface is instantiated like a module but is connected to ports like a signal.

4.1.1   Communication Between the Testbench and DUT

The next few sections show a testbench connected to an arbiter, using individual signals and again using interfaces. Figure 4.2 is a diagram of the top level design including a testbench, arbiter, clock generator, and the signals that connect them. This DUT (Design Under Test) is a trivial design, so you can concentrate on the SystemVerilog concepts and not get bogged down in the design. At the end of the chapter, an ATM router is shown.

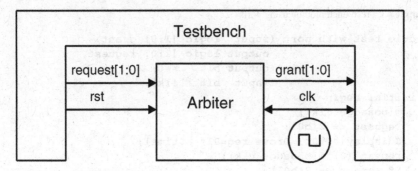

Fig. 4.2   Testbench – Arbiter without interfaces

4.1.2   Communication with Ports

The following code shows the steps needed to connect an RTL block to a testbench. First is the header for the arbiter model, shown in Sample 4.1. This uses the Verilog-2001 style port declarations where the type and direction are in the header. Some code has been left out for clarity.

As discussed in Section 2.1.1, SystemVerilog has expanded the classic `reg` type so that you can use it like a `wire` to connect blocks. In recognition of its new capabilities, the `reg` type has the new name of `logic`. The only place where you cannot use a `logic` variable is a net with multiple structural drivers, where you must use a net such as `wire`.

Sample 4.1   Arbiter model using ports

```
module arb_with_port (output logic [1:0] grant,
                      input  logic [1:0] request,
                      input  bit         rst, clk);

  always @(posedge clk or posedge rst) begin
    if (rst)
      grant <= 2'b00;
    else if (request[0])     // High priority
      grant <= 2'b01;
    else if (request[1])     // Low priority
      grant <= 2'b10;
    else
      grant <= '0;
  end
endmodule
```

The testbench in Sample 4.2 is kept in a module to separate it from the design. Typically, it connects to the design with ports.

Sample 4.2 Testbench module using ports

```
module test_with_port (input   logic [1:0] grant,
                        output logic [1:0] request,
                        output bit    rst,
                        input  bit    clk);
  initial begin
    @(posedge clk);
    request <= 2'b01;
    $display("@%0t: Drove req=01", $time);
    repeat (2) @(posedge clk);
    if (grant == 2'b01)
      $display("@%0t: Success: grant == 2'b01", $time);
    else
      $display("@%0t: Error: grant != 2'b01", $time);
    $finish;
  end
endmodule
```

The top module connects the testbench and DUT, and includes a simple clock generator.

Sample 4.3 Top-level module with ports

```
module top;
  logic [1:0] grant, request;
  bit   clk;
  always #50 clk = ~clk;

  arb_with_port  a1 (grant, request, rst, clk);   // Sample 4-1
  test_with_port t1 (grant, request, rst, clk);   // Sample 4-2
endmodule
```

In Sample 4.3, the modules are simple, but real designs with hundreds of pins require pages of signal and port declarations. All these connections can be error prone. As a signal moves through several layers of hierarchy, it has to be declared and connected over and over. Worst of all, if you just want to add a new signal, it has to be declared and connected in multiple files. SystemVerilog interfaces can help in each of these cases.

4.2 The Interface Construct

Designs have become so complex that even the communication between blocks may need to be separated out into separate entities. To model this, SystemVerilog uses the interface construct that you can think of as an intelligent bundle of wires.

It contains the connectivity, synchronization, and optionally, the functionality of the communication between two or more blocks and, optionally, error checking. They connect design blocks and/or testbenches.

Design-level interfaces are covered in Sutherland (2006). This book concentrates on interfaces that connect design blocks and testbenches.

4.2.1   Using an Interface to Simplify Connections

The first improvement to the arbiter example is to bundle the wires together into an interface. Figure 4.3 shows the testbench and arbiter, communicating using an interface. Note how the interface extends into the two blocks, representing the drivers and receivers that are functionally part of both the test and the DUT. The clock can be part of the interface or a separate port.

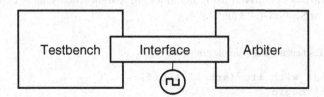

Fig. 4.3 An interface straddles two modules

The simplest interface is just a bundle of bidirectional signals as shown in Sample 4.4. Use the `logic` data type so you can drive the signals from procedural statements.

Sample 4.4 Simple interface for arbiter

```
interface arb_if(input bit clk);
  logic [1:0] grant, request;
  bit rst;
endinterface
```

Sample 4.5 is the device under test, the arbiter, that uses an interface instead of ports.

Sample 4.5   Arbiter using a simple interface

```
module arb_with_ifc (arb_if arbif);
  always @(posedge arbif.clk or posedge arbif.rst)
    begin
    if (arbif.rst)
      arbif.grant <= '0;
    else if (arbif.request[0])      // High priority
      arbif.grant <= 2'b01;
    else if (arbif.request[1])      // Low priority
      arbif.grant <= 2'b10;
    else
      arbif.grant <= '0;
  end
endmodule
```

Sample 4.6 shows the testbench. You refer to a signal in an interface by making a hierarchical reference using the instance name: `arbif.request`. Interface signals should always be driven using nonblocking assignments. This is explained in more detail in Section 4.4.3 and 4.4.4.

Sample 4.6   Testbench using a simple arbiter interface

```
module test_with_ifc (arb_if arbif);
  initial begin
    @(posedge arbif.clk);
    arbif.request <= 2'b01;
    $display("@%0t: Drove req=01", $time);
    repeat (2) @(posedge arbif.clk);
    if (arbif.grant != 2'b01)
      $display("@%0t: Error: grant != 2'b01", $time);
    $finish;
  end
endmodule
```

All these blocks are instantiated and connected in the `top` module as shown in Sample 4.7.

Sample 4.7   Top module with a simple arbiter interface

```
module top;
  bit clk;
  always #50 clk = ~clk;

  arb_if         arbif(clk);      // From Sample 4-4
  arb_with_ifc   a1 (arbif);      // From Sample 4-5
  test_with_ifc  t1(arbif);       // From Sample 4-6
endmodule : top
```

You can see an immediate benefit, even on this small device: the connections become cleaner and less prone to mistakes. If you wanted to put a new signal in an interface, you would just have to add it to the interface definition and the modules that actually used it. You would not have to change any module such as top that just passes the interface through. This language feature greatly reduces the chance for wiring errors.

This book only shows interfaces with a single clock that is connected to a generator at the top level. If your interface requires multiple clocks, treat them like the other signals inside the interface, and connect the interface to a clock generator. You are more productive if you work at a high level and treat the interface as a cycle based construct. The next level up is transaction-based, which is beyond typical RTL code.

Make sure you declare your interfaces outside of modules and program blocks. If you forget, expect all sorts of trouble. Some compilers may not support defining an interface inside a module. If allowed, the interface would be local to the module and thus not visible to the rest of the design. Sample 4.8 shows the common mistake of including the interface definition right after other include statements.

Sample 4.8 Bad test module includes interface

```
module bad_test(arb_if arbif);
'include "MyTest.sv"    // Legal include
'include "arb_if.sv"    // BAD:Interface hidden in module
...
```

4.2.2 Connecting Interfaces and Ports

If you have a Verilog-2001 legacy design with ports that cannot be changed to use an interface, you can just connect the interface's signals to the individual ports. Sample 4.9 connects the original arbiter from Sample 4.1 to the interface in Sample 4.4.

Sample 4.9 Connecting an interface to a module that uses ports

```
module top;
  bit  clk;
  always #50 clk = ~clk;

  arb_if arbif(clk);
  arb_with_port a1 (.grant  (arbif.grant), // .port (ifc.signal)
                    .request (arbif.request),
                    .rst    (arbif.rst),
                    .clk    (arbif.clk));
  test_with_ifc t1(arbif);    // From Sample 4-6
endmodule : top
```

4.2.3 Grouping Signals in an Interface Using Modports

Sample 4.5 uses a point-to-point connection scheme with no signal directions in the interface. The original modules using ports had this information that the compiler uses to check for wiring mistakes. The `modport` construct in an interface lets you group signals and specify directions. The MONITOR modport in Sample 4.10 allows you to connect a monitor module to the interface.

Sample 4.10 Interface with modports

```
interface arb_if(input bit clk);
  logic [1:0] grant, request;
  bit rst;

  modport TEST (output request, rst,
                input  grant, clk);

  modport DUT (input request, rst, clk,
               output grant);

  modport MONITOR (input request, grant, rst, clk);

endinterface
```

Sample 4.11 shows the arbiter model and testbench, with the modport in their port connection list. Note that you put the modport name, DUT or TEST, after the interface name, arb_if. Other than the modport name, these are identical to the previous examples.

Sample 4.11 Arbiter model with interface using modports

```
module arb_with_mp (arb_if.DUT arbif);
  ...
endmodule
```

Sample 4.12 Testbench with interface using modports

```
module test_with_mp (arb_if.TEST arbif);
  ...
endmodule
```

Even though the code didn't change much (except that the interface grew larger), this interface more accurately represents the real design, especially the signal direction.

There are two ways to use these modport names in your design. You can specify them in the modules that connect to the interface signals. In this case, the top model does not change from Sample 4.7, except for the module names. This book recommends this style, as the modport is an implementation detail that should not clutter the top level module.

The alternative is to specify the modport when you instantiate the module as shown in Sample 4.13.

Sample 4.13 Top level module with modports

```
module top;
   logic [1:0] grant, request;
   bit    clk;
   always #50 clk = ~clk;

   arb_if        arbif(clk);        // Sample 4-10
   arb_with_mp   a1 (arbif.DUT);    // Sample 4-11
   test_with_mp  t1 (arbif.TEST);   // Sample 4-12
endmodule
```

With this style, you have the flexibility to instantiate a module more than once, with each instance connected to a different modport, that is, a different subset of interface signals. For example, a byte-wide RAM model could connect to one of four slots on a 32-bit bus. In this case, you would need to specify the modport when you instantiate the module, not in the module itself.

Note that modports are defined in the interface, and specified in the module port list, but never in the signal name. The name arb_if.TEST.grant is illegal!

4.2.4 Using Modports with a Bus Design

Not every signal needs to go in every modport. Consider a CPU – memory bus modeled with an interface. The CPU is the bus master and drives a subset of the signals, such as request, command, and address. The memory is a slave and receives those signals and drives ready. Both master and slave drive data. The bus arbiter only looks at request and grant, and ignores all other signals. So your interface would have three modports for master, slave, and arbiter, plus an optional monitor modport.

4.2.5 Creating an Interface Monitor

You can create a bus monitor using the MONITOR modport. Sample 4.14 shows a trivial monitor for the arbiter. For a real bus, you could decode the commands and print the status: completed, failed, etc.

Sample 4.14   Arbiter monitor with interface using modports

```
module monitor (arb_if.MONITOR arbif);

  always @(posedge arbif.request[0]) begin
    $display("@%0t: request[0] asserted", $time);
    @(posedge arbif.grant[0]);
    $display("@%0t: grant[0] asserted", $time);
  end

  always @(posedge arbif.request[1]) begin
    $display("@%0t: request[1] asserted", $time);
    @(posedge arbif.grant[1]);
    $display("@%0t: grant[1] asserted", $time);
  end
endmodule
```

4.2.6   Interface Trade-Offs

An interface cannot contain module instances, only instances of other interfaces. There are trade-offs in using interfaces with modports as compared with traditional ports connected with signals.

The advantages to using an interface are as follows.

- An interface is ideal for design reuse. When two blocks communicate with a specified protocol using more than two signals, consider using an interface. If groups of signals are repeated over and over, as in a networking switch, you should additionally use virtual interfaces, as described in Chapter 10.
- The interface takes the jumble of signals that you declare over and over in every module or program and puts it in a central location, reducing the possibility of misconnecting signals.
- To add a new signal, you just have to declare it once in the interface, not in higher-level modules, once again reducing errors.
- Modports allow a module to easily tap a subset of signals from an interface. You can specify signal direction for additional checking.

The disadvantages of using an interface are as follows.

- For point-to-point connections, interfaces with modports are almost as verbose as using ports with lists of signals. Interfaces have the advantage that all the declarations are still in one central location, reducing the chance for making an error.
- You must now use the interface name in addition to the signal name, possibly making the modules more verbose, but more readable for debugging.
- If you are connecting two design blocks with a unique protocol that will not be reused, interfaces may be more work than just wiring together the ports.

• It is difficult to connect two different interfaces. A new interface (bus_if) may contain all the signals of an existing one (arb_if), plus new signals (address, data, etc.). You may have to break out the individual signals and drive them appropriately.

4.2.7   More Information and Examples

The SystemVerilog LRM specifies many other ways for you to use interfaces. See Sutherland (2006) for more examples of using interfaces for design.

4.2.8   Logic vs. Wire in an Interface

This book recommends declaring the signals in your interface as logic while the VMM has a rule that says to use a wire. The difference is ease-of-use vs. reusability.

If your testbench drives an asynchronous signal in an interface with a procedural assignment, the signal must be a logic type. A wire can only be driven with a continuous assignment statement. Signals in a clocking block are always synchronous and can be declared as logic or wire. Sample 4.15 shows how the logic signal can be driven directly, whereas the wire requires additional code.

Sample 4.15 Driving logic and wires in an interface

```
interface asynch_if();
  logic l;
  wire w;
endinterface


module test(asynch_if ifc);
  logic local_wire;
  assign ifc.w = local_wire;

  initial begin
    ifc.l <= 0;        // Drive asych logic directly ...
    local_wire <= 1; // but drive wire through assign
    ...
  end
endmodule
```

Another reason to use logic for interface signals is that the compiler will give an error if you unintentionally use multiple structural drivers.

The VMM takes a more long-term approach. Take the case where you have created test code that works well on the current project and is later used in a new design.

What if your interface with all its `logic` signals is connected such that now a signal has multiple structural drivers? The engineers will have to change that `logic` to a `wire`, and, if the signal does not go through a clocking block, change the procedural assignment statements. Now there are two versions of the interface, and existing tests must be modified before they can be reused. Rewriting good code goes against the VMM principles.

4.3   Stimulus Timing

The timing between the testbench and the design must be carefully orchestrated. At a cycle level, you need to drive and receive the synchronous signals at the proper time in relation to the clock. Drive too late or sample too early, and your testbench is off a cycle. Even within a single time slot (for example, everything that happens at time 100ns), mixing design and testbench events can cause a race condition, such as when a signal is both read and written at the same time. Do you read the old value, or the one just written? In Verilog, nonblocking assignments help when a test module drives the DUT, but the test could not always be sure it sampled the last value driven by the design. SystemVerilog has several constructs to help you control the timing of the communication.

4.3.1   Controlling Timing of Synchronous Signals
with a Clocking Block

An interface should contain a clocking block to specify the timing of synchronous signals relative to the clocks. Clocking blocks are mainly used by testbenches but also allow you to create abstract synchronous models. Signals in a clocking block are driven or sampled synchronously, ensuring that your testbench interacts with the signals at the right time. Synthesis tools do not support clocking blocks, so your RTL code can not take advantage of them. The chief benefit of clocking blocks is that you can put all the detailed timing information in here, and not clutter your testbench.

An interface can contain multiple clocking blocks, one per clock domain, as there is a single clock expression in each block. Typical clock expressions are @ (posedge clk) for a single edge clock and @ (clk) for a DDR (double data rate) clock.

You can specify a clock skew in the clocking block using the `default` statement, but the default behavior is that input signals are sampled just before the design executes, and the outputs are driven back into the design during the current time slot. The next section provides more details on the timing between the design and testbench.

Once you have defined a clocking block, your testbench can wait for the clocking expression with `@arbif.cb` rather than having to spell out the exact clock and edge. Now if you change the clock or edge in the clocking block, you do not have to change your testbench.

Sample 4.16 is similar to Sample 4.10 except that the TEST modport now treats request and grant as synchronous signals. The clocking block cb declares that the signals are active on the positive edge of the clock. The signal directions are relative to the modport where they are used. So request is a synchronous output in the TEST modport, and grant is an synchronous input. The signal rst is asynchronous in the TEST modport.

Sample 4.16 Interface with a clocking block

```
interface arb_if(input bit clk);
  logic [1:0] grant, request;
  bit rst;

  clocking cb @(posedge clk);      // Declare cb
    output request;
    input grant;
  endclocking

  modport TEST (clocking cb,       // Use cb
                output rst);

  modport DUT (input request, rst, clk,
               output grant);
endinterface

// Trivial test, see Sample 4-21 for a better one
module test_with_cb(arb_if.TEST arbif);
  initial begin
    @arbif.cb;
    arbif.cb.request <= 2'b01;
    @arbif.cb;
    $display("@%0t: Grant = %b", $time, arbif.cb.grant);
    @arbif.cb;
    $display("@%0t: Grant = %b", $time, arbif.cb.grant);
    $finish;
  end
endmodule
```

4.3.2 Timing Problems in Verilog

Your testbench needs to be separate from the design, not just logically but also temporally. Consider how a hardware tester interacts with a chip for synchronous signals. In a real hardware design, the DUT's storage elements latch their inputs from the tester at the active clock edge. These values propagate through the storage element outputs, and then the logic clouds to the inputs of the next storage element. The time from the input of the first storage to the next must be less than a clock cycle.

So a hardware tester needs to drive the chip's inputs at the clock edge, and read the outputs just before the following edge.

A testbench has to mimic this tester behavior. It should drive on or after the active clock edge, and should sample as late as possible as allowed by the protocol timing specification, just before the active clock edge.

If the DUT and testbench are made of Verilog modules only, this outcome is nearly impossible to achieve. If the testbench drives the DUT at the clock edge, there could be race conditions. What if the clock propagates to some DUT inputs before the testbench stimulus, but is a little later to other inputs? From the outside, the clock edges all arrive at the same simulation time, but in the design, some inputs get the value driven during the last cycle, whereas other inputs get values from the current cycle.

One way around this problem is to add small delays to the system, such as #0. This forces the thread of Verilog code to stop and be rescheduled after all other code. Invariably though, a large design has several sections that all want to execute last. Whose #0 wins out? It could vary from run to run and be unpredictable between simulators. Multiple threads using #0 delays cause indeterministic behavior. Avoid using #0 as it will make your code unstable and not portable.

The next solution is to use a larger delay, #1. RTL code has no timing, other than clock edges, so one time unit after the clock, the logic has settled. However, what if one module uses a time precision of 1ns, whereas another used a resolution of just 10ps? Does that #1 mean 1ns, 10ps, or something else? You want to drive as soon as possible after the clock cycle with the active clock edge, but not during that time, and before anything else can happen. Worse yet, your DUT may contain a mix of RTL code with no delays and gate code with delays. Just as you should avoid using #0, stay away from #1 delays to fix timing problems. See Cummings (2000) and other papers by him for additional guidelines.

4.3.3   Testbench – Design Race Condition

Sample 4.17 shows a potential race condition between the testbench and design. The race condition occurs when the test drives the start signal and then the other ports. The memory is waiting on the start signal and could wake up immediately, whereas the write signal still has its old value, while addr and data have new values. This behavior is perfectly legal according to the LRM. You could delay all these signals slightly by using nonblocking assignments, as recommended by Cummings (2000), but remember that the testbench and the design are both using these assignments. It is still possible to get a race condition between the testbench and design.

Sampling the design outputs has a similar problem. You want to grab the values at the last possible moment, just before the active clock edge. Perhaps you know the next clock edge is at 100ns. You can't sample right at the clock edge at 100ns, as some design values may have already changed. You should sample at Tsetup just before the clock edge.

Sample 4.17 Race condition between testbench and design

```
module memory(input wire start, write,
              input wire [7:0] addr,
              inout wire [7:0] data);
  logic [7:0] mem[256];
  always @(posedge start) begin
    if (write)
      mem[addr] <= data;
    ...
  end
endmodule

module test(output logic start, write,
            output logic [7:0] addr, data);
  initial begin
    start = 0;              // Initialize signals
    write = 0;
    #10;                    // Short delay
    addr = 8'h42;           // Start first command
    data = 8'h5a;
    start = 1;
    write = 1;
    ...
  end
endmodule
```

4.3.4 The Program Block and Timing Regions

The root of the problem is the mixing of design and testbench events during the same time slot, though even in pure RTL the same problem can happen. Good coding guidelines such as proper use of nonblocking assignments can reduce these race conditions, but improperly coded assignments have the habit of creeping in. What if there were a way you could separate these events temporally, just as you separated the code? At 100ns, your testbench could sample the design outputs before the clock has had a chance to change and any design activity has occurred. By definition, these values would be the last possible ones from the previous time slot. Then, after all the design events are done, your testbench would start.

How does SystemVerilog know to schedule the testbench events separately from the design events? In SystemVerilog, your testbench code is in a program block, which is similar to a module in that it can contain code and variables and be instantiated in other modules. However, a program cannot have any hierarchy such as instances of modules, interfaces, or other programs.

A new region of the time slot was introduced in SystemVerilog as shown in Fig. 4.4. In Verilog, most events are executed in the Active region. There are dozens of other regions for nonblocking assignments, PLI execution, etc., but they can be

Fig. 4.4 Main regions inside
a SystemVerilog time step

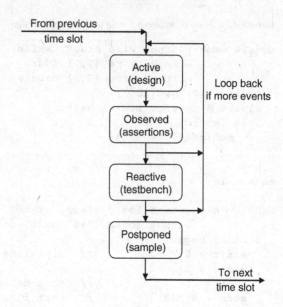

ignored for the purposes of this book. See Table 4.1, the LRM, and Cummings (2006) for more details on the SystemVerilog event regions.

First to execute during a time slot is the Active region, where design events run. These include your traditional RTL and gate code plus the clock generator. The second region is the Observed region, where SystemVerilog Assertions are evaluated. Following that is the Reactive region where the testbench code in a program executes. Note that time does not strictly flow forwards — events in the Observed and Reactive regions can trigger further design events in the Active region in the current cycle. Last is the Postponed region, which samples signals at the end of the time slot, in the readonly period, after design activity has completed.

Table 4.1 Primary SystemVerilog scheduling regions

Name	Activity
Active	Simulation of design code in modules
Observed	Evaluation of SystemVerilog Assertions
Reactive	Execution of testbench code in programs
Postponed	Sampling design signals for testbench input

Sample 4.18 shows part of the testbench code for the arbiter. Note that the statement @arbif.cb waits for the active edge of the clocking block, @(posedge clk), as shown in Sample 4.16. This sample shows that your testbench code is

written at a slightly higher level of abstraction, using cycle-by-cycle timing instead of worrying about individual clock edges.

Section 4.4 explains more about the driving and sampling of interface signals.

Sample 4.18 Testbench using interface with clocking block

```
program automatic test (arb_if.TEST arbif);
  initial begin
    @arbif.cb;
    arbif.cb.request <= 2'b01;
    $display("@%0t: Drove req=01", $time);
    repeat (2) @arbif.cb;
    if (arbif.cb.grant == 2'b01)
      $display("@%0t: Success: grant == 2'b01", $time);
    else
      $display("@%0t: Error: grant != 2'b01", $time);
    end
endprogram : test
```

Your test should be contained in a single program. You should use OOP to build a dynamic, hierarchical testbench from objects instead of modules. A simulation may have multiple program blocks if you are using code from other people or combining several tests.

As discussed in Section 3.6.1, you should always declare your program block as `automatic` so that it behaves more like the routines in stack-based languages you may have worked with, such as C.

Note that not all vendors regard program blocks equally. See Rich (2009) for an alternate opinion.

4.3.5 Specifying Delays Between the Design and Testbench

The default timing of the clocking block is to sample inputs with a skew of #1step and to drive the outputs with a delay of #0. The 1step delay specifies that signals are sampled in the Postponed region of the previous time slot, before any design activity. So you get the output values just before the clock changes. The testbench outputs are synchronous by virtue of the clocking block, so they flow directly into the design. The program block, running in the Reactive region, generates the stimulus that is applied to the DUT, which is then evaluated in the Active region during the same time slot. The DUT evaluates its logic and drives its outputs, which are the inputs to the testbench through the clocking blocks. These are then sampled in the Postponed region and the cycle repeats. If you have a design background, you can remember this by imagining that the clocking block inserts a synchronizer between the design and testbench, as shown in Fig. 4.5. With proper use of program and clocking blocks, race conditions between the testbench and DUT can be all but eliminated.

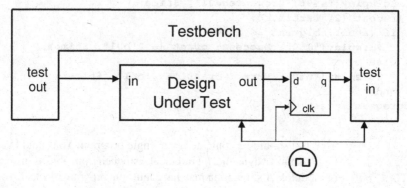

Fig. 4.5 A clocking block synchronizes the DUT and testbench

4.4 Interface Driving and Sampling

Your testbench needs to drive and sample signals from the design, primarily through interfaces with clocking blocks. The next section uses the arbiter interface from Sample 4.16 and the top-level module from Sample 4.9.

Asynchronous signals such as rst pass through the interface with no delays. The signals in the clocking block get synchronized as shown in the sections below.

4.4.1 Interface Synchronization

You can use the Verilog @ and wait constructs to synchronize the signals in a testbench. Sample 4.19 shows the various constructs.

Sample 4.19 Signal synchronization

```
program automatic test(bus_if.TB bus);
  initial begin
    @bus.cb;                      // Continue on active edge
                                  // in clocking block
    repeat (3) @bus.cb;           // Wait for 3 active edges
    @bus.cb.grant;                // Continue on any edge
    @(posedge bus.cb.grant);      // Continue on posedge
    @(negedge bus.cb.grant);      // Continue on negedge
    wait (bus.cb.grant==1);       // Wait for expression
                                  // No delay if already true
    @(posedge bus.cb.grant or
      negedge bus.rst);           // Wait for several signals
  end
endprogram
```

4.4.2 Interface Signal Sample

When you read a signal from a clocking block, you get the value sampled from just before the last clock edge, i.e., from the Postponed region. Sample 4.20 shows a program block that reads the synchronous grant signal from the DUT. The arb module drives grant to 1 & 2 in the middle of the 100ns cycle, and then to 3 exactly at the clock edge. This code is for illustration only and is not real, synthesizable RTL.

Sample 4.20 Synchronous interface sample and drive from module

```
program automatic test(arb_if.TEST arbif);
  initial begin
    $monitor("@%0t: grant=%h", $time, arbif.cb.grant);
    #500ns $display("End of test");
  end
endprogram

module arb_dummy(arb_if.DUT arbif);
  initial
    fork
      # 70ns arbif.grant = 1;
      #170ns arbif.grant = 2;
      #250ns arbif.grant = 3;
    join
endmodule
```

The waveforms in Fig. 4.6 show that in the program, arbif.cb.grant gets the value from just before the clock edge. When the interface input changes right at a clock edge, such as 250ns, the value does not propagate to the testbench until the next cycle, which starts at 350ns.

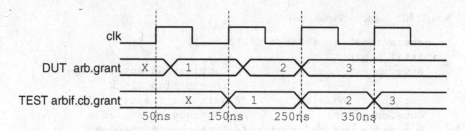

Fig. 4.6 Sampling a synchronous interface

4.4.3   Interface Signal Drive

Sample 4.21 has an abbreviated version of the arbiter test program, which uses the
arbiter interface defined in Sample 4.16.

Sample 4.21 Testbench using interface with clocking block

```
program automatic test_with_cb (arb_if.TEST arbif);

  initial begin
    @arbif.cb;
    arbif.cb.request <= 2'b01;
    $display("@%0t: Drove req=01", $time);
    repeat (2) @arbif.cb;
    if (arbif.cb.grant == 2'b01)
      $display("@%0t: Success: grant == 2'b01", $time);
    else
      $display("@%0t: Error: grant != 2'b01", $time);
  end

endprogram : test_with_cb
```

 When using modports with clocking blocks, a synchronous
interface signal such as request must be prefixed with both the
interface name, arbif, and the clocking block name, cb. So in
Sample 4.21, arbif.cb.request is legal, but arbif.
request is not. This is the most common coding mistake with interfaces and clocking blocks.

4.4.4   Driving Interface Signals Through a Clocking Block

You should always drive interface signals in a clocking block with a synchronous
drive using a nonblocking assignment. This is because the design signal does not
change immediately after your assignment – remember that your testbench executes

in the Reactive region while design code is in the Active region. If your testbench drives `arbif.cb.request` at 100ns, the same time as `arbif.cb` (which is `@(posedge clk)` according to the clocking block), `request` changes in the design at 100ns. However, if your testbench tries to drive `arbif.cb.request` at time 101ns, between clock edges, the change does not propagate until the next clock edge. In this way, your drives are always synchronous. In Sample 4.20, `arbif.grant` is driven by a module and can use a blocking assignment.

If the testbench drives the synchronous interface signal at the active edge of the clock, as shown in Sample 4.22, the value propagates immediately to the design. This is because the default output delay is `#0` for a clocking block. If the testbench drives the output just after the active edge, the value is not seen in the design until the next active edge of the clock.

Sample 4.22   Interface signal drive

```
busif.cb.request <= 1;        // Synchronous drive
busif.cb.cmd <= cmd_buf;      // Synchronous drive
```

Sample 4.23 shows what happens if you drive a synchronous interface signal at various points during a clock cycle. This uses the interface from Sample 4.16 and the top module and clock generator from Sample 4.9.

Sample 4.23   Driving a synchronous interface

```
program automatic test_with_cb(arb_if.TEST arbif);
   initial fork
     # 70ns arbif.cb.request <= 3;
     #170ns arbif.cb.request <= 2;
     #250ns arbif.cb.request <= 1;
     #500ns finish;
   join
endprogram

module arb(arb_if.DUT arbif);
   initial
     $monitor("@%0t: req=%h", $time, arbif.request);
endmodule
```

Note that in Fig. 4.7, the value 3, driven in the middle of the first cycle, is seen by the DUT at the start of the second cycle. The value 2 is driven in the middle of the second cycle. It is never seen by the DUT as the testbench drives a 1 at the end of the second cycle.

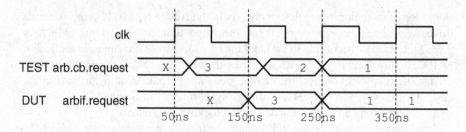

Fig. 4.7 Driving a synchronous interface

Driving clocking block signals asynchronously can lead to dropped values. Instead, drive at the clock edge by using a cycle delay prefix on your drives as shown in Sample 4.24.

Sample 4.24 Interface signal drive

```
##2 arbif.cb.request <= 0;   // Wait 2 cycles then assign
##3;         // Illegal - must be used with an assignment
```

If you want to wait for two clock cycles before driving a signal, you can either use "repeat (2) @arbif.cb;" or use the cycle delay ##2. This latter delay only works as a prefix to a drive of a signal in a clocking block, as it needs to know which clock to use for the delay.

The cycle delay of ##0 in an assignment that drives the value immediately if the clock was asserted in this time slot, according to the clocking block. If the clock was not just asserted, the signal is driven at the next active edge of the clock. The cycle delay of ##1 always waits for the next active edge of the clock, even if the clock was asserted in the current time slot.

The naked cycle delay statement ##3; works if you have a default clocking block for your program or module. This book only recommends putting a clocking block in an interface and not creating a default clocking block. You should always be specific about which clock is being referenced.

4.4.5   Bidirectional Signals in the Interface

In Verilog-1995, if you want to drive a bidirectional signal such as a port from procedural code, you need a continuous assignment to connect the reg to the wire. In SystemVerilog, synchronous bidirectional signals in interfaces are easier to use as the continuous assignment is added for you, as shown in Sample 4.25. When you write to the net from a program, SystemVerilog actually writes to a temporary variable that drives the net. Your program reads directly from the wire, seeing the value that is resolved from all the drivers. Design code in a module still uses the classic register plus continuous assignment statement.

Sample 4.25 Bidirectional signals in a program and interface

```
interface bidir_if (input bit clk);
  wire [7:0] data;    // Bidirectional signal
  clocking cb @(posedge clk);
    inout data;
  endclocking
  modport TEST (clocking cb);
endinterface

program automatic test(bidir_if.TEST mif);
  initial begin
    mif.cb.data <= 'z;        // Tri-state the bus
    @mif.cb;
    $displayh(mif.cb.data);   // Read from the bus
    @mif.cb;
    mif.cb.data <= 7'h5a;     // Drive the bus
    @mif.cb;
    $displayh(mif.cb.data);   // Read from the bus
    mif.cb.data <= 'z;        // Release the bus
  end
endprogram
```

The SystemVerilog LRM is not clear on driving an asynchronous bidirectional signal using an interface. Two possible solutions are to use a cross-module reference and continuous assignment or to use a virtual interface as shown in Chapter 10.

4.4.6 Specifying Delays in Clocking Blocks

A clocking block ensures that your signals are driven and sampled at the specified clock edge. You can skew these times with either a `default` statement, or by specifying the delays for individual signals. This can be useful when simulating netlists with real delays. Sample 4.26 shows a clocking block with a default statement that has the skews for all signals. In this example, the inputs are sampled 15ns before the posedge of the clock and the outputs are driven 10ns after the posedge of the clock.

Sample 4.26 Clocking block with default statement

```
clocking cb @(posedge clk);
  default input #15ns output #10ns;
  output request;
  input grant;
endclocking
```

Sample 4.27 shows the equivalent clocking block, but with the delays specified on the individual signals.

Sample 4.27 Clocking block with delays on individual signals

```
clocking cb @(posedge clk);
  output #10ns request;
  input  #15ns grant;
endclocking
```

4.5 Program Block Considerations

4.5.1   The End of Simulation

In Verilog, simulation continues while there are scheduled events, or until a $finish
is executed. SystemVerilog adds an additional way to end simulation. A program
block is treated as if it contains a test. If there is only a single program, simulation
ends when you complete the last statement in every initial block in the program,
as this is considered the end of the test. Simulation ends even if there are threads
still running in the program or modules. As a result, you don't have to shut down
every monitor and driver when a test is done.

If there are several program blocks, simulation ends when the last program completes. This way simulation ends when the last test completes. You can terminate
any program block early by executing $exit. Of course you can still explicitly call
$finish to end simulation, but this might cause issues if you have multiple
programs.

However, simulation is not yet over. A module or program can have a final
block that contains code to be run just before the simulator terminates, as shown in
Sample 4.28. This is a great place to perform clean up work such as closing files,
and printing a report of the number of errors and warnings encountered. You cannot
schedule any events, or have any delays in a final block that could cause time to
elapse. You do not have to worry about freeing any memory that was allocated as
this will be done automatically.

Sample 4.28 A final block

```
program automatic test;
  int errors, warnings;

  initial begin
    ... // Main program activity
  end

  final
    $display("Test completed with %0d errors and %0d warnings",
             errors, warnings);
endprogram
```

4.5.2   Why are `Always` Blocks not Allowed in a Program?

In SystemVerilog you can put `initial` blocks in a program, but not `always` blocks. This may seem odd if you are used to Verilog modules, but there are several reasons SystemVerilog programs are closer to a program in C, with one (or more) entry points, than Verilog's many small blocks of concurrently executing hardware. In a design, an `always` block might trigger on every positive edge of a clock from the start of simulation. In contrast, a testbench has the steps of initialization, stimulate and respond to the design, and then wrap up simulation. An `always` block that runs continuously would not work.

When the last `initial` block completes in the program, simulation implicitly ends just as if you had executed `$finish`. If you had an always block, it would run for ever, so you would have to explicitly call `$exit` to signal that the program completed. But don't despair. If you really need an `always` block, you can use `initial` `forever` to accomplish the same thing.

4.5.3   The Clock Generator

Now that you have seen the program block, you may wonder if the clock generator should be in a module. The clock is more closely tied to the design than the testbench, and so the clock generator should remain in a module. The generator should be instantiated at the same level as the DUT so it can drive both the DUT and testbench As you refine the design, you create clock trees, and you have to carefully control the skews as the clocks enter the system and propagate through the blocks.

The testbench is much less picky. It just wants a clock edge to know when to drive and sample signals. Functional verification is concerned with providing the right values at the right cycle, not with fractional nanosecond delays and relative clock skews.

The program block is not the place to put a clock generator. Sample 4.29 tries to put the generator in a program block but just causes a race condition. The `clk` and `data` signals both propagate from the Reactive region to the design in the Active region and could cause a race condition depending on which one arrived first.

Sample 4.29   Bad clock generator in program block

```
program automatic bad_generator (output bit clk, data);
  initial
    forever #5 clk <= ~clk ;

  initial
    forever @(posedge clk)
      data <= ~data;
endprogram
```

Avoid race conditions by always putting the clock generator in a module. If you want to randomize the generator's properties, create a class with random variables for skew, frequency, and other characteristics, as shown in Chapter 6. You can use this class in the generator module, or in the testbench.

Sample 4.30 shows a good clock generator in a module. It deliberately avoids an edge at time 0 to prevent race conditions. All clock edges are generated with a blocking assignment to trigger events during the Active region. If you must generate a clock edge at time 0, use a nonblocking assignment to set the initial value so all clock sensitive logic such as `always` blocks will have started before the clock changes value.

Sample 4.30   Good clock generator in module

```
module clock_generator (output bit clk);
  bit local_clk = 0;
  assign clk = local_clk;   // Drive port from local signal
  always #50 local_clk = ~local_clk;
endmodule
```

Lastly, don't try to verify the low-level timing with functional verification. The testbenches described in this book check the behavior of the DUT but not the timing, which is better done with a static timing analysis tool. Your testbenches should be flexible enough to be compatible with gate-level simulations run with back-annotated timing.

4.6   Connecting It All Together

Now you have a design described in a module, a testbench in a program block, and interfaces that connect them together. Sample 4.31 has the top-level module that instantiates and connects all the pieces.

Sample 4.31   Top module with implicit port connections

```
module top;
  bit  clk;
  always #50 clk = ~clk;

  arb_if arbif(.*);        // ... arbif(clk)   From Sample 4-4
  arb_with_ifc  a1(.*);   // ... a1(arbif)    From Sample 4-5
  test_with_ifc t1(.*);   // ... t1(arbif)    From Sample 4-6
endmodule : top
```

This is almost identical to Sample 4.7. It uses a shortcut notation.* (implicit port connection) that automatically connects module instance ports to signals at the current level if they have the same name and data type.

4.6.1   An Interface in a Port List Must Be Connected

The SystemVerilog compiler won't let you compile a single module or program that uses an interface in the port list. Why not? After all, a module or program with ports made of individual signals can be compiled without being instantiated, as shown in Sample 4.32.

Sample 4.32   Module with just port connections

```
module uses_a_port(inout bit not_connected);
  ...
endmodule
```

The compiler creates wires and connects them to the dangling signals. However, a module or program with an interface in its port list must be connected to an instance of the interface.

Sample 4.33   Module with an interface

```
// This will not compile without interface declaration
module uses_an_interface(arb_if.DUT arbif);
  initial arbif.grant = 0;
endmodule
```

For Sample 4.33, the compiler is not able to build even a simple interface. If you have modports or a program block using clocking blocks in an interface, the compiler has an even more difficult time. Even if you are just looking to wring out syntax bugs, you must complete the connections. This can be done as shown in Sample 4.34.

Sample 4.34   Top module connecting DUT and interface

```
module top;
  bit clk;
  always #50 clk = !clk;

  arb_if arbif(clk);              // Interface with modport
  uses_an_interface u1(arbif);  //    needed to compile this
endmodule
```

4.7 Top-Level Scope

Sometimes you need to create things in your simulation that are outside of a program or module so that they are seen by all blocks. In Verilog, only macros extend across module boundaries, and so are used for creating global constants. SystemVerilog introduces the *compilation unit*, that is a group of source files that are compiled together. The scope outside the boundaries of any `module`, `macromodule`, `interface`, `program`, `package`, or `primitive` is known as the *compilation-unit scope*, also referred to as $unit. Anything such as a parameter defined in this scope is similar to a global because it can be seen by all lower-level blocks. However, it is not truly global as the `parameter` cannot be seen during compilation of other files.

This leads to some confusion. Some simulators compile all the SystemVerilog code together, so $unit is global. Other simulators and synthesis tools compile a single module or group of modules at a time, so $unit may be just the contents of one or a few files. As a result, $unit is not portable. Packages allow you to have code outside of a program or module while eliminating the requirement of compiling all the modules at the same time.

This book calls the scope outside blocks the "top-level scope." You can define variables, parameters, data types and even routines in this space. Sample 4.35 declares a top-level parameter, TIMEOUT, that can be used anywhere in the hierarchy. This example also has a `const` string that holds an error message. You can declare top-level constants either way.

Sample 4.35 Top-level scope for arbiter design

```
// root.sv
`timescale 1ns/1ns
parameter int TIMEOUT = 1_000_000;
const string time_out_msg = "ERROR: Time out";
module top;
  test t1();
endmodule

program automatic test;
  ...
  initial begin
    #TIMEOUT;
    $display("%s", time_out_msg);
    $finish;
    end
endprogram
```

The instance name $root allows you to unambiguously refer to names in the system, starting with the top-level scope. In this respect, $root is similar to "/" in the Unix file system. For tools such as VCS that compile all files at once, $root and $unit are equivalent. The name $root also solves an old Verilog problem.

When your code refers to a name in another module, such as i1.var, the compiler first looks in the local scope, then looks up to the next higher scope, and so on until it reaches the top. You may have wanted to use i1.var in the top module, but an instance named i1 in an intermediate scope may have sidetracked the search, giving you the wrong variable. You use $root to make unambiguous cross module references by specifying the absolute path.

Sample 4.36 shows a program that is instantiated in the module top that is implicitly instantiated in the top-level scope. The program can use a relative or absolute reference to the clk signal in the module. You can use a macro to hold the hierarchical path so that when the path changes, you only have to change one piece of code. The LRM does not allow modules to be explicitly instantiated in the top-level scope.

Sample 4.36 Cross-module references with $root

```
module top;
  bit clk;
  test t1(.*);
endmodule

`define TOP $root.top
program automatic test;
  initial begin
    // Absolute reference
    $display("clk=%b", $root.top.clk);
    $display("clk=%b", `TOP.clk);     // With macro

    // Relative reference
    $display("clk=%b", top.clk);
    end
endprogram
```

4.8   Program–Module Interactions

The program block can read and write all signals in modules, and can call routines in modules, but a module has no visibility into a program. This is because your testbench needs to see and control the design, but the design should not depend on anything in the testbench.

 A program can call a routine in a module to perform various actions. The routine can set values on internal signals, also known as "backdoor load." Next, because the current SystemVerilog standard does not define how to force signals from a program block, you need to write a task in the design to do the force, and then call it from the program.

Lastly, it is a good practice for your testbench to use a function to get information from the DUT. Reading signal values can work most of the time, but if the design code changes, your testbench may interpret the values incorrectly. A function in the module can encapsulate the communication between the two and make it easier for your testbench to stay synchronized with the design. Chapter 10 shows how to embed functions and SystemVerilog Assertions in an interface.

4.9 SystemVerilog Assertions

You can create temporal assertions about signals in your design to check their behavior and temporal relationship with SystemVerilog Assertions (SVA). The simulator keeps track of what assertions have triggered, so you can gather functional coverage data on them.

4.9.1 Immediate Assertions

An immediate assertion checks if an expression is true when the statement is executed. Your testbench procedural code can check the values of design signals and testbench variables and take action if there is a problem. For example, if you have asserted the bus request, you expect that grant will be asserted two cycles later. You could use an `if`-statement as shown in Sample 4.37.

Sample 4.37 Checking a signal with an `if`-statement

```
arbif.cb.request <= 2'b01;
repeat (2) @arbif.cb;
if (arbif.cb.grant != 2'b01)
  $display("Error, grant != 2'b01");
```

An assertion is more compact than an `if`-statement. However, note that the logic is reversed compared to the `if`-statement above. You want the expression inside the parentheses to be true; otherwise, print an error as shown in Sample 4.38.

Sample 4.38 Simple immediate assertion

```
arbif.cb.request <= 2'b01;
repeat (2) @arbif.cb;
a1: assert (arbif.cb.grant == 2'b01);
```

If the `grant` signal is asserted correctly, the test continues. If the signal does not have the expected value, the simulator produces a message similar Sample 4.39.

Sample 4.39 Error from failed immediate assertion

```
"test.sv", 7: top.t1.a1: started at 55ns failed at 55ns
Offending '(arbif.cb.grant == 2'b1)'
```

This says that on line 7 of the file `test.sv`, the assertion `top.t1.a1` started at 55ns to check the signal `arbif.cb.grant`, but failed immediately. The label `a1` should be unique so that you can quickly locate the failing assertion.

You may be tempted to use the full SystemVerilog Assertion syntax to check an elaborate sequence over a range of time, but use care as they can be hard to debug. Assertions are declarative code, and execute very differently than the surrounding procedural code. In just a few lines of assertions, you can verify temporal relations; the equivalent procedural code would be more complicated and verbose, but easier for the next person to understand when they have to read your code.

If you are a VHDL programmer, you may be tempted at this point to start sprinkling immediate assertions across your code. Resist the temptation! Your code will work correctly for weeks or months until someone decides to improve simulation performance by disabling assertions. The simulator will no longer execute the expression in the assertion. If the expression has a side effect such as incrementing a value or calling a function, it will no longer occur.

4.9.2 Customizing the Assertion Actions

An immediate assertion has optional then- and else-clauses. If you want to augment the default message, you can add your own as shown in Sample 4.40.

Sample 4.40 Creating a custom error message in an immediate assertion

```
a40: assert (arbif.cb.grant == 2'b01)
else $error("Grant not asserted");
```

If `grant` does not have the expected value, you'll see an error message similar to Sample 4.41.

Sample 4.41 Error from failed immediate assertion

```
"test.sv", 7: top.t1.a40: started at 55ns failed at 55ns
Offending '(arbif.cb.grant == 2'b01)'
Error: "test.sv", 7: top.t1.a40: at time 55 ns
Grant not asserted
```

SystemVerilog has four functions to print messages: `$info`, `$warning`, `$error`, and `$fatal`. These are allowed only inside an assertion, not in procedural code, though future versions of SystemVerilog may allow this.

You can use the then-clause to record when an assertion completed successfully as shown in Sample 4.42.

Sample 4.42 Creating a custom error message

```
a42: assert (arbif.cb.grant == 2'b01)
  grants_received++;            // Another succesful result!
else
  $error("Grant not asserted");
```

4.9.3   Concurrent Assertions

The other type of assertion is the concurrent assertion that you can think of as a small
model that runs continuously, checking the values of signals for the entire simulation.
These are instantiated similarly to other design blocks and are active for the entire
simulation. You need to specify a sampling clock in the assertion. Sample 4.43 has
a small assertion to check that the arbiter request signal does not have X or Z values
except during reset. This code is placed outside of procedural blocks such as `initial`
and `always`. Sample 4.43 is for illustration only. See one of the books listed below
for a more information.

Sample 4.43 Concurrent assertion to check for X/Z

```
interface arb_if(input bit clk);
  logic [1:0] grant, request;
  bit rst;

  property request_2state;
    @(posedge clk) disable iff (rst)
    $isunknown(request) == 0;        // Make sure no Z or X found
  endproperty
  assert_request_2state: assert property (request_2state);

endinterface
```

4.9.4   Exploring Assertions

There are many other uses for assertions. For example, you can put assertions in an
interface. Now your interface not only transmits signal values but also checks the
protocol.

This Section has provided a brief introduction to SystemVerilog Assertions. For
more information, see Vijayaraghhavan and Ramanathan (2005) and Haque et al.
(2007).

4.10   The Four-Port ATM Router

The arbiter example is a good introduction to interfaces, but real designs have more than a single input and output. This section discusses a four-port ATM (Asynchronous Transfer Mode) router, shown in Fig. 4.8.

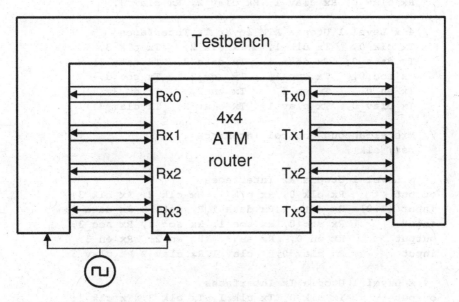

Fig. 4.8 Testbench – ATM router diagram without interfaces

4.10.1   ATM Router with Ports

The following code fragments show the tangle of wires you would have to endure to connect an RTL block to a testbench. First is the header for the ATM router model. This uses the Verilog-1995 style port declarations, where the type and direction are separate from the header.

The actual code for the router in Sample 4.44 is crowded out by nearly a page of port declarations.

Sample 4.44 ATM router model header with ports

```
module atm_router_ports(
  // 4 x Level 1 Utopia ATM layer Rx Interfaces
     Rx_clk_0,   Rx_clk_1,   Rx_clk_2,   Rx_clk_3,
     Rx_data_0,  Rx_data_1,  Rx_data_2,  Rx_data_3,
     Rx_soc_0,   Rx_soc_1,   Rx_soc_2,   Rx_soc_3,
     Rx_en_0,    Rx_en_1,    Rx_en_2,    Rx_en_3,
     Rx_clav_0,  Rx_clav_1,  Rx_clav_2,  Rx_clav_3,

  // 4 x Level 1 Utopia ATM layer Tx Interfaces
     Tx_clk_0,   Tx_clk_1,   Tx_clk_2,   Tx_clk_3,
     Tx_data_0,  Tx_data_1,  Tx_data_2,  Tx_data_3,
     Tx_soc_0,   Tx_soc_1,   Tx_soc_2,   Tx_soc_3,
     Tx_en_0,    Tx_en_1,    Tx_en_2,    Tx_en_3,
     Tx_clav_0,  Tx_clav_1,  Tx_clav_2,  Tx_clav_3,

  // Miscellaneous control interfaces
     rst, clk);

// 4 x Level 1 Utopia Rx Interfaces
  output        Rx_clk_0, Rx_clk_1, Rx_clk_2, Rx_clk_3;
  input [7:0]   Rx_data_0,Rx_data_1,Rx_data_2,Rx_data_3;
  input         Rx_soc_0, Rx_soc_1, Rx_soc_2, Rx_soc_3;
  output        Rx_en_0,  Rx_en_1,  Rx_en_2,  Rx_en_3;
  input         Rx_clav_0,Rx_clav_1,Rx_clav_2,Rx_clav_3;

// 4 x Level 1 Utopia Tx Interfaces
  output        Tx_clk_0, Tx_clk_1, Tx_clk_2, Tx_clk_3;
  output [7:0]  Tx_data_0,Tx_data_1,Tx_data_2,Tx_data_3;
  output        Tx_soc_0, Tx_soc_1, Tx_soc_2, Tx_soc_3;
  output        Tx_en_0,  Tx_en_1,  Tx_en_2,  Tx_en_3;
  input         Tx_clav_0,Tx_clav_1,Tx_clav_2,Tx_clav_3;

// Miscellaneous control interfaces
  input         rst, clk;
  ...
endmodule
```

So what sort of synthesizable code goes in the "…" at the end of Sample 4.44 ?
See Sutherland (2006) for more information and examples of using interfaces in
modules and other SystemVerilog design constructs.

4.10.2   ATM Top-Level Module with Ports

Sample 4.45 contains the top-level module.

Sample 4.45 Top-level module without an interface

```
module top;
  bit clk, rst;
  always #5 clk = !clk;
  wire Rx_clk_0,  Rx_clk_1,  Rx_clk_2,  Rx_clk_3,
       Rx_soc_0,  Rx_soc_1,  Rx_soc_2,  Rx_soc_3,
       Rx_en_0,   Rx_en_1,   Rx_en_2,   Rx_en_3,
       Rx_clav_0, Rx_clav_1, Rx_clav_2, Rx_clav_3,
       Tx_clk_0,  Tx_clk_1,  Tx_clk_2,  Tx_clk_3,
       Tx_soc_0,  Tx_soc_1,  Tx_soc_2,  Tx_soc_3,
       Tx_en_0,   Tx_en_1,   Tx_en_2,   Tx_en_3,
       Tx_clav_0, Tx_clav_1, Tx_clav_2, Tx_clav_3;

  wire [7:0] Rx_data_0, Rx_data_1, Rx_data_2, Rx_data_3,
             Tx_data_0, Tx_data_1, Tx_data_2, Tx_data_3;

  atm_router_ports
              a1(Rx_clk_0, Rx_clk_1, Rx_clk_2, Rx_clk_3,
                 Rx_data_0,Rx_data_1,Rx_data_2,Rx_data_3,
                 Rx_soc_0, Rx_soc_1, Rx_soc_2, Rx_soc_3,
                 Rx_en_0,  Rx_en_1,  Rx_en_2,  Rx_en_3,
                 Rx_clav_0,Rx_clav_1,Rx_clav_2,Rx_clav_3,
                 Tx_clk_0, Tx_clk_1, Tx_clk_2, Tx_clk_3,
                 Tx_data_0,Tx_data_1,Tx_data_2,Tx_data_3,
                 Tx_soc_0, Tx_soc_1, Tx_soc_2, Tx_soc_3,
                 Tx_en_0,  Tx_en_1,  Tx_en_2,  Tx_en_3,
                 Tx_clav_0,Tx_clav_1,Tx_clav_2,Tx_clav_3,
                 rst, clk);

  test_ports t1 (Rx_clk_0, Rx_clk_1, Rx_clk_2, Rx_clk_3,
                 Rx_data_0,Rx_data_1,Rx_data_2,Rx_data_3,
                 Rx_soc_0, Rx_soc_1, Rx_soc_2, Rx_soc_3,
                 Rx_en_0,  Rx_en_1,  Rx_en_2,  Rx_en_3,
                 Rx_clav_0,Rx_clav_1,Rx_clav_2,Rx_clav_3,
                 Tx_clk_0, Tx_clk_1, Tx_clk_2, Tx_clk_3,
                 Tx_data_0,Tx_data_1,Tx_data_2,Tx_data_3,
                 Tx_soc_0, Tx_soc_1, Tx_soc_2, Tx_soc_3,
                 Tx_en_0,  Tx_en_1,  Tx_en_2,  Tx_en_3,
                 Tx_clav_0,Tx_clav_1,Tx_clav_2,Tx_clav_3,
                 rst, clk);
endmodule
```

Sample 4.46 shows the top of the testbench module. Once again, note that the ports and wires take up the majority of the module.

Sample 4.46 Verilog-1995 testbench using ports

```verilog
module test_ports(
    // 4 x Level 1 Utopia ATM layer Rx Interfaces
    Rx_clk_0,  Rx_clk_1,  Rx_clk_2,  Rx_clk_3,
    Rx_data_0, Rx_data_1, Rx_data_2, Rx_data_3,
    Rx_soc_0,  Rx_soc_1,  Rx_soc_2,  Rx_soc_3,
    Rx_en_0,   Rx_en_1,   Rx_en_2,   Rx_en_3,
    Rx_clav_0, Rx_clav_1, Rx_clav_2, Rx_clav_3,

    // 4 x Level 1 Utopia ATM layer Tx Interfaces
    Tx_clk_0,  Tx_clk_1,  Tx_clk_2,  Tx_clk_3,
    Tx_data_0, Tx_data_1, Tx_data_2, Tx_data_3,
    Tx_soc_0,  Tx_soc_1,  Tx_soc_2,  Tx_soc_3,
    Tx_en_0,   Tx_en_1,   Tx_en_2,   Tx_en_3,
    Tx_clav_0, Tx_clav_1, Tx_clav_2, Tx_clav_3,

    rst, clk);     // Miscellaneous control signals

// 4 x Level 1 Utopia Rx Interfaces
   input         Rx_clk_0, Rx_clk_1, Rx_clk_2, Rx_clk_3;
   output [7:0]  Rx_data_0,Rx_data_1,Rx_data_2,Rx_data_3;
   reg    [7:0]  Rx_data_0,Rx_data_1,Rx_data_2,Rx_data_3;
   output        Rx_soc_0, Rx_soc_1, Rx_soc_2, Rx_soc_3;
   reg           Rx_soc_0, Rx_soc_1, Rx_soc_2, Rx_soc_3;
   input         Rx_en_0,  Rx_en_1,  Rx_en_2,  Rx_en_3;
   output        Rx_clav_0,Rx_clav_1,Rx_clav_2,Rx_clav_3;
   reg           Rx_clav_0,Rx_clav_1,Rx_clav_2,Rx_clav_3;

// 4 x Level 1 Utopia Tx Interfaces
   input         Tx_clk_0,  Tx_clk_1,  Tx_clk_2,  Tx_clk_3;
   input [7:0]   Tx_data_0, Tx_data_1,Tx_data_2,Tx_data_3;
   input         Tx_soc_0,  Tx_soc_1,  Tx_soc_2,  Tx_soc_3;
   input         Tx_en_0,  Tx_en_1,  Tx_en_2,   Tx_en_3;
   output        Tx_clav_0, Tx_clav_1,Tx_clav_2,Tx_clav_3;
   reg           Tx_clav_0, Tx_clav_1,Tx_clav_2,Tx_clav_3;

// Miscellaneous control interfaces
   output        rst;
   reg           rst;
   input         clk;

   initial begin
     // Reset the device
     rst <= 1;
     Rx_data_0 <= 0;
     ...
     end
endmodule
```

You just saw three pages of code, and it was all just connectivity — no testbench, no design! Interfaces provide a better way to organize all this information and eliminate the repetitive parts that are so error prone.

4.10.3   Using Interfaces to Simplify Connections

Figure 4.9 shows the ATM router connected to the testbench, with the signals grouped into interfaces.

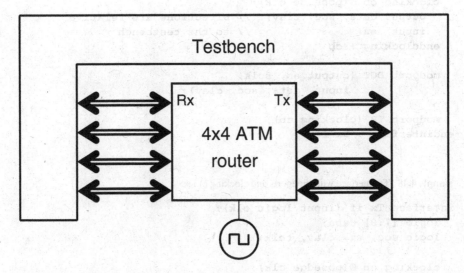

Fig. 4.9 Testbench - router diagram with interfaces

4.10.4   ATM Interfaces

Sample 4.47 and 4.48 show the Rx and Tx interfaces with modports and clocking blocks.

Sample 4.47   Rx interface with modports and clocking block

```
interface Rx_if (input logic clk);
  logic [7:0] data;
  logic soc, en, clav, rclk;

  clocking cb @(posedge clk);
    output data, soc, clav;  // Directions are relative
    input  en;               // to the testbench
  endclocking : cb

  modport DUT (output en, rclk,
               input  data, soc, clav);

  modport TB (clocking cb);
endinterface : Rx_if
```

Sample 4.48   Tx interface with modports and clocking block

```
interface Tx_if (input logic clk);
  logic [7:0] data;
  logic soc, en, clav, tclk;

  clocking cb @(posedge clk);
      input  data, soc, en;
      output clav;
  endclocking : cb

  modport DUT (output data, soc, en, tclk,
               input  clk, clav);

  modport TB (clocking cb);
endinterface : Tx_if
```

4.10.5   ATM Router Model Using an Interface

Sample 4.49 contains the ATM router model and testbench, which need to specify the modport in their port connection list. Note that you put the modport name after the interface name, Rx_if.

Sample 4.49 ATM router model with interface using modports

```
module atm_router(Rx_if.DUT Rx0, Rx1, Rx2, Rx3,
                  Tx_if.DUT Tx0, Tx1, Tx2, Tx3,
                  input logic clk, rst);
  ...
endmodule
```

4.10.6 ATM Top Level Module with Interfaces

The top module, shown in Sample 4.50, has shrunk considerably, along with the chances of making a mistake.

Sample 4.50 Top-level module with interface

```
module top;
  bit clk, rst;
  always #5 clk = !clk;

  Rx_if Rx0 (clk), Rx1 (clk), Rx2 (clk), Rx3 (clk);
  Tx_if Tx0 (clk), Tx1 (clk), Tx2 (clk), Tx3 (clk);

  atm_router a1 (Rx0, Rx1, Rx2, Rx3,                // or just (.*)
                 Tx0, Tx1, Tx2, Tx3, clk, rst);

  test       t1 (Rx0, Rx1, Rx2, Rx3,                // or just (.*)
                 Tx0, Tx1, Tx2, Tx3, clk, rst);
endmodule : top
```

4.10.7 ATM Testbench with Interface

Sample 4.51 shows the part of the testbench that captures cells coming in from the TX port of the router. Note that the interface names are hard-coded, so you have to duplicate the same code four times for the 4×4 ATM router. For example, only the task receive_cell0 is shown, and the final code would also have receive_cell1, receive_cell2, and receive_cell30. Chapter 10 shows how to simplify the code by using virtual interfaces.

Sample 4.51 Testbench using an interface with a clocking block

```
program automatic test(Rx_if.TB Rx0, Rx1, Rx2, Rx3,
                       Tx_if.TB Tx0, Tx1, Tx2, Tx3,
                       input logic clk, output bit rst);

  bit [7:0] bytes[ATM_CELL_SIZE];

  initial begin
    // Reset the device
    rst <= 1;
    Rx0.cb.data <= 0;
    ...
    receive_cell0();
    ...
  end

  task receive_cell0();
    @(Tx0.cb);
    Tx0.cb.clav <= 1;          // Assert ready to receive
    wait (Tx0.cb.soc == 1);    // Wait for Start of Cell

    foreach (bytes[i]) begin
      wait (Tx0.cb.en == 0);   // Wait for enable
        @(Tx0.cb);

      bytes[i] = Tx0.cb.data;
      @(Tx0.cb);
      Tx0.cb.clav <= 0;        // Deassert flow control
    end
  endtask : receive_cell0

endprogram : test
```

4.11   The Ref Port Direction

SystemVerilog introduces a new port direction for connecting modules: `ref`. You should be familiar with the `input`, `output`, and `inout` directions. The last is for modeling bidirectional connections. If you drive a signal with multiple `inout` ports, SystemVerilog will calculate the value of the signal by combining the values of all drivers, taking in to account driver strengths and Z values.

A `ref` port is a different beast. It is essentially a way to make two names that both reference the same variable. There is only one storage location, but multiple aliases. Ref ports can only connect to variables, not signals. See Section 3.4.3 for information on the `ref` direction for routine arguments.

In Sample 4.52, the incr module has two ref ports, c and d. These two variables share storage with the c and d variables in the top module. When top changes the value of c, it is seen immediately by incr. Then incr increments c and the result is seen back in the top module. If the port c was declared as inout, you would have had to build tristate drivers such as continuous assignment statements, and make sure you properly drove an enable signal and Z values. Don't consider ref ports as a convenient replacement for inout ports as only the latter are supported for synthesis.

Sample 4.52 Ref ports

```
module incr(ref int c, d);  // c variable is read and written
  always @(c)
    #1 d = c++; // d=c; c=c+1;
endmodule

module top;
  int c, d;
  incr i1(c, d);
  initial begin
    $monitor("@%0d: c=%0d, d=%0d", $time, c, d);
    c = 2;
    #10;
    c = 8;
    #10;
  end
endmodule
```

4.12   Conclusion

In this chapter you have learned how to use SystemVerilog's interfaces to organize the communication between design blocks and your testbench. With this design construct, you can replace dozens of signal connections with a single interface, making your code easier to maintain and improve, and reducing the number of wiring mistakes.

SystemVerilog also introduces the program block to hold your testbench and to reduce race conditions between the device under test and the testbench. With a clocking block in an interface, your testbenches will drive and sample design signals correctly relative to the clock.

4.13   Exercises

1. Design an interface and testbench for the ARM Advanced High-performance Bus (AHB). You are provided a bus master as verification IP that can initiate AHB transactions. You are testing a slave design. The testbench instantiates the interface, slave, and master. Your interface will display an error if the transaction type is not IDLE or NONSEQ on the negative edge of HCLK. The AHB signals are described in Table 4.2.

Table 4.2 AHB Signal Description

Signal	Width	Direction	Description
HCLK	1	Output	Clock
HADDR	21	Output	Address
HWRITE	1	Output	Write flag: 1=write, 0=read
HTRANS	2	Output	Transaction type: 2′b00=IDLE, 2′b10=NONSEQ
HWDATA	8	Output	Write data
HRDATA	8	Input	Read data

2. For the following interface, add the following code.

   a. A clocking block that is sensitive to the negative edge of the clock, and all I/O that are synchronous to the clock.
   b. A modport for the testbench called `master`, and a modport for the DUT called `slave`
   c. Use the clocking block in the I/O list for the `master` modport.

   ```
   interface my_if(input bit clk);
     bit write;
     bit [15:0] data_in;
     bit [7:0] address;
     logic [15:0] data_out;
   endinterface
   ```

3. For the clocking block in Exercise 2, fill in the `data_in` and `data_out` signals in the following timing diagram.

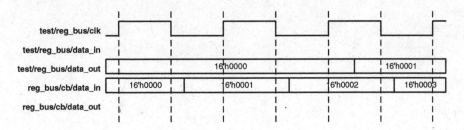

4. Modify the clocking block in Exercise 2 to have:

   a. output skew of 25ns for output write and address
   b. input skew of 15ns
   c. restrict `data_in` to only change on the positive edge of the clock

5. For the clocking block in Exercise 4, fill in the following timing diagram, assuming a clock period of 100ns.

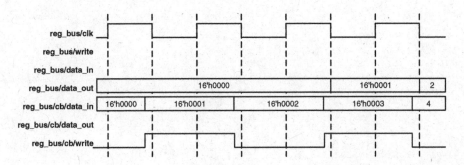

Chapter 5
Basic OOP

5.1 Introduction

With procedural programming languages such as Verilog and C, there is a strong division between data structures and the code that uses them. The declarations and types of data are often in a different file than the algorithms that manipulate them. As a result, it can be difficult to understand the functionality of a program, as the two halves are separate.

Verilog users have it even worse than C users, as there are no structures in Verilog, only bit vectors and arrays. If you wanted to store information about a bus transaction, you would need multiple arrays: one for the address, one for the data, one for the command, and more. Information about transaction N is spread across all the arrays. Your code to create, transmit, and receive transactions is in a module that may or may not be actually connected to the bus. Worst of all, the arrays are all static, so if your testbench only allocated 100 array entries, and the current test needed 101, you would have to edit the source code to change the size and recompile. As a result, the arrays are sized to hold the greatest conceivable number of transactions, but during a normal test, most of that memory is wasted.

Object-Oriented Programming (OOP) lets you create complex data types and tie them together with the routines that work with them. You can create testbenches and system-level models at a more abstract level by calling routines to perform an action rather than toggling bits. When you work with transactions instead of signal transitions, you are working at a higher level, and your code is more easily written and understood. As a bonus, your testbench is decoupled from the design details, making it more robust and easier to maintain and reuse on future projects.

C. Spear and G. Tumbush, *SystemVerilog for Verification: A Guide to Learning the Testbench Language Features*, DOI 10.1007/978-1-4614-0715-7_5,
© Springer Science+Business Media, LLC 2012

If you already are familiar with OOP, skim this chapter, as SystemVerilog follows OOP guidelines fairly closely. Be sure to read Section 5.18 to learn how to build a testbench. Chapter 8 presents advanced OOP concepts such as inheritance and more testbench techniques; it should be read by everyone.

5.2   Think of Nouns, not Verbs

Grouping data and code together helps you in creating and maintaining large testbenches. How should data and code be brought together? You can start by thinking of how you would perform the testbench's job.

The goal of a testbench is to apply stimulus to a design and then check the result to see if it is correct. The data that flows into and out of the design is grouped together into transactions such as bus cycles, opcodes, packets, or data samples. The best way to organize the testbench is around the transactions, and the operations that you perform on them. In OOP, the transaction is the focus of your testbench.

You can think of an analogy between cars and testbenches. Early cars required detailed knowledge about their internals (nouns) to operate. You had to advance or retard the spark, open and close the choke, keep an eye on the engine speed and be aware of the traction of the tires if you drove on a slippery surface such as a wet road. Today your interactions with the car are at a high level. When you get into a car, you perform discrete actions (verbs), such as starting, moving forward, turning, stopping, and listening to music while you drive. If you want to start a car, just turn the key in the ignition, and you are done. Get the car moving by pressing the gas pedal; stop it with the brakes. Are you driving on snow? Don't worry: the anti-lock brakes help you stop safely and in a straight line. You don't have to think about the low level details.

Your testbench should be structured the same way. Traditional testbenches were oriented around the operations that had to happen: create a transaction, transmit it, receive it, check it, and make a report. Instead, you should think about the structure of the testbench, and what each part does. The generator creates transactions and passes them to the next level. The driver talks with the design that responds with transactions that are received by a monitor. The scoreboard checks these against the expected data. You should divide your testbench into blocks, and then define how they communicate. This chapter shows many examples of these components.

How do you represent these blocks in SystemVerilog? A class may describe a data-centric block such as a bus transaction, network packet, or CPU instruction. Or a class might represent a control block such as a driver or scoreboard. Either way, a class encapsulates the data together with the routines that manipulate it. The details of how the class implements actions such as data generation or checking is hidden from the outside, making the class more reusable.

5.3   Your First Class

Sample 5.1 shows a class for a generic transaction. It contains an address, a checksum, and an array of data values. There are two routines in the `Transaction` class: one that displays the address, and another that computes a checksum of the data.

 To make it easier to match the beginning and end of a named block, you can put a label on the end of it. In Sample 5.1 these end labels may look redundant, but in complex code, with many nested blocks, the labels help you find the mate for a simple `end`, `endtask`, `endfunction`, or `endclass`.

Sample 5.1 Simple transaction class

```
class Transaction;
  bit [31:0] addr, csm, data[8];

  function void display();
    $display("Transaction: %h", addr);
  endfunction : display

  function void calc_csm();
    csm = addr ^ data.xor;
  endfunction : calc_csm

endclass : Transaction
```

 Every company has its own naming style. This book uses the following convention: Class names start with a capital letter and avoid using underscores, as in `Transaction` or `Packet`. Constants are all upper case, as in `CELL_SIZE`, and variables are lower case, as in `count` or `opcode`. You are free to use whatever style you want.

5.4   Where to Define a Class

You can define and use classes in SystemVerilog in a `program`, `module`, `package`, or outside of any of these.

When you start a project, you might store a single class per file. When the number of files gets too large, you can group a set of related classes and type definitions into a SystemVerilog `package` as shown in Sample 5.2. For instance, you might group together all USB3 transactions and BFMs into a single `package`. Now you can compile the package separately from the rest of the system. Unrelated classes, such as those for other transactions, scoreboards, or different protocols, should remain in separate files.

Code samples in this book leave out the packages to keep the text more compact.

Sample 5.2 Class in a package

```
// File abc.svh
package abc;
  class Transaction;
    // Class body
  endclass
endpackage
```

Sample 5.3 shows how to import a package into a program.

Sample 5.3 Importing a package in a program

```
program automatic test;
  import abc::*;
  Transaction tr;

  // Test code

endprogram
```

5.5   OOP Terminology

What separates you, an OOP novice, from an expert? The first thing is the words you use. You already know some OOP concepts from working with Verilog. Here are some OOP terms, definitions, and rough equivalents in Verilog 2001.

- Class – a basic building block containing routines and variables. The analogue in Verilog is a module.
- Object – an instance of a class. In Verilog, you need to instantiate a module to use it.
- Handle – a pointer to an object. In Verilog, you use the name of an instance when you refer to signals, and routines from outside the module. A handle is like the address of the object, but stored in a pointer that can only refer to one type. A handle is similar to a reference in other OOP languages.
- Property – a variable that holds data. In Verilog, this is a signal such as a register or wire.
- Method – the procedural code that manipulates variables, contained in tasks and functions. Verilog modules have tasks and functions plus `initial` and `always` blocks.
- Prototype – the header of a routine that shows the name, type, and argument list plus return type. The body of the routine contains the executable code. See Section 5.10 for more on prototypes and out-of-body experiences.

   This book uses the more traditional terms from Verilog of "variable" and "routine" when discussing non-OOP code, and "property" and "method" for classes.

   In Verilog you build complex designs by creating modules and instantiating them hierarchically. In OOP you create classes and construct them (creating objects) to create a similar hierarchy. Modules are instantiated during compilation while classes are constructed at run time.

   Here is an analogy to explain these OOP terms. Think of a class as the blueprint for a house. This plan describes the structure of the house, but you cannot live in a blueprint; you need to build the physical house. An object is the actual house. Just as one set of blueprints can be used to build a whole subdivision of houses, a single class can be used to build many objects. The house address is like a handle in that it uniquely identifies your house. Inside your house you have things such as lights (on or off), with switches to control them. A class has variables that hold values, and routines that control the values. A class for the house might have many lights. A single call to `turn_on_porch_light()` sets the light variable ON in a single house.

5.6   Creating New Objects

Both Verilog and OOP have the concept of instantiation, but there are some differences in the details. A Verilog module, such as a counter, is instantiated when you compile your design. A SystemVerilog class, such as a network packet, is instantiated during simulation, when needed by the testbench. Verilog instances are static, as the hardware does not change during simulation; only signal values change. SystemVerilog stimulus objects are constantly being created and used to drive the DUT and check the results. Later, the objects may be freed so their memory can be used by new ones. Back to the house analogy: the address is normally static, unless your house burns down, causing you to construct a new one. Garbage collection at home is not automatic, especially if you have teenagers.

   The analogy between OOP and Verilog has a few other exceptions. The top-level Verilog module is implicitly instantiated. However, a SystemVerilog class must be instantiated before it can be used. Next, a Verilog instance name only refers to a single instance, whereas a SystemVerilog handle can refer to many objects, though only one at a time.

5.6.1   Handles and Constructing Objects

In Sample 5.4, `tr` is a handle that points to an object of type `Transaction`. For brevity, you can just say `tr` is a `Transaction` handle.

Sample 5.4 Declaring and using a handle

```
Transaction tr;   // Declare a handle
tr = new();       // Allocate a Transaction object
```

When you declare the handle `tr`, it is initialized to the special value `null`. On the next line, call the `new()` function to construct the `Transaction` object.

This special `new` function allocates space for the `Transaction`, initializes the variables to their default value (0 for 2-state variables and X for 4-state ones), and returns the address where the object is stored. For example, the `Transaction` class has two 32-bit registers (`addr` and `csm`) and an array with eight values (data), for a total of 10 longwords, or 40 bytes. So when you call `new`, SystemVerilog allocates at least 40 bytes of storage. If you have used C, this step is similar to calling the `malloc` function. Note that SystemVerilog requires additional memory for 4-state variables and housekeeping information such as the object's type.

This process is called instantiation as you are making an instance of the class. The `new` function is sometimes called the constructor, as it builds the object, just as your carpenter constructs a house from wood and nails. For every class, SystemVerilog creates a default `new` function to allocate and initialize an object.

5.6.2   *Custom Constructor*

You can define your own `new()` function to set your own values. Note that you must not give a return value type as the constructor is a special function and automatically returns a handle to an object of the same type as the class.

Sample 5.5 Simple user-defined new() function

```
class Transaction;
  logic [31:0] addr, csm, data[8];

  function new();
    addr = 3;
    data = '{default:5};
  endfunction

endclass
```

In Sample 5.5, first SystemVerilog allocates the space for the object automatically. Next it sets `addr` and `data` to fixed values but leaves `csm` at its default value of X. You can use arguments with default values to make a more flexible constructor, as shown in Sample 5.6. Now you can specify the value for `addr` and `data` when you call the constructor, or use the default values.

Sample 5.6   A new() function with arguments

```
class Transaction;
  logic [31:0] addr, csm, data[8];

  function new(input logic [31:0] a=3, d=5);
    addr = a;
    data = '{default:d};
  endfunction
endclass

initial begin
  Transaction tr;
  tr = new(.a(10));   // a=10, d uses default of 5
end
```

How does SystemVerilog know which new() function to call? It looks at the type of the handle on the left side of the assignment. In Sample 5.7, the call to new inside the Driver constructor calls the new() function for Transaction, even though the one for Driver is closer. Since tr is a Transaction handle, SystemVerilog does the right thing and creates an object of type Transaction.

Sample 5.7   Calling the right new() function

```
class Transaction;
  logic [31:0] addr, csm, data[8];
endclass : Transaction

class Driver;
  Transaction tr;
  function new();        // Driver's new function
    tr = new();          // Call the Transaction new function
  endfunction
endclass : Driver
```

5.6.3   *Separating the Declaration and Construction*

You should avoid declaring a handle and calling the constructor, new, all in one statement. While this is legal syntax and less verbose, it can create ordering problems, as the constructor is called before the first procedural statement. You might need to initialize objects in a certain order, but if you call new() in the declaration, you won't have the same control. Additionally, if you forget to use automatic storage, the constructor is called at the start of simulation, not when the block is entered.

5.6.4 The Difference Between New() and New[]

You may have noticed that this `new()` function looks a lot like the `new[]` operator described in Section 2.3, used to set the size of dynamic arrays. They both allocate memory and initialize values. The big difference is that the `new()` function is called to construct a single object, whereas the `new[]` operator is building an array with multiple elements. `new()` can take arguments for setting object values, whereas `new[]` only takes a single value for the number of elements in the array. Just remember that the new with square brackets `[]` is for arrays, while the one with parentheses `()` is for classes, which usually contain methods.

5.6.5 Getting a Handle on Objects

 New OOP users often confuse an object with its handle. The two are very distinct. You **declare** a handle and **construct** an object. Over the course of a simulation, a handle can point to many objects. This is the dynamic nature of OOP and SystemVerilog. Don't get the handle confused with the object.

In Sample 5.8, `t1` first points to one object, then another. Fig. 5.1 shows the resulting handles and objects.

Sample 5.8 Allocating multiple objects

```
Transaction t1, t2;   // Declare two handles
t1 = new();           // Allocate first Transaction object
t2 = t1;              // t1 & t2 point to it
t1 = new();           // Allocate second Transaction object
```

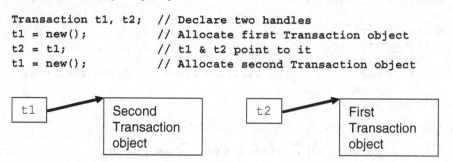

Fig. 5.1 Handles and objects after allocating multiple objects

Why would you want to create objects dynamically? During a simulation you may need to create hundreds or thousands of transactions. SystemVerilog lets you create objects automatically, when you need them. In Verilog, you would have to use a fixed-size array large enough to hold the maximum number of transactions.

Note that this dynamic creation of objects is different from anything else offered before in the Verilog language. An instance of a Verilog module and its name are bound together statically during compilation. Even with `automatic` variables, which come and go during simulation, the name and storage are always tied together.

An analogy for handles is people who are attending a conference. Each person is similar to an object. When you arrive, a badge is "constructed" by writing your name on it. This badge is a handle that can be used by the organizers to keep track of each person. When you take a seat for the lecture, space is allocated. You may have multiple badges for attendee, presenter, or organizer. When you leave the conference, your badge may be reused by writing a new name on it, just as a handle can point to different objects through assignment. Lastly, if you lose your badge and there is nothing to identify you, you will be asked to leave. The space you take, your seat, is reclaimed for use by someone else.

5.7 Object Deallocation

Now you know how to create an object — but how do you get rid of it? For example, your testbench creates and sends thousands of transactions, such as packets, instructions, frames, interrupts, etc. into your DUT. Once you know the transaction has completed successfully, you don't need to keep it around. You should reclaim the memory; otherwise, a long simulation might run out of memory.

Garbage collection is the process of automatically freeing objects that are no longer referenced. One way SystemVerilog can tell if an object is no longer being used is by keeping track of the number of handles that point to it. When the last handle no longer references an object, SystemVerilog releases the memory for it. (The actual algorithm to find unused objects varies between simulators. This section describes reference counting, which is the easiest to understand).

Sample 5.9 Creating multiple objects

```
Transaction t;    // Create a handle
t = new();        // Allocate a new Transaction
t = new();        // Allocate a second one, free the first
t = null;         // Deallocate the second
```

The second line in Sample 5.9 calls new() to construct an object and store the address in the handle t. The next call to new() constructs a second object and stores its address in t, overwriting the previous value. Since there are no handles pointing to the first object, SystemVerilog can deallocate it. The object may be deleted immediately, or after a short wait. The last line explicitly clears the handle so that now the second object can be deallocated.

If you are familiar with C++, these concepts of objects and handles are familiar, but there are some important differences. A SystemVerilog handle can only point to objects of one type, so they are called "type-safe." In C, a typical void pointer is only an address in memory, and you can set it to any value or modify it with operators such as pre-increment. You cannot be sure that a pointer is valid. A C++ typed pointer is much safer, but you may be tempted by C's flexibility. SystemVerilog does not allow any modification of a handle or using a handle of one type to refer to an object of another type. (SystemVerilog's OOP specification is closer to Java than C++).

Since SystemVerilog garbage collects an object when no more handles refer to it, you can be sure your code always uses valid handles. In C / C++, a pointer can refer to an object that no longer exists. Garbage collection in those languages is manual, so your code can suffer from "memory leaks" when you forget to deallocate objects.

SystemVerilog cannot garbage collect an object that is still referenced somewhere by a handle. For example, if you keep objects in a linked list, SystemVerilog cannot deallocate the objects until you manually clear all handles by setting them to `null`. If an object contains a routine that forks off a thread, the object is not deallocated while the thread is running. Likewise, any objects that are used by a spawned thread may not be deallocated until the thread terminates. See Chapter 7 for more information on threads.

5.8   Using Objects

Now that you have allocated an object, how do you use it? Going back to the Verilog module analogy, you can refer to variables and routines in an object with the "." notation, as shown in Sample 5.10.

Sample 5.10 Using variables and routines in an object

```
Transaction t;        // Declare a handle to a Transaction
t = new();            // Construct a Transaction object
t.addr = 32'h42;      // Set the value of a variable
t.display();          // Call a routine
```

In strict OOP, the only access to variables in an object should be through accessor functions such as `get()` and `put()`. This is because accessing variables directly limits your ability to change the underlying implementation in the future. If a better (or simply different) algorithm comes along in the future, you may not be able to adopt it because you would also need to modify all of the references to the variables.

The problem with this methodology is that it was written for large software applications with lifetimes of a decade or more. With dozens of programmers making modifications, stability is paramount. However, you are creating a testbench, where the goal is maximum control of all variables to generate the widest range of stimulus values.

One of the ways to accomplish this is with constrained-random stimulus generation, which cannot be done if a variable is hidden behind a screen of methods. While the `get()` and `put()` methods are fine for compilers, GUIs, and APIs, you should stick with public variables that can be directly accessed anywhere in your testbench.

The exception to this rule is for verification IP that is created and maintained by a group such as a company that has no direct relationship to the end user. For example, if you purchase a PCI transactor from another company, they will restrict access to the internals, forcing you to treat it as a black box. The developer must give you enough methods to generate both good transactions and inject all flavors of errors.

5.9   Class Methods

A method in a class is just a `task` or `function` defined inside the scope of the class. Sample 5.11 defines `display()` methods for the `Transaction` and `PCI_Tran`. SystemVerilog calls the correct one, based on the handle type.

Sample 5.11   Routines in the class

```
class Transaction;
  bit [31:0] addr, csm, data[8];
  function void display();
    $display("@%0t: TR addr=%h, csm=%h, data=%p",
             $time, addr, csm, data);
  endfunction
endclass

class PCI_Tran;
  bit [31:0] addr, data;  // Use realistic names
  function void display();
    $display("@%0t: PCI: addr=%h, data=%h", $time, addr, data);
  endfunction
endclass

Transaction t;
PCI_Tran pc;

initial begin
  t = new();        // Construct a Transaction
  t.display();      // Display a Transaction
  pc = new();       // Construct a PCI transaction
  pc.display();     // Display a PCI Transaction
end
```

A method in a class uses automatic storage by default, so you don't have to worry about remembering the `automatic` modifier.

5.10   Defining Methods Outside of the Class

A good rule of thumb is you should limit a piece of code to one "page" or screen in your favorite editor to keep it understandable. You may be familiar with this rule for routines, but it also applies to classes. If you can see everything in a class on the screen at one time, it is easier to understand.

However, if each method takes a page, how can the whole class fit on a page? In SystemVerilog you can break a method into the prototype (method name and arguments) inside the class, and the body (the procedural code) that goes after the class.

Here is how you create out-of-block declarations. Copy the first line of the method, with the name and arguments, and add the `extern` keyword at the beginning. Now take the entire method and move it after the class body, and add the class name and two colons (`::` the scope operator) before the method name. The above classes could be defined as shown in Sample 5.12.

Sample 5.12   Out-of-block method declarations

```
class Transaction;
  bit [31:0] addr, csm, data[8];
  extern function void display();
endclass

function void Transaction::display();
  $display("@%0t: Transaction addr=%h, csm=%h, data=%p",
           $time, addr, csm, data);
endfunction

class PCI_Tran;
  bit [31:0] addr, data;   // Use realistic names
  extern function void display();
endclass

function void PCI_Tran::display();
  $display("@%0t: PCI: addr=%h, data=%h",
           $time, addr, data);
endfunction
```

A common coding mistake is when the prototype does not match the out-of-body. SystemVerilog requires that the prototype be identical to the out-of-block method declaration, except for the class name and scope operator, `::`. The prototype can have qualifiers such as `local`, `protected`, or `virtual`, but not the out-of-body. If any arguments have default values, they must be given in the prototype, but they are optional in the out-of-body.

Another common mistake is to leave out the class name when you declare the method outside of the class. As a result, it is defined at the next higher scope (probably the program or package scope), and the compiler gives an error when the task tries to access class-level variables and methods. This is shown in Sample 5.13.

Sample 5.13   Out-of-body method missing class name

```
class Bad_OOB;
  bit [31:0] addr, csm, data[8]; // Class-level variable
  extern function void display();
endclass

function void display();          // Missing "Bad_OOB::"
  $display("addr=%0d", addr);     // Error, addr not found
endfunction
```

5.11   Static Variables vs. Global Variables

Every object has its own local variables that are not shared with any other object. If you have two Transaction objects, each has its own addr, csm, and data variables. Sometimes though, you need a variable that is shared by all objects of a certain type. For example, you might want to keep a running count of the number of transactions that have been created. Without OOP, you would probably create a global variable. Then you would have a global variable that is used by one small piece of code, but is visible to the entire testbench. This "pollutes" the global name space and makes variables visible to everyone, even if you want to keep them local.

5.11.1   A Simple Static Variable

In SystemVerilog you can create a static variable inside a class. This variable is shared amongst all instances of the class, but its scope is limited to the class. In Sample 5.14, the static variable count holds the number of objects created so far. It is initialized to 0 in the declaration because there are no transactions at the beginning of the simulation. Each time a new object is constructed, it is tagged with a unique value, and count is incremented.

Sample 5.14 Class with a static variable

```
class Transaction;
  static int count = 0; // Number of objects created
  int id;              // Unique instance ID
  function new();
    id = count++;      // Set ID, bump count
  endfunction
endclass

Transaction t1, t2;
initial begin
  t1 = new();          // 1st instance, id=0, count=1
  $display("First id=%0d, count=%0d", t1.id, t1.count);
  t2 = new();          // 2nd instance, id=1, count=2
  $display("Second id=%0d, count=%0d", t2.id, t2.count);
end
```

In Sample 5.14, there is only one copy of the static variable count, regardless of how many Transaction objects are created. You can think that count is stored with the class and not the object. The variable id is not static, so every Transaction has its own copy, as shown in Fig. 5.2. Now you don't need to make a global variable for the count.

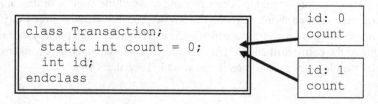

Fig. 5.2 Static variables in a class

 Using the ID field is a good way to track objects as they flow through a design. When debugging a testbench, you often need a unique value. SystemVerilog does not let you print the address of an object, but you can make an ID field. Whenever you are tempted to make a global variable, consider making a class-level static variable. A class should be self-contained, with as few outside references as possible.

5.11.2 Accessing Static Variables Through the Class Name

Sample 5.14 showed how you can reference a static variable using a handle. You don't need a handle; you could use the class name followed by : :, the class scope resolution operator, shown in Sample 5.15.

Sample 5.15 The class scope resolution operator

```
class Transaction;
  static int count = 0;              // Number of objects created
endclass

initial begin
  run_test();
  $display("%0d transactions were created",
           Transaction::count);  // Reference static w/o handle
end
```

5.11.3   Initializing Static Variables

A static variable is usually initialized in the declaration. You can't easily initialize it in the class constructor, as this is called for every single new object. You would need another static variable to act as a flag, indicating whether the original variable had been initialized. If you have a more elaborate initialization, you could use an initial block. Make sure static variables are initialized before the first object is constructed.

Another use for a static variable is when every instance of a class needs information from a single object. For example, a transaction class may refer to a configuration object that has the number of transactions. If you have a non-static handle in the Transaction class, every object will have its own copy, wasting space. Sample 5.16 shows how to use a static variable instead.

Sample 5.16 Static storage for a handle

```
class Transaction;
  static Config cfg;       // A handle with static storage
endclass

initial begin
  Transaction::cfg = new(.num_trans(42));
end
```

5.11.4   Static Methods

As you employ more static variables, the code to manipulate them may grow into a full fledged routine. In SystemVerilog you can create a static method inside a class that can read and write static variables, even before the first instance has been created.

Sample 5.17 has a simple static function to display the values of the static variables. SystemVerilog does not allow a static method to read or write non-static variables, such as id. You can understand this restriction based on the code below. When the function display_statics is called at the end of the example, no Transaction objects have been constructed, so no storage has been created for id variables.

Sample 5.17 Static method displays static variable

```
class Transaction;
  static Config cfg;
  static int count = 0;
  int id;

  // Static method to display static variables.
  static function void display_statics();
    if (cfg == null)
      $display("ERROR: configuration not set");
    else
      $display("Transaction cfg.num_trans=%0d, count=%0d",
               cfg.num_trans, count);
  endfunction
endclass

Config cfg;
initial begin
  cfg = new(.num_trans(42));            // Pass argument by name
  Transaction::cfg = cfg;
  Transaction::display_statics();  // Static method call
end
```

5.12   Scoping Rules

When writing your testbench, you need to create and refer to many variables.
SystemVerilog follows the same basic rules as Verilog, with a few helpful
improvements.

A scope is a block of code such as a module, program, task, function, class, or
begin/end block. The for and foreach loops automatically create a block so
that an index variable can be declared or created local to the scope of the loop.

You can only define new variables in a block. New in SystemVerilog is the ability to declare a variable in an unnamed begin-end block.

A name can be relative to the current scope or absolute starting with $root. For
a relative name, SystemVerilog looks up the list of scopes until it finds a match. If
you want to be unambiguous, use $root at the start of a name. Variables can not be
declared in $root, that is, outside of any module, program or package.

Sample 5.18 uses the same name in several scopes. Note that in actual code, you
would use more meaningful names! The name limit is used for a global variable,
a program variable, a class variable, a function variable, and a local variable in an
initial block. The latter is in an unnamed block, so the label created is tool dependent, along with the signal's hierarchical name.

Sample 5.18   Name scope

```
program automatic top;
  int limit;                    // $root.top.limit

  class Foo;
    int limit, array[];         // $root.top.Foo.limit

    // $root.top.Foo.print.limit
    function void print (input int limit);
      for (int i=0; i<limit; i++)
        $display("%m: array[%0d]=%0d", i, array[i]);
    endfunction
  endclass

  initial begin
    int limit = 3;
    Foo bar;

    bar = new();
    bar.array = new[limit];
    bar.print (limit);
  end
endprogram
```

For testbenches, you can declare variables in the `program` or in the `initial` block. If a variable is only used inside a single `initial` block, such as a counter, you should declare it there to avoid possible name conflicts with other blocks. Note that if you declare a variable in an unnamed block, such as the `initial` in Sample 5.18, there is no hierarchical name that works consistently across all tools.

Declare your classes outside of any `program` or `module` in a `package`. This approach can be shared by all the testbenches, and you can declare temporary variables at the innermost possible level. This style also eliminates a common bug that happens when you forget to declare a variable inside a class. SystemVerilog looks for that variable in higher scopes.

If a block uses an undeclared variable, and another variable with that name happens to be declared in the program block, the class uses it instead, with no warning. In Sample 5.19, the function `Bad::display` did not declare the loop variable i, so SystemVerilog uses the program level i instead. Calling the function changes the value of `test.i`, probably not what you want!

Sample 5.19   Class uses wrong variable

```
program automatic test;
  int i;  // Program-level variable

  class Bad;
    logic [31:0] data[];

    // Calling this function changes the program variable i
    function void display();
      // Forgot to declare i in next statement
      for (i=0; i<data.size(); i++)
        $display("data[%0d]=%x", i, data[i]);
    endfunction
  endclass
endprogram
```

If you move the class into a package, the class cannot see the program-level variables, and thus won't use them unintentionally as shown in Sample 5.20.

Sample 5.20   Move class into package to find bug

```
package Better;
  class Bad;
    logic [31:0] data[];

    // ** Will not compile because of undeclared i
    function void display();
      for (i = 0; i<data.size(); i++)
        $display("data[%0d]=%x", i, data[i]);
    endfunction
  endclass
endpackage


program automatic test;
  int i;  // Program-level variable
  import Better::*;
  //...
endprogram
```

5.12.1 . What is This?

When you use a variable name, SystemVerilog looks in the current scope for it, and then in the parent scopes until the variable is found. This is the same algorithm used by Verilog. What if you are deep inside a class and want to unambiguously refer to a class-level object? This style code is most commonly used in constructors, where the programmer uses the same name for a class variable and an argument. In Sample 5.21, the keyword "this" removes the ambiguity to let SystemVerilog know that you are assigning the local variable, name, to the class variable, name.

Sample 5.21 Using this to refer to class variable

```
class Scoping;
  string name;

  function new(input string name);
    this.name = name;      // class name = local name
  endfunction

endclass
```

Some people think this argument naming style makes the code easier to read; others think it is a shortcut by a lazy programmer.

5.13   Using One Class Inside Another

A class can contain an instance of another class, using a handle to an object. This is just like Verilog's concept of instantiating a module inside another module to build up the design hierarchy. A common reason for using containment are code reuse and controlling complexity. For example, every one of your transactions may have a statistics block, including timestamps indicating when the transaction started and ended transmission, and information about all transactions, as shown in Fig. 5.3 and Sample 5.22.

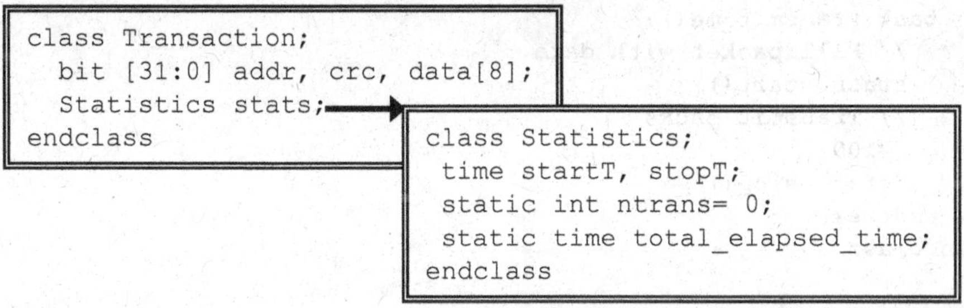

```
class Transaction;
  bit [31:0] addr, crc, data[8];
  Statistics stats;
endclass
```
→
```
class Statistics;
  time startT, stopT;
  static int ntrans= 0;
  static time total_elapsed_time;
endclass
```

Fig. 5.3 Contained objects Sample 5.22

Sample 5.22 Statistics class declaration

```
class Statistics;
  time startT;                    // Transaction start time
  static int ntrans = 0;          // Transaction count
  static time total_elapsed_time = 0;

  function void start();
    startT = $time;
  endfunction

  function void stop();
    time how_long = $time - startT;
    ntrans++;                     // Another trans completed
    total_elapsed_time += how_long;
  endfunction

endclass
```

Now you can use the Statistics class inside another class such as a transaction as can been seen in Sample 5.23.

Sample 5.23 Encapsulating the Statistics class

```
class Transaction;
  bit [31:0] addr, csm, data[8];
  Statistics stats;              // Statistics handle

  function new();
    stats = new();               // Make instance of Statistics
  endfunction

  task transmit_me();
    // Fill packet with data
    stats.start();
    // Transmit packet
    #100;
    stats.stop();
  endtask
endclass
```

The outermost class, Transaction, can refer to things in the Statistics class using the usual hierarchical syntax, such as stats.startT.

Remember to instantiate the object; otherwise, the handle stats is null and the call to start fails. This is best done in the constructor of the outer class, Transaction.

As your classes become larger, they may become hard to manage. When your variable declarations and method prototypes grow larger than a page, you should see if there is a logical grouping of items in the class so that it can be split into several smaller ones.

This is also a potential sign that it's time to refactor your code, i.e., split it into several smaller, related classes. See Chapter 8 for more details on class inheritance. Look at what you're trying to do in the class. Is there something you could move into one or more base classes, i.e., decompose a single class into a class hierarchy? A classic indication is similar code appearing at various places in the class. You need to factor that code out into a function in the current class, one of the current class's parent classes, or both.

5.13.1   How Big or Small Should My Class Be?

 Just as you may want to split up classes that are too big, you should also have a lower limit on how small a class should be. A class with just one or two members makes the code harder to understand as it adds an extra layer of hierarchy and forces you to constantly jump back and forth between the parent class and all the children to understand what it does. In addition, look at how often it is used. If a small class is only instantiated once, you might want to merge it into the parent class.

One Synopsys customer put each transaction variable into its own class for fine control of randomization. The transaction had a separate object for the address, checksum, data, etc. In the end, this approach only made the class hierarchy more complex. On the next project they flattened the hierarchy.

See Section 8.4 for more ideas on partitioning classes.

5.13.2   Compilation Order Issue

Sometimes you need to compile a class that includes another class that is not yet defined. The declaration of the handle causes an error, as the compiler does not recognize the new type. Declare the class name with a `typedef` statement, as shown in Sample 5.24.

Sample 5.24 Using a typedef class statement

```
typedef class Statistics;   // Define a lower level class

class Transaction;
  Statistics stats;          // Use Statistics class
  ...
endclass

class Statistics;            // Define Statistics class
  ...
endclass
```

5.14   Understanding Dynamic Objects

In a statically allocated language such as Verilog, every signal has a unique variable associated with it. For example, there may be a wire called grant, the integer count, and a module instance i1. In OOP, there is not the same one-to-one correspondence. There can be many objects, but only a few named handles. A testbench may allocate a thousand transaction objects during a simulation, but may only have a few handles to manipulate them. This situation takes some getting used to if you have only written Verilog code.

In reality, there is a handle pointing to every active object. Some handles may be stored in arrays or queues, or in another object, like a linked list. For objects stored in a mailbox, the handle is in an internal SystemVerilog structure. See Section 7.6 for more information on mailboxes. Remember that as soon as you assign a new value to the last handle pointing to an object, that object can be garbage collected.

5.14.1   Passing Objects and Handles to Methods

What happens when you pass an object into a method? Perhaps the method only needs to read the values in the object, such as transmit above. Or, your method may modify the object, like a method to create a packet. Either way, when you call the method, you pass a handle to the object, not the object itself.

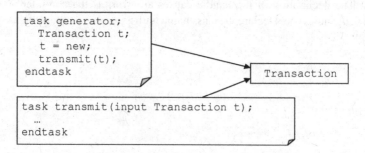

Fig. 5.4 Handles and objects across methods

In Fig. 5.4, the `generator` task has just called `transmit`. There are two handles, `generator.t` and `transmit.t`, that both refer to the same object.

When you call a method, if you pass a scalar variable such as a handle into a `ref` argument, SystemVerilog passes the address of the variable so the method can modify it. If you don't use `ref`, SystemVerilog copies the scalar's value into the argument variable, so any change to the argument in the method does not affect the original value.

Sample 5.25   Passing objects

```
// Transmit a packet onto a 32-bit bus
task transmit(input Transaction tr);
   tr.data[0] = ~tr.data[0];  // Corrupt the first word
   CBbus.rx_data <= tr.data[0];
   ...
endtask

Transaction tr;
initial begin
   tr = new();          // Allocate the object
   tr.addr = 42;        // Initialize values
   transmit(tr);        // Pass object to task
end
```

In Sample 5.25, the initial block allocates a `Transaction` object and calls the `transmit` task with the handle that points to the object. Using this handle, `transmit` can read and write values in the object. However, if `transmit` tries to modify the handle, the result won't be seen in the initial block, as the `t` argument was not declared as `ref`.

A method can modify an object, even if the handle argument does not have a `ref` modifier. This frequently causes confusion for new users, as they mix up the handle with the object. As shown above, `transmit` can modify `data[0]` in the object without changing the value of `t`. If you don't want an object modified in a method, pass a copy of it so that the original object is untouched. See Section 5.15 for more on copying objects.

5.14.2   Modifying a Handle in a Task

A common coding mistake is to forget to use `ref` on method arguments that you want to modify, especially handles. In Sample 5.26, the argument `tr` is not declared as `ref`, so any change to it is not be seen by the calling code. The argument `tr` has the default direction of `input`.

Sample 5.26 Bad transaction creator task, missing `ref` on handle

```
function void create(Transaction tr); // Bug, missing ref
  tr = new();
  tr.addr = 42;
  // Initialize other fields
  ...
endfunction

Transaction t;
initial begin
  create(t);               // Create a transaction
  $display(t.addr);        // Fails because t=null
end
```

Even though `create` modified the argument `tr`, the handle `t` in the calling block remains `null`. You need to declare the argument `tr` as `ref` as can be seen in Sample 5.27.

Sample 5.27 Good transaction creator task with ref on handle

```
function void create(ref Transaction tr);
  ...
endfunction : create
```

If a method is only going to modify the properties of the object, the method should declare the handle as an input argument. If a method is going to modify the handle, for example to make it point to a new object, the method must declare the handle as a ref argument.

5.14.3 Modifying Objects in Flight

A very common mistake is forgetting to create a new object for each transaction in the testbench. In Sample 5.28, the `generate_bad` task creates a `Transaction` object with random values, and transmits it into the design over several cycles.

Sample 5.28 Bad generator creates only one object

```
task generator_bad(input int n);
  Transaction t;
  t = new();                      // Create one new object
  repeat (n) begin
    t.addr = $random();           // Initialize variables
    $display("Sending addr=%h", t.addr);
    transmit(t);                  // Send it into the DUT
  end
endtask
```

What are the symptoms of this mistake? The code above creates only one Transaction, so every time through the loop, generator_bad changes the object at the same time it is being transmitted. When you run this, the $display shows many addr values, but all transmitted Transaction objects have the same value of addr. The bug becomes visible when transmit spawns off a thread that takes several cycles to send the transaction, and so the values in the object are re-randomized in the middle of transmission. If your transmit task makes a copy of the object, you can recycle the same object over and over. This bug can also happen with mailboxes as shown in Sample 7.32

To avoid this bug, you need to create a new Transaction during each pass through the loop as seen in Sample 5.29.

Sample 5.29 Good generator creates many objects

```
task generator_good(input int n);
  Transaction t;
  repeat (n) begin
    t = new();                    // Create one new object
    t.addr = $random();           // Initialize variables
    $display("Sending addr=%h", t.addr);
    transmit(t);                  // Send it into the DUT
  end
endtask
```

5.14.4 Arrays of Handles

As you write testbenches, you need to be able to store and reference many objects. You can make arrays of handles, each of which refers to an object. Sample 5.30 shows storing ten bus transaction handles in an array.

Sample 5.30 Using an array of handles

```
task generator();
  Transaction tarray[10];
  foreach (tarray[i]) begin
    tarray[i] = new();      // Construct each object
    transmit(tarray[i]);
  end
endtask
```

The array `tarray` is made of handles, not objects. So you need to construct each object in the array before using it, just as you would for a normal handle. There is no way to call **new** on an entire array of handles.

There is no such thing as an "array of objects", though you may use this term as a shorthand for an array of handles that points to objects. Keep in mind that some handles may be set to `null`, or that multiple handles could point to a single object.

5.15　Copying Objects

You may want to make a copy of an object to keep a method from modifying the original, or in a generator to preserve the constraints. You can either use the simple, built-in copy available with `new` operator or you can write your own for more complex classes. See Section 8.2 for more reasons why you should make a copy method.

5.15.1　Copying an Object with the New Operator

Copying an object with the `new` operator is easy and reliable as shown in Sample 5.31. Memory for the new object is allocated and all variables from the existing object are copied. However any `new()` function that you may have defined is not called.

Sample 5.31 Copying a simple class with `new`

```
class Transaction;
  bit [31:0] addr, csm, data[8];
  function new();
    $display("In %m");
  endfunction
endclass

Transaction src, dst;
initial begin
  src = new();          // Create first object
  dst = new src;        // Make a copy with new operator
end
```

This is a shallow copy, similar to a photocopy of the original, blindly transcribing values from source to destination. If the class contains a handle to another class, only the handle's value is copied by the new operator, not a full copy of the lower level object. In Sample 5.32, the Transaction class contains a handle to the Statistics class, originally shown in Sample 5.22.

Sample 5.32 Copying a complex class with new operator

```
class Transaction;
  bit [31:0] addr, csm, data[8];
  static int count = 0;
  int id;
  Statistics stats;            // Handle points to Statistics object

  function new();
    stats = new();             // Construct a new Statistics object
    id = count++;
  endfunction
endclass

Transaction src, dst;
initial begin
  src = new();           // Create a Transaction object
  src.stats.startT = 42; // Results in Figure 5-5
  dst = new src;         // Copy src to dst with new operator
                         // Results in Figure 5-6
  dst.stats.startT = 96; // Changes stats for dst & src
  $display(src.stats.startT); // 96, see Figure 5-7
end
```

The initial block creates the first **Transaction** object and modifies a variable in the contained object **stats** as shown in Fig. 5.5.

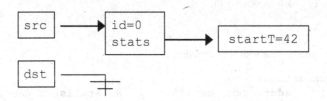

Fig. 5.5 Objects and handles before copy with the new operator

When you use the new operator to make a copy, the **Transaction** object is copied, but not the **Statistics** one. This is because the new operator does not call your own new() function. Instead, the values of variables and handles are copied. So now both **Transaction** objects have the same id as shown in Fig. 5.6.

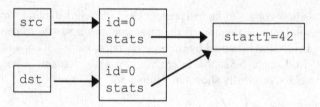

Fig. 5.6 Objects and handles after copy with the new operator

Worse yet, both **Transaction** objects point to the same **Statistics** object so modifying startT with the src handle affects what is seen with the dst handle as you can see in Figure 5.7.

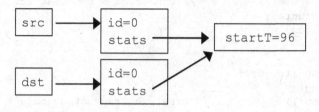

Fig. 5.7 Both src and dst objects refer to a single statistics object and see updated startT value

5.15.2 *Writing Your Own Simple Copy Function*

If you have a simple class that does not contain any references to other classes, writing a copy function is easy as you can see in Samples 5.33 and 5.34. Instead of calling the new() function and copying each individual variable, the copy function could have instead used the new operator, but then it would need to replicate any processing done in new(), such as setting the id.

Sample 5.33 Simple class with copy function

```
class Transaction;
  bit [31:0] addr, csm, data[8];   // No Statistic handle

  function Transaction copy();
    copy = new();                  // Construct destination
    copy.addr = addr;              // Fill in data values
    copy.csm  = csm;
    copy.data = data;              // Array copy
  endfunction
endclass
```

Sample 5.34 Using a copy function

```
Transaction src, dst;
initial begin
  src = new();                    // Create first object
  dst = src.copy();               // Make a copy of the object
end
```

5.15.3   Writing a Deep Copy Function

For nontrivial classes, you should always create your own copy function as seen in
Sample 5.35. You can make it a deep copy by calling the copy functions of all the
contained objects. Your own copy function makes sure all your user fields (such as
id) remain consistent. The downside of making your own copy function is that you
need to keep it up to date as you add new variables – forget one and you could spend
hours debugging to find the missing value.

Sample 5.35 Complex class with deep copy function

```
class Transaction;
  bit [31:0] addr, csm, data[8];
  Statistics stats;              // Handle points to Statistics object
  static int count = 0;
  int id;

  function new();
    stats = new();
    id = count++;
  endfunction

  function Transaction copy();
    copy = new();                // Construct destination object
    copy.addr = addr;            // Fill in data values
    copy.csm  = csm;
    copy.data = data;
    copy.stats = stats.copy();   // Call Statistics::copy
  endfunction
endclass
```

The new() constructor is called by copy so every object gets a unique id. Add
a copy() method for the Statistics class as shown in Sample 5.36, and every
other class in the hierarchy.

Sample 5.36 Statistics class declaration

```
class Statistics;
  time startT;         // Transaction times
  ...                  // See Sample 5-22 for rest of class
  function Statistics copy();
    copy = new();
    copy.startT = startT;
  endfunction
endclass
```

Now when you make a copy of the Transaction object, it will have its own Statistics object as shown in Sample 5.37.

Sample 5.37 Copying a complex class with new operator

```
Transaction src, dst;
initial begin
  src = new();               // Create first object
  src.stats.startT = 42;     // Set start time
  dst = src.copy();          // Deep copy src to dst
  dst.stats.startT = 96;     // Changes stats for dst only
  $display(src.stats.startT); // "42", See Figure 5-8
end
```

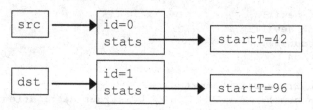

Fig. 5.8 Objects and handles after deep copy

The good news is that the UVM data macros create the copy function automatically, so you are spared from having to write them by hand. Manually creating these is very error prone, especially when you add new variables.

5.15.4   *Packing Objects to and from Arrays Using Streaming Operators*

Some protocols, such as ATM, transmit control and data values one byte at a time. Before you send out a transaction, you need to pack together the variables in the object to a byte array. Likewise, after receiving a string of bytes, you need to unpack them back into a transaction object. For both of these functions, use the streaming operators, as shown in Section 2.12.

You can't just stream the entire object as this would gather all properties, including both data and also meta-data such as timestamps and self-checking information that you may not want packed. You need to write your own `pack` function like the one in Samples 5.38 and 5.39 that only uses the properties that you choose.

More good news - the UVM data macros create the pack and unpack methods.

Sample 5.38   Transaction class with pack and unpack functions

```
class Transaction;
  bit [31:0] addr, csm, data[8];  // Real data
  static int count = 0;           // Meta-data does not
  int id;                         //      get packed

  function new();
    id = count++;
  endfunction

  function void display();
    $write("Tr: id=%0d, addr=%x, csm=%x", id, addr, csm);
    foreach(data[i]) $write(" %x", data[i]);
    $display;
  endfunction

  function void pack(ref byte bytes[$]);
    bytes = { >> {addr, csm, data}};
  endfunction

  function Transaction unpack(ref byte bytes[$]);
    { >> {addr, csm, data}} = bytes;
  endfunction
endclass : Transaction
```

Sample 5.39 Using the pack and unpack functions

```
Transaction tr, tr2;
byte b[$];                     // Queue of bytes

initial begin
  tr = new();
  tr.addr = 32'ha0a0a0a0;     // Fill object with values
  tr.csm = '1;
  foreach (tr.data[i])
    tr.data[i] = i;

  tr.pack(b);                  // Pack object into byte array
  $write("Pack results: ");
  foreach (b[i])
    $write("%h", b[i]);
   $display;

  tr2 = new();
  tr2.unpack(b);
  tr2.display();
end
```

5.16 Public vs. Local

The core concept of OOP is encapsulating data and related methods into a class. Variables are kept local to the class by default to keep one class from poking around inside another. A class provides a set of accessor methods to access and modify the data. This would also allow you to change the implementation without needing to let the users of the class know. For instance, a graphics package could change its internal representation from Cartesian coordinates to polar as long as the user interface (accessor methods) have the same functionality.

Consider the `Transaction` class that has a payload and a checksum so that the hardware can detect errors. In conventional OOP, you would make a method to set the payload also set the checksum so they would stay synchronized. Thus your objects would always be filled with correct values.

However, testbenches are not like other programs, such as a web browser or word processor. A testbench needs to create errors. You want to have a bad checksum so you can test how the hardware reacts to errors.

OOP languages such as C++ and Java allow you to specify the visibility of variables and methods. By default, everything in a class is local unless labeled otherwise.

In SystemVerilog, everything is public unless labeled `local` or `protected`. You should stick with this default so you have the greatest control over the operation of the DUT, which is more important than long-term software stability. For example, making the checksum visible allows you to easily inject errors into the

DUT. If the checksum was local, you would have to write extra code to bypass the data-hiding mechanisms, resulting in a larger and more complex testbench.

5.17   Straying Off Course

As a new OOP student, you may be tempted to skip the extra thought needed to group items into a class, and just store data in a few variables. Avoid the temptation! A basic DUT monitor samples several values from an interface. Don't just store them in some integers and pass them to the next stage. This saves you a few minutes at first, but eventually you need to group these values together to form a complete transaction. Several of these transactions may need to be grouped to create a higher-level transaction such as a DMA transfer. Instead, immediately put those interface values into a transaction class. Now you can store related information (port number, receive time) along with the data, and easily pass this object to the rest of your testbench.

5.18   Building a Testbench

Now that you have seen the basics of OOP, you can see how to create a layered test-bench from a set of classes. Figure 5.9 is the diagram from Chapter 1. Obviously, the transactions flowing between the blocks are objects, but each block is also modeled with a class.

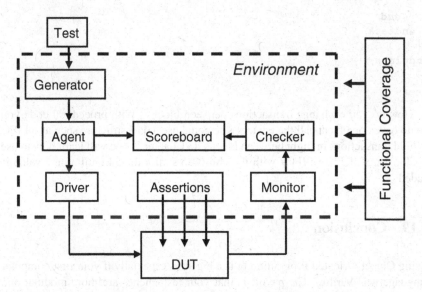

Fig. 5.9 Layered testbench

The Generator, Agent, Driver, Monitor, Checker, and Scoreboard are all classes, modeled as transactors (described below). They are instantiated inside the Environment class. For simplicity, the test is at the top of the hierarchy, as is the program that instantiates the Environment class. The Functional coverage definitions can be put inside or outside the Environment class. See Section 1.10 for a description of the layered verification environment and its components.

A transactor is made of a simple loop that receives a transaction object from a previous block, makes some transformations, and sends it to the following one as you can see in Sample 5.40. Some, such as the Generator, have no upstream block, so this transactor constructs and randomizes every transaction, while others, such as the Driver, receive a transaction and send it into the DUT as signal transitions.

Sample 5.40 Basic Transactor

```
class Transactor;   // Generic class
  Transaction tr;

  task run();
    forever begin
      // Get transaction from upstream block
      ...
      // Do some processing
      ...
      // Send it to downstream block
      ...
    end
  endtask

endclass
```

How do you exchange transactions between blocks? With procedural code you could have one object call the next, or you could use a data structure such as a FIFO to hold transactions in flight between blocks. In Chapter 7, you will learn how to use mailboxes, which are FIFOs with the ability to stall a thread until a new value is added.

5.19   Conclusion

Using Object-Oriented Programming is a big step, especially if your first computer language was Verilog. The payoff is that your testbenches are more modular and thus easier to develop, debug, and reuse.

Have patience — your first OOP testbench may look more like Verilog with a few classes added. As you get the hang of this new way of thinking, you begin to create and manipulate classes for both transactions and the transactors in the testbench that manipulate them.

In Chapter 8 you will learn more OOP techniques so your test can change the behavior of the underlying testbench without having to change any of the existing code.

5.20 Exercises

1. Create a class called MemTrans that contains the following members, then construct a MemTrans object in an initial block.

   a. An 8-bit data_in of logic type
   b. A 4-bit address of logic type
   c. A void function called print that prints out the value of data_in and address

2. Using the MemTrans class from Exercise 1, create a custom constructor, the new function, so that data_in and address are both initialized to 0.

3. Using the MemTrans class from Exercise 1, create a custom constructor so that data_in and address are both initialized to 0 but can also be initialized through arguments passed into the constructor. In addition, write a program to perform the following tasks.

   a. Create two new MemTrans objects.
   b. Initialize address to 2 in the first object, passing arguments by name.
   c. Initialize data_in to 3 and address to 4 in the second object, passing arguments by name.

4. Modify the solution from Exercise 3 to perform the following tasks.

   a. After construction, set the address of the first object to 4'hF.
   b. Use the print function to print out the values of data_in and address for the two objects.
   c. Explicitly deallocate the 2nd object.

5. Using the solution from Exercise 4, create a static variable last_address that holds the initial value of the address variable from the most recently created object, as set in the constructor. After allocating objects of class MemTrans (done in Exercise 4) print out the current value of last_address.

6. Using the solution from Exercise 5, create a static method called print_last_address that prints out the value of the static variable last_address. After allocating objects of class MemTrans, call the method print_last_address to print out the value of last_address.

7. Given the following code, complete the function print_all in class MemTrans to print out data_in and address using the class PrintUtilities. Demonstrate using the function print_all.

```
class PrintUtilities;

  function void print_4(input string name,
                        input [3:0] val_4bits);
    $display("%t: %s = %h", $time, name, val_4bits);
  endfunction

  function void print_8(input string name,
                        input [7:0] val_8bits);
    $display("%t: %s = %h", $time, name, val_8bits);
  endfunction

endclass


class MemTrans;
  bit [7:0] data_in;
  bit [3:0] address;
  PrintUtilities print;

  function new();
    print = new();
  endfunction

  function void print_all;
    // Fill in function body
  endfunction
endclass
```

8. Complete the following code where indicated by the comments starting with //.

```
program automatic test;
   import my_package::*;   // Define class Transaction

   initial begin
     // Declare an array of 5 Transaction handles
     // Call a generator task to create the objects
   end

   task generator(...);      // Complete the task header
     // Create objects for every handle in the array
     // and transmit the object.
   endtask

   task transmit(Transaction tr);
     .......
   endtask : transmit

endprogram
```

9. For the following class, create a copy function and demonstrate its use. Assume the Statistics class has its own copy function.

```
package automatic my_package;
   class MemTrans;
     bit [7:0] data_in;
     bit [3:0] address;
     Statistics stats;
     function new();
       data_in = 3;
       address = 5;
       stats = new();
     endfunction
   endclass;
endpackage
```

Chapter 6
Randomization

6.1 Introduction

As designs grow larger, it becomes more difficult to create a complete set of stimuli needed to check their functionality. You can write a directed testcase to check a certain set of features, but you cannot write enough directed testcases when the number of features keeps doubling on each project. Worse yet, the interactions between all these features are the source for the most devious bugs and are the least likely to be caught by going through a laundry list of features.

The solution is to create test cases automatically using constrained-random tests (CRT). A directed test finds the bugs you think are there, but a CRT finds bugs you never thought about, by using random stimulus. You restrict the test scenarios to those that are both valid and of interest by using constraints.

Creating the environment for a CRT takes more work than creating one for directed tests. A simple directed test just applies stimulus, and then you manually check the result. These results are captured as a golden log file and compared with future simulations to see whether the test passes or fails. A CRT requires an environment to predict the result, using a reference model, transfer function, or other techniques, plus functional coverage to measure the effectiveness of the stimulus. However, once this environment is in place, you can run hundreds of tests without having to hand-check the results, thereby improving your productivity. This trade-off of test-authoring time (your work) for CPU time (machine work) is what makes CRT so valuable.

A CRT is made of two parts: the test code that uses a stream of random values to create input to the DUT, and a seed to the pseudo-random number generator (PRNG), shown in Section 6.16.1 at the end of this chapter. You can make a CRT behave differently just by using a new seed. This feature allows you to leverage each test so each is the functional equivalent of many directed tests, just by changing seeds. You are able to create more equivalent tests using these techniques than with directed testing.

C. Spear and G. Tumbush, *SystemVerilog for Verification: A Guide to Learning the Testbench Language Features*, DOI 10.1007/978-1-4614-0715-7_6, © Springer Science+Business Media, LLC 2012

You may feel that these random tests are like throwing darts. How do you know when you have covered all aspects of the design? The stimulus space is too large to generate every possible input, so you need to generate a useful subset. In Chapter 9 you will learn how to measure verification progress by using functional coverage.

There are many ways to use randomization, and this chapter gives many examples. It highlights the most useful techniques, but choose what works best for you.

6.2   What to Randomize

When you think of randomizing the stimulus to a design, the first thing you may think of are the data fields. These are the easiest to create – just call $random. The problem is that this approach has a very low payback in terms of bugs found: you only find data-path bugs, perhaps with bit-level mistakes. The test is still inherently directed. The challenging bugs are in the control logic. As a result, you need to randomize all decision points in your DUT. Everywhere control paths diverge, randomization increases the probability that you'll take a different path in each test case.

You need to think broadly about all design input such as the following items.

- Device configuration
- Environment configuration
- Primary input data
- Encapsulated input data
- Protocol exceptions
- Delays
- Transaction status
- Errors and violations

6.2.1   Device Configuration

What is the most common reason why bugs are missed during testing of the RTL design? Not enough different configurations have been tried! Most tests just use the design as it comes out of reset, or apply a fixed set of initialization vectors to put it into a known state. This is like testing a PC's operating system right after it has been installed, and without any applications; of course the performance is fine, and there are no crashes.

Over time, in a real world environment, the DUT's configuration becomes more and more random. In a real world example, a verification engineer had to verify a timedivision multiplexor switch that had 600 input channels and 12 output channels. When the device was installed in the end-customer's system, channels would be allocated and deallocated over and over. At any point in time, there was little correlation between adjacent channels. In other words, the configuration would seem random.

To test this device, the verification engineer had to write several dozen lines of Tcl code to configure each channel. As a result, she was never able to try configurations with more than a handful of channels enabled. Using a CRT methodology, she wrote a testbench that randomized the parameters for a single channel, and then put this in a loop to configure the whole device. Now she had confidence that her tests would uncover bugs that previously would have been missed.

6.2.2   Environment Configuration

The device that you are designing operates in an environment containing other devices. When you are verifying the DUT, it is connected to a testbench that mimics this environment. You should randomize the entire environment, including the number of objects and how they are configured.

Another company was creating a PCI switch that connected multiple buses to an internal memory bus. At the start of simulation the customer used randomization to choose the number of PCI buses (1–4), the number of devices on each bus (1–8), and the parameters for each device (master or slave, CSR addresses, etc.). Even though there were many possible combinations, this company knew all had been covered.

6.2.3   Primary Input Data

This is what you probably thought of first when you read about random stimulus: take a transaction such as a bus write or ATM cell and fill it with some random values. How hard can that be? Actually it is fairly straightforward as long as you carefully prepare your transaction classes. You should anticipate any layered protocols and error injection.

6.2.4   Encapsulated Input Data

Many devices process multiple layers of stimulus. For example, a device may create TCP traffic that is then encoded in the IP protocol, and finally sent out inside Ethernet packets. Each level has its own control fields that can be randomized to try new combinations. So you are randomizing the data and the layers that surround it. You need to write constraints that create valid control fields but that also allow injecting errors.

6.2.5 Protocol Exceptions, Errors, and Violations

Anything that can go wrong, will, eventually. The most challenging part of design and verification is how to handle errors in the system. You need to anticipate all the cases where things can go wrong, inject them into the system, and make sure the design handles them gracefully, without locking up or going into an illegal state. A good verification engineer tests the behavior of the design to the edge of the functional specification and sometimes even beyond.

When two devices communicate, what happens if the transfer stops partway through? Can your testbench simulate these breaks? If there are error detection and correction fields, you must make sure all combinations are tried.

The random component of these errors is that your testbench should be able to send functionally correct stimuli and then, with the flip of a configuration bit, start injecting random types of errors at random intervals.

6.2.6 Delays

Many communication protocols specify ranges of delays. The bus grant comes one to three cycles after request. Data from the memory is valid in the fourth to tenth bus cycle. However, many directed tests, optimized for the fastest simulation, use the shortest latency, except for that one test that only tries various delays. Your testbench should always use random, legal delays during every test to try to find that (hopefully) one combination that exposes a design bug.

Below the cycle level, some designs are sensitive to clock jitter. By sliding the clock edges back and forth by small amounts, you can make sure your design is not overly sensitive to small changes in the clock cycle.

The clock generator should be in a module outside the testbench so that it creates events in the Active region along with other design events. However, the generator should have parameters such as frequency and offset that can be set by the testbench during the configuration phase.

Note that the methodology described in this book is for finding functional errors, not timing errors. Your constrained random testbench should not purposefully violate setup and hold and hold requirements. These are better validated using timing analysis tools.

6.3 Randomization in SystemVerilog

The random stimulus generation in SystemVerilog is most useful when used with OOP. You first create a class to hold a group of related random variables, and then have the random-solver fill them with random values. You can create constraints to limit the random values to legal values, or to test specific features.

You can randomize individual variables, but this case is the least interesting. True constrained-random stimuli is created at the transaction level, not one value at a time.

6.3.1   Simple Class with Random Variables

Sample 6.1 shows a packet class with random variables and constraints, plus testbench code that constructs and randomizes a packet.

Sample 6.1   Simple random class

```
class Packet;
  // The random variables
  rand  bit [31:0] src, dst, data[8];
  randc bit [ 7:0] kind;
  // Limit the values for src
  constraint c {src > 10;
                 src < 15;}
endclass

Packet p;
initial begin
  p = new();// Create a packet
  if (!p.randomize())
    $finish;
  transmit(p);
end
```

This class has four random variables. The first three use the `rand` modifier, so that every time you randomize the class, the variables are assigned a value. Think of rolling dice where each roll could be a new value or repeat the current one. The `kind` variable is `randc`, which means random cyclic, so that the random solver does not repeat a random value until every possible value has been assigned. Think of dealing cards from a deck where you deal out every card in the deck in random order, then shuffle the deck, and deal out the cards in a different order. Note that the cyclic pattern is for a single variable. A `randc` array with five elements has five different patterns, like five decks of cards, dealt in parallel. Simulators are only required to implement `randc` variables up to 8 bits wide with 256 different values, but most support much larger ranges.

A constraint is just a set of relational expressions that must be true for the chosen value of the variables. In this example, the `src` variable must be greater than 10 and less than 15. Note that the constraint expression is grouped using curly braces: {}. This is because this code is declarative, not procedural, which uses `begin...end`.

The `randomize` function returns 0 if a problem is found with the constraints. The code checks the result and stops simulation with $finish if there is a problem. Alternatively, you might want to call a special routine to end simulation, after doing some housekeeping chores like printing a summary report. The rest of the book uses a macro instead of this extra code.

You should not randomize an object in the class constructor. Your test may need to turn constraints on or off, change weights, or even add new constraints before randomization. The constructor is for initializing the object's variables, and if you called `randomize` at this early stage, you might end up throwing away the results.

Variables in your classes should be random and public. This gives your test the most control over the DUT's stimulus and control. You can disable randomization of a variable, as shown in Section 6.11.2. If you forget to make a variable random, you must edit the environment, which you want to avoid. The exception is that configuration variables such as weights and limits should not be random in transaction classes as their values are chosen at the start of simulation and do not change.

6.3.2   Checking the Result from Randomization

The `randomize` function assigns random values to any variable in the class that has been labeled as `rand` or `randc`, and also makes sure that all active constraints are obeyed. Randomization can fail if your code has conflicting constraints (see next section), so you should always check the status. If you don't check, the variables may get unexpected values, causing your simulation to fail.

The remaining code samples in this book employ the macro in Sample 6.2 to check for the result of randomization. If you adopt this style, you can easily add code to give meaningful error messages and gracefully wind down simulation. The macro shows off several coding tricks, including wrapping the generated code in a do...while statement so it can be used like a normal statement terminated with a semicolon, including in an if-else statement, something that VMM log macros did right, but not OVM.

Sample 6.2 Randomization check macro and example

```
`define SV_RAND_CHECK(r) \
   do begin \
     if (!(r)) begin \
       $display("%s:%0d: Randomization failed \"%s\"", \
                `__FILE__, `__LINE__, `"r`"); \
       $finish; \
     end \
   end while (0)

initial begin
  Packet p = new();                    // Create a packet
  `SV_RAND_CHECK(p.randomize());       // Randomize it
end
```

6.3.3   The Constraint Solver

The process of solving constraint expressions is handled by the SystemVerilog constraint solver. The solver chooses values that satisfy the constraints. The values come from SystemVerilog's PRNG, that is started with an initial seed. If you give a SystemVerilog simulator the same seed and the same testbench, it should always produce the same results. Note that changing the tool version or switches such as debug level can change results. See the exercises at the end of this chapter to see how to specify the initial seed.

The solver is specific to the simulation vendor, and a constrained-random test may not give the same results when run on different simulators, or even on different versions of the same tool. The SystemVerilog standard specifies the meaning of the expressions, and the legal values that are created, but does not detail the precise order in which the solver should operate. See Section 6.16 for more details on random number generators.

6.3.4   What can be Randomized?

SystemVerilog allows you to randomize integral variables, that is, variables that contain a simple set of bits. This includes 2-state and 4-state types, though randomization only generates 2-state values. You can have integers, bit vectors, etc. You cannot have a random string, or refer to a handle in a constraint. Randomizing `real` variables is not yet defined in the LRM. '

6.4   Constraint Details

Useful stimulus is more than just random values — there are relationships between the variables. Otherwise, it may take too long to generate interesting stimulus values, or the stimulus might contain illegal values. You define these interactions in SystemVerilog using constraint blocks that contain one or more constraint expressions. SystemVerilog chooses random values so that the expressions are true.

 At least one variable in each expression should be random, either `rand` or `randc`. The following class fails when randomized, unless `age` happens to be in the right range. The solution is to add the modifier `rand` or `randc` before `age`.

Sample 6.3 Constraint without random variables

```
class Child;
  bit [7:0] age;  // Error - should be rand or randc
  constraint c_teenager {age > 12;
                         age < 20;}
endclass
```

The `randomize` function tries to assign new values to random variables and to make sure all constraints are satisfied. In Sample 6.3, since there are no random variables, `randomize` just checks the value of `age` to see if it is in the bounds specified by the constraint `c_teenager`. Unless the variable falls in the range of 13:19, `randomize` fails. While you can use a constraint to check that a non-random variable has a valid value, use an `assert` or `if`-statement instead. It is much easier to debug your procedural checker code than read through an error message from the random solver.

6.4.1   Constraint Introduction

Sample 6.4 shows a simple class with random variables and constraints. The specific constructs are explained in the following sections. Notice that in constraint blocks, you use curly braces, `{ }`, to group together multiple expressions. The `begin…end` keywords are for procedural code.

Sample 6.4   Constrained-random class

```
class Stim;
  const bit [31:0] CONGEST_ADDR = 42;
  typedef enum {READ, WRITE, CONTROL} stim_e;
  randc stim_e kind;    // Enumerated var
  rand bit [31:0] len, src, dst;
  rand bit congestion_test;

  constraint c_stim {
    len < 1000;
    len > 0;
    if (congestion_test) {
      dst inside { [CONGEST_ADDR-10:CONGEST_ADDR+10] };
      src == CONGEST_ADDR;
    }
    else
      src inside {0, [2:10], [100:107]};
  }
endclass
```

6.4.2   Simple Expressions

Sample 6.4 showed a constraint block with several expressions. The first two control the values for the `len` variable. As you can see, a variable can be used in multiple expressions.

There should be a maximum of only one operator in an expression, such as <, <=, ==, >=, or >. Sample 6.5 shows a SystemVerilog gotcha in that it incorrectly tries to generate three variables in a fixed order.

Sample 6.5 Bad ordering constraint

```
class Order_bad;
  rand bit [7:0] lo, med, hi;
  constraint bad  {lo < med < hi;} // Gotcha!
endclass
```

Sample 6.6 Result from incorrect ordering constraint

```
lo =   20, med = 224, hi = 164
lo = 114, med =  39, hi = 189
lo = 186, med = 148, hi = 161
lo = 214, med = 223, hi = 201
```

Sample 6.6 shows the results, which are not what was intended. The constraint bad in Sample 6.5 is broken down into multiple binary relational expressions, going from left to right: ((lo < med) < hi). First, the expression (lo < med) is evaluated, which gives 0 or 1. Then hi is constrained to be greater than the result. The variables lo and med are randomized but not constrained. The correct constraint is shown in Sample 6.7. For more examples, see Sutherland (2007).

Sample 6.7 Constrain variables to be in a fixed order

```
class Order_good;
  rand bit [7:0] lo, med, hi;
  constraint good {lo < med;    // Only use binary constraints
                   med < hi;}
endclass
```

6.4.3   Equivalence Expressions

The most common mistake with constraints is trying to make an assignment in a constraint block, which can only contain expressions. Instead, use the equivalence operator to set a random variable to a value, e.g., len==42. You can build complex relationships between one or more random variables: len == (header.addr_mode * 4 + payload.size()).

6.4.4   Weighted Distributions

A bug in the DUT may be found with constrained random stimulus if you apply enough patterns. However, it may take a long time for a particular corner case to be generated. When reviewing functional coverage result, see if corner cases are being generated. If not, you can use a weighted distribution to skew the stimulus in a particular direction, and thus accelerate finding bugs. The dist operator allows you to create weighted distributions so that some values are chosen more often than others.

The dist operator takes a list of values and weights, separated by the : = or the : / operator. The values and weights can be constants or variables. The values can be a single value or a range such as [lo:hi]. The weights are not percentages and do not have to add up to 100. The : = operator specifies that the weight is the same for every specified value in the range, whereas the : / operator specifies that the weight is to be equally divided between all the values.

Sample 6.8   Weighted random distribution with dist

```
class Transaction;
  rand bit [1:0] src, dst;
  constraint c_dist {
    src dist {0:=40, [1:3]:=60};
    // src = 0, weight = 40/220
    // src = 1, weight = 60/220
    // src = 2, weight = 60/220
    // src = 3, weight = 60/220

    dst dist {0:/40, [1:3]:/60};
    // dst = 0, weight = 40/100
    // dst = 1, weight = 20/100
    // dst = 2, weight = 20/100
    // dst = 3, weight = 20/100
  }
endclass
```

In Sample 6.8, src gets the value 0, 1, 2, or 3. The weight of 0 is 40, whereas, 1, 2, and 3 each have the weight of 60, for a total of 220. The probability of choosing 0 is 40/220, and the probability of choosing 1, 2, or 3 is 60/220 each.

Next, dst gets the value 0, 1, 2, or 3. The weight of 0 is 40, whereas 1, 2, and 3 share a total weight of 60, for a total of 100. The probability of choosing 0 is 40/100, and the probability of choosing 1, 2, or 3 is only 20/100 each.

Once again, the values and weights can be constants or variables. You can use variable weights to change distributions on the fly or even to eliminate choices by setting the weight to zero, as shown in Sample 6.9.

Sample 6.9 Dynamically changing distribution weights

```
// Bus operation, byte, word, or longword
class BusOp;
  // Operand length
  typedef enum {BYTE, WORD, LWRD } length_e;
  rand length_e len;

  // Weights for dist constraint
  bit [31:0] w_byte=1, w_word=3, w_lwrd=5;

  constraint c_len {
    len dist {BYTE := w_byte,      // Choose a random
              WORD := w_word,      // length using
              LWRD := w_lwrd};     // variable weights
  }
endclass
```

In Sample 6.9, the len enumerated variable has three values. With the default weighting values, longword lengths are chosen more often, as w_lwrd has the largest value. Don't worry, you can change the weights on the fly during simulation to get a different distribution.

6.4.5   Set Membership and the Inside Operator

You can create sets of values with the inside operator. The SystemVerilog solver chooses between the values in the set with equal probability, unless you have other constraints on the variable. As always, you can use variables in the sets.

Sample 6.10 Random sets of values

```
class Ranges;
  rand bit [31:0] c;        // Random variable
  bit [31:0] lo, hi;        // Non-random variables used as limits
  constraint c_range {
    c inside {[lo:hi]};    // lo <= c && c <= hi
  }
endclass
```

In Sample 6.10, SystemVerilog uses the values for lo and hi to determine the range of possible values. You can use the variables as parameters for your constraints so that the testbench can alter the behavior of the stimulus generator without rewriting the constraints. Note that if lo > hi, an empty set is formed, and the constraint fails.

If you want any value, as long as it is not inside a set, invert the constraint with the NOT operator: ! as shown in Sample 6.11.

Sample 6.11 Inverted random set constraint

```
constraint c_range {
  !(c inside {[lo:hi]});  // c < lo or c > hi
}
```

6.4.6 Using an Array in a Set

Sample 6.12 shows how you can choose from a set of values by storing them in an array.

Sample 6.12 Random set constraint for an array

```
class Fib;
  rand bit [7:0] f;
  bit [7:0] vals[] = '{1,2,3,5,8};
  constraint c_fibonacci {
    f inside vals;
  }
endclass
```

This is expanded into the constraints in Sample 6.13.

Sample 6.13 Equivalent set of constraints

```
constraint c_fibonacci {
  (f == vals[0]) ||      // f==1
  (f == vals[1]) ||      // f==2
  (f == vals[2]) ||      // f==3
  (f == vals[3]) ||      // f==5
  (f == vals[4]);        // f==8
}
```

Likewise, you can use the NOT operator to tell SystemVerilog to choose any value except those in an array as shown in Sample 6.14.

Sample 6.14 Choose any value except those in an array

```
class NotFib;
  rand bit [7:0] notf;
  bit [7:0] vals[] = '{1,2,3,5,8};
  constraint c_fibonacci {
    !(notf inside vals);
  }
endclass
```

Always make sure your constraints work as you expect. You could create functional coverage groups and generate reports, or print a histogram of values with the code in Sample 6.15, with the output in Sample 6.16.

Sample 6.15 Printing a histogram

```
initial begin
  Fib fib;
  int count[9], maxx[$];

  fib = new();
  repeat (20_000) begin
    `SV_RAND_CHECK(fib.randomize());
    count[fib.f]++;          // Count the number of hits
  end
  maxx = count.max();        // Get largest value in count

  // Print histogram of count
  foreach(count[i])
    if (count[i]) begin
      $write("count[%0d]=%5d ", i, count[i]);
      repeat (count[i]*40/maxx[0]) $write("*");
      $display;
    end
end
```

Sample 6.16 Histogram for inside constraint

```
count[1]= 3980 ************************************
count[2]= 3924 ************************************
count[3]= 3922 ************************************
count[5]= 4175 *************************************
count[8]= 3999 ************************************
```

Samples 6.17 and 6.18 choose a day of the week from a list of enumerated values. You can change the list of choices on the fly. If you make choice a randc variable, the simulator tries every possible value before repeating.

Sample 6.17 Class to choose from an array of possible values

```
class Days;
  typedef enum {SUN, MON, TUE, WED,
                THU, FRI, SAT} days_e;
  days_e choices[$];
  rand days_e choice;
  constraint cday {choice inside choices;}
endclass
```

Sample 6.18 Choosing from an array of values

```
initial begin
  Days days;
  days = new();

  days.choices = {Days::SUN, Days::SAT};
  `SV_RAND_CHECK(days.randomize());
  $display("Random weekend day %s\n", days.choice.name());

  days.choices = {Days::MON, Days::TUE, Days::WED,
                  Days::THU, Days::FRI};
  `SV_RAND_CHECK(days.randomize());
  $display("Random week day %s", days.choice.name());
end
```

The name function returns a string with the name of an enumerated value.

If you want to dynamically add or remove values from a set, think twice before using the inside operator because of its performance. Perhaps you have a set of values that you want to choose just once. You could use inside to choose values from a queue, and delete them to slowly shrink the queue. This requires the solver to solve N constraints, where N is the number of elements left in the queue. Instead, use a randc variable that is an index into an array of choices as shown in Samples 6.19 and 6.20. Choosing a randc value takes a short, constant time, whereas solving a large number of constraints is more expensive, especially if your array has more than a few dozen elements.

Sample 6.19 Using randc to choose array values in random order

```
class RandcInside;
  int array[];              // Values to choose
  randc bit [15:0] index;   // Index into array

  function new(input int a[]); // Construct & initialize
    array = a;
  endfunction

  function int pick();        // Return most recent pick
    return array[index];
  endfunction

  constraint c_size {index < array.size();}
endclass
```

Sample 6.20   Testbench for randc choosing array values in random order

```
initial begin
  RandcInside ri;

  ri = new('{1,3,5,7,9,11,13});
  repeat (ri.array.size()) begin
    `SV_RAND_CHECK(ri.randomize());
    $display("Picked %2d [%0d]", ri.pick(), ri.index);
  end
end
```

Note that constraints and routines can be mixed in any order.

6.4.7   Bidirectional Constraints

By now you may have realized that constraint blocks are not procedural code, executing from top to bottom. They are declarative code, all active at the same time. If you constrain a variable with the `inside` operator with the set [10:50] and have another expression that constrains the variable to be greater than 20, SystemVerilog solves both constraints simultaneously and only chooses values between 21 and 50.

SystemVerilog constraints are solved bidirectionally, which means that the constraints on all random variables are solved concurrently. Adding or removing a constraint on any one variable affects the value chosen for all variables that are related directly or indirectly. Consider the constraint in Sample 6.21.

Sample 6.21   Bidirectional constraints

```
class Bidir;
  rand bit [15:0] r, s, t;
  constraint c_bidir {        // All are solved in parallel
    r < t;                    // A value for r affects s, t
    s == r;
    t < 10;
    s > 5;
  }
endclass
```

The SystemVerilog solver looks at all four constraints simultaneously. The variable r has to be less than t, which has to be less than 10. However, r is also constrained to be equal to s, which is greater than 5. Even though there is no direct constraint on the lower value of `t`, the constraint on `s` restricts the choices. Table 6.1 shows the possible values for these three variables.

Table 6.1 Solutions for bidirectional constraint

Solution	r	s	t
A	6	6	7
B	6	6	8
C	6	6	9
D	7	7	8
E	7	7	9
F	8	8	9

6.4.8 Implication Constraints

Normally, all constraint expressions are active in a block. What if you want to have an expression active only some of the time? Set the highest address, but only for IO space mode. SystemVerilog supports two implication operators, `->` and `if`.

Sample 6.22 Constraint block with implication operator

```
class BusOp;
  rand bit [31:0] addr;
  rand bit io_space_mode;
  constraint c_io {
    io_space_mode ->
      addr[31] == 1'b1;
  }
```

The expression `A->B` is equivalent to the expression (`!A || B`). When the implication operator appears in a constraint, the solver picks values for `A` and `B` so the expression is true. Truth Table 6.2 shows the value of the expression for the logical values of `A` and `B`.

Table 6.2 Implication operator truth table

A->B	B=false	B=true
A=false	true	true
A=true	false	true

When A is true, B must be true, but when A is false, B can be true or false. Note that this is a partly bidirectional constraint, but that A->B does not imply that B->A. The two expressions produce different results.

In Sample 6.23, when d==1, the variable e must be 1, but when e==1, d can be 0 or 1.

Sample 6.23 Implication operator

```
class LogImp;
  rand bit d, e;
  constraint c {
    (d==1) -> (e==1);
  }
endclass
```

If you add the constraint {e==0;}, the variable d must be 0; But if you add a constraint {e==1;} the values of d are not constrained, it can still be 0 or 1.

Sample 6.24 shows how Sample 6.22 could be written with an if implication constraint.

Sample 6.24 Constraint block with if implication operator

```
class BusOp;
  rand bit [31:0] addr;
  rand bit io_space_mode;
  constraint c_io {
    if (io_space_mode)
      addr[31] == 1'b1;
  }
```

The if-else operator is a great way to choose between multiple expressions. For example, the bus defined in Sample 6.9 might support byte, word, and longword reads, but only longword writes if written like Sample 6.25.

Sample 6.25 Constraint block with if-else operator

```
class BusOp;
  rand operand_e op;
  rand length_e len;

  constraint c_len_rw {
    if (op == READ) {
      len inside { [BYTE:LWRD] };
    }
    else {
      len == LWRD;
    }
  }
```

The constraint `if (A) B else C;` is equivalent to the two constraints `(A && B);` and `(!A && C);`. Sample 6.26 shows how you can chain together multiple choices.

Sample 6.26 Constraint block with multiple if-else operator

```
class BusOp;
  ...
  constraint c_addr_space {
    if (addr_space == MEM)
      addr inside {[0:32'h0FFF_FFFF]};
    else if (addr_space == IO)
      addr inside {[32'1000_0000:32'h7FFF_FFFF]};
    else
      addr inside {[32'h8000_0000:32'hFFFF_FFFF]};
  }
```

6.4.9   Equivalence Operator

The equivalence operator `<->` is bidirectional. `A<->B` is defined as `((A->B) && (B->A))`. Table 6.3 is the truth table for the logical values of A and B as constrained in Sample 6.27.

Table 6.3 Equivalence operator truth table

A<->B	B=false	B=true
A=false	true	false
A=true	false	true

Sample 6.27 Equivalence constraint

```
rand bit d, e;
constraint c { (d==1) <-> (e==1); }
```

When d is true, e must also be true, and when d is false, e must also be false. So this operator is the same as a logical XNOR. If you start with the constraint d<->e, and add a constraint such as d==1, e is set to 1 by the solver. The constraint d<->e and e==0 cause d to be set to 0 by the solver. If your class has all three of the constraints, d<->e, d==1, and e==0, the solver will not be able to choose values for d and e.

6.5   Solution Probabilities

Whenever you deal with random values, you need to understand the probability of the outcome. SystemVerilog does not guarantee the exact solution found by the random constraint solver, but you can influence the distribution. Any time you work with random numbers, you have to look at thousands or millions of values to average out the noise. Some simulators, such as Synopsys VCS, have multiple solvers to allow you to trade memory usage vs. performance. The distributions will vary between different simulators. The tables were generated with Synopsys VCS 2011.03.

6.5.1   Unconstrained

Start with two random variables in a class with no constraints as shown in Sample 6.28.

Sample 6.28   Class Unconstrained

```
class Unconstrained;
  rand bit x;          // 0 or 1
  rand bit [1:0] y;    // 0, 1, 2, or 3
endclass
```

Table 6.4 shows the eight possible solutions. Since there are no constraints, each has the same probability. You have to run thousands of randomizations to see the actual results approach the listed probabilities.

Table 6.4   Solutions for Unconstrained class

Solution	x	y	Probability
A	0	0	1/8
B	0	1	1/8
C	0	2	1/8
D	0	3	1/8
E	1	0	1/8
F	1	1	1/8
G	1	2	1/8
H	1	3	1/8

6.5.2   Implication

In Sample 6.29, the value of y depends on the value of x. This is indicated with the implication operator in the following constraint. This example and the rest in this section also behave in the way same with the if implication operator.

Sample 6.29 Class with implication constraint

```
class Impl;
  rand bit x;           // 0 or 1
  rand bit [1:0] y;     // 0, 1, 2, or 3
  constraint c_xy {
    (x==0) -> (y==0);
  }
endclass
```

Table 6.5 shows the possible solutions and probability. You can see that the random solver recognizes that there are eight combinations of x and y, but all the solutions where x==0 (solutions A–D) have been merged together.

Table 6.5 Solutions for Impl class

Solution	x	y	Probability
A	0	0	1/2
B	0	1	0
C	0	2	0
D	0	3	0
E	1	0	1/8
F	1	1	1/8
G	1	2	1/8
H	1	3	1/8

6.5.3 Implication and Bidirectional Constraints

Note that the implication operator says that when x==0, y is forced to 0, but when y==0, there is no constraint on x. However, implication is bidirectional in that if y were forced to a nonzero value, x would have to be 1. Sample 6.30 has the constraint y>0, so x can never be 0 and Table 6.6 shows the solutions.

Sample 6.30 Class with implication constraint and additional constraint

```
class Imp2;
  rand bit x;          // 0 or 1
  rand bit [1:0] y;    // 0, 1, 2, or 3
  constraint c_xy {
    y > 0;             // Force y = 1, 2, or 3
    (x==0) -> (y==0);
  }
endclass
```

Table 6.6 Solutions for `Imp2` class

Solution	x	y	Probability
A	0	0	0
B	0	1	0
C	0	2	0
D	0	3	0
E	1	0	0
F	1	1	1/3
G	1	2	1/3
H	1	3	1/3

6.5.4 Guiding Distribution with Solve...Before

You can guide the SystemVerilog solver using the "solve...before" constraint as seen in Sample 6.31.

Sample 6.31 Class with implication and solve...before

```
class SolveXBeforeY;
  rand bit x;          // 0 or 1
  rand bit [1:0] y;    // 0, 1, 2, or 3
  constraint c_xy {
    (x==0) -> y==0;
    solve x before y;
  }
endclass
```

The `solve…before` constraint does not change the solution space, just the probability of the results. The solver chooses values of x (0, 1) with equal probability. In 1000 calls to `randomize`, x is 0 about 500 times, and 1 about 500 times. When x is 0, y must be 0. When x is 1, y can be 0, 1, 2, or 3 with equal probability as shown in Table 6.7.

Table 6.7 Solutions for `solve x before y` constraint

Solution	x	y	Probability
A	0	0	1/2
B	0	1	0
C	0	2	0
D	0	3	0
E	1	0	1/8
F	1	1	1/8
G	1	2	1/8
H	1	3	1/8

If you use the constraint `solve y before x`, you get a very different distribution as shown in Table 6.8.

Table 6.8 Solutions for `solve y before x` constraint

Solution	x	y	Probability
A	0	0	1/8
B	0	1	0
C	0	2	0
D	0	3	0
E	1	0	1/8
F	1	1	1/4
G	1	2	1/4
H	1	3	1/4

 Only use `solve…before` if you are dissatisfied with how often some values occur. Excessive use can slow the constraint solver and make your constraints difficult for others to understand.

For the simple class in Sample 6.31, the equivalence operator, `<->`, gives the same solution as the implication operator `->`. Try adding additional constraints and plot the results for your favorite simulator.

6.6 Controlling Multiple Constraint Blocks

A class can contain multiple constraint blocks. One block might ensure you have a valid transaction, as described in Section 6.7, but you might need to disable this when testing the DUT's error handling. Or you might want to have a separate constraint for each test. Perhaps one constraint would restrict the data length to create small transactions (great for testing congestion), whereas another would make long transactions.

You can turn constraints on and off with the `constraint_mode` function. You can control a single constraint with `handle.constraint.constraint_mode(arg)`. To control all constraints in an object, use `handle.constraint_mode(arg)`, as shown in Sample 6.32. When the argument for `constraint_mode` is 0, the constraint is turned off, and when it is 1, the constraint is turned on.

Sample 6.32 Using `constraint_mode`

```
class Packet;
  rand bit [31:0] length;
  constraint c_short {length inside {[1:32]}; }
  constraint c_long  {length inside {[1000:1023]}; }
endclass

Packet p;
initial begin
  p = new();

  // Create a long packet by disabling short constraint
  p.c_short.constraint_mode(0);
  `SV_RAND_CHECK(p.randomize());

  transmit(p);

  // Create a short packet by disabling all constraints
  // then enabling only the short constraint
  p.constraint_mode(0);
  p.c_short.constraint_mode(1);
  `SV_RAND_CHECK(p.randomize());
  transmit(p);
end
```

While many small constraints may give you more flexibility, the process of turning them on and off is more complex. For example, when you turn off all constraints that create data, you are also disabling all the ones that check the data's validity.

If you just want to make a random variable non-random, use `rand_mode` as described in Section 6.11.2.

6.7 Valid Constraints

A good randomization technique is to create several constraints to ensure the correctness of your random stimulus, known as "valid constraints." In Sample 6.33, a bus read-modify-write command is only allowed for a longword data length.

Sample 6.33 Checking write length with a valid constraint

```
class Transaction;
  typedef enum {BYTE, WORD, LWRD, QWRD} length_e;
  typedef enum {READ, WRITE, RMW, INTR} access_e;
  rand length_e length;
  rand access_e access;

  constraint valid_RMW_LWRD {
    (access == RMW) -> (length == LWRD);
  }
endclass
```

Now you know the bus transaction obeys the rule. Later, if you want to violate the rule, use `constraint_mode` to turn off this one constraint. You can turn these off with `constraint_mode` when you want to generate errors. For example, what if a packet has a zero-length payload? You should have a naming convention to make these constraints stand out, such as using the prefix `valid` as shown above.

6.8 In-Line Constraints

As you write more tests, you can end up with many constraints. They can interact with each other in unexpected ways, and the extra code to enable and disable them adds to the test complexity. Additionally, constantly adding and editing constraints to a class could cause problems in a team environment.

Many tests only randomize objects at one place in the code. SystemVerilog allows you to add an extra constraint using `randomize with`. This is equivalent to adding an extra constraint to any existing ones in effect. Sample 6.34 shows a base class with constraints, then two `randomize with` statements.

Sample 6.34 The randomize() with statement

```
class Transaction;
  rand bit [31:0] addr, data;
  constraint c1 {addr inside{[0:100],[1000:2000]};}
endclass

initial begin
  Transaction t;
  t = new();

  // addr is 50-100, 1000-1500, data < 10
  `SV_RAND_CHECK(t.randomize() with {addr >= 50; addr <= 1500;
                                     data < 10;});

  driveBus(t);

  // force addr to a specific value, data > 10
  `SV_RAND_CHECK(t.randomize() with {addr == 2000; data > 10;});

  driveBus(t);
end
```

The extra constraints are added to the existing ones in effect. Use constraint_ mode if you need to disable a conflicting constraint. Note that inside the with{} statement, SystemVerilog uses the scope of the class. That is why Sample 6.34 used just addr, not t.addr.

A common mistake is to surround your in-line constraints with parenthesis instead of curly braces { }. Just remember that constraint blocks use curly braces, so your in-line constraint must use them too. Braces are for declarative code.

6.9   The pre_randomize and post_randomize Functions

Sometimes you need to perform an action immediately before every randomize call or immediately afterwards. For example, you may want to set some nonrandom class variables (such as limits or weights) before randomization starts, or you may need to calculate the error correction bits for random data. SystemVerilog lets you do this with two functions, pre_randomize and post_randomize that are created automatically in any class with random variables.

6.9.1   Building a Bathtub Distribution

For some applications, you want a nonlinear random distribution. For instance, small and large packets are more likely to find a design bug such as buffer overflow than medium-sized packets. So you want a bathtub shaped distribution; high on both ends, and low in the middle. You could build an elaborate `dist` constraint, but it might require lots of tweaking to get the shape you want. Verilog has several functions for nonlinear distribution, such as `$dist_exponential`, but none for a bathtub. The graph in Fig. 6.1 shows how you can combine two exponential curves to make a bathtub curve. The `pre_randomize` method in Sample 6.35 calculates a point on an exponential curve, then randomly chooses to put this on the left curve, or right. As you pick points on either the left and right curves, you gradually build a distribution of the combined values.

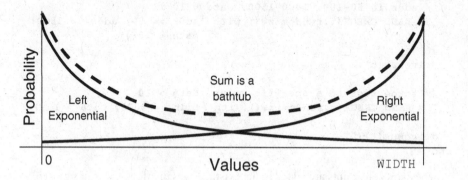

Fig. 6.1 Building a bathtub distribution

Sample 6.35 Building a bathtub distribution

```
class Bathtub;
  int value;   // Random variable with bathtub dist
  int WIDTH = 50, DEPTH=6, seed=1;

  function void pre_randomize();
    // Calculate an expontal curve
    value = $dist_exponential(seed, DEPTH);
    if (value > WIDTH) value = WIDTH;

    // Randomly put this point on the left or right curve
    if ($urandom_range(1))      // Random 0 or 1
      value = WIDTH - value;
  endfunction

endclass
```

Every time this object is randomized, the variable `value` gets updated. Across many randomizations, you will see the desired nonlinear distribution. Since the variable is calculated procedurally, not through the random constraint solver, it does not need the `rand` modifier.

See Sample 6.64 for another example of `post_randomize`.

6.9.2   *Note on Void Functions*

The functions `pre_randomize` and `post_randomize` can only call other functions, not tasks that could possibly consume time. You cannot have a delay in the middle of a call to `randomize`. When you are debugging a randomization problem, you can call your display routines if you planned ahead and made them void functions.

Chapter 8 describes advanced OOP concepts including extended classes and virtual methods. The `pre_randomize` and `post_randomize` functions are not virtual and so they are called based on the type of the handle, not the object. Additionally, if your extended class's `pre_randomize or post_randomize` need functionality in the base class's `pre_randomize` and `post_randomize` functions, they should call these methods using the super prefix, as in `super.pre_randomize`.

6.10   Random Number Functions

You can use all the Verilog-1995 distribution functions, plus several that are new for SystemVerilog. Consult a statistics book for more details on the "dist" functions. Some of the useful functions include the following.

- `$random` — Flat distribution, returning signed 32-bit random
- `$urandom` — Flat distribution, returning unsigned 32-bit random
- `$urandom_range` — Flat distribution over a range
- `$dist_exponential` — Exponential decay, as shown in Fig. 6.1
- `$dist_normal` — Bell-shaped distribution
- `$dist_poisson` — Bell-shaped distribution
- `$dist_uniform` — Flat distribution

The `$urandom_range` function takes two arguments, an optional low value, and a high value as shown in Sample 6.36.

Sample 6.36 $urandom range usage

```
a = $urandom_range(3, 10); // Pick a value from 3 to 10
a = $urandom_range(10, 3); // Pick a value from 3 to 10
b = $urandom_range(5);     // Pick a value from 0 to 5
```

6.11   Constraints Tips and Techniques

How can you create constrained-random tests that can be easily modified? There are
several tricks you can use. The most general technique is to use OOP to extend the
original class as described in sections 6.11.8 and 8.2.4 but this requires more planning. So first learn some simple techniques, but keep your mind open to other ways.

6.11.1   Constraints with Variables

Most constraint examples in this book use constants to make them more readable.
In Sample 6.37, length is randomized over a range that uses a variable for the
upper bound.

Sample 6.37   Constraint with a variable bound

```
class Packet;
  rand bit [31:0] length;
  bit [31:0] max_length = 100;  // Configuration, not rand
  constraint c_length {
    length inside {[1:max_length]};
  }
endclass
```

By default, this class creates random lengths between 1 and 100, but by changing
the variable max_length, you can vary the upper limit.

You can use variables in the dist constraint to turn on and off values and ranges.
In Sample 6.38, each bus command has a different weight variable.

Sample 6.38   dist constraint with variable weights

```
typedef enum {READ8, READ16, READ32} read_e;
class ReadCommands;
  rand read_e read_cmd;
  int read8_wt=1, read16_wt=1, read32_wt=1;
  constraint c_read {
    read_cmd dist {READ8   := read8_wt,
                   READ16  := read16_wt,
                   READ32  := read32_wt};
  }
endclass
```

By default, this constraint produces each command with equal probability. If you
want to have a greater number of READ8 commands, increase the read8_wt weight
variable. Most importantly, you can turn off generation of a command by dropping
its weight to 0.

6.11.2 Using Nonrandom Values

If you have a set of constraints that produces stimulus that is almost what you want, but not quite, you could call `randomize`, and then set a variable to the value you want — you don't have to use the random value. However, your stimulus values may not be correct according to the constraints you created to check validity.

If there are just a few random variables that you want to override, use the `rand_mode` function to make them nonrandom. When you call this method with the argument 0 for a random variable, the `rand` or `randc` qualifier is disabled and the variable's value is no longer changed by the random solver, but the value is still checked in if it appears in a constraint. Setting the random mode to 1 turns the qualifier back on so the variable can changed by the solver.

Sample 6.39 `rand_mode` disables randomization of variables

```
// Packet with variable length payload
class Packet;
  rand bit [7:0] length, payload[];
  constraint c_valid {length > 0;
                      payload.size() == length;}

  function void display(input string msg);
    $display("\n%s:", msg);
    $write("\tPacket len=%0d, bytes = ", length);
    for(int i=0; (i<4 && i<payload.size()); i++)
      $write(" %0d", payload[i]);
    $display;
  endfunction
endclass

Packet p;
initial begin
  p = new();
  `SV_RAND_CHECK(p.randomize());  // Randomize all variables
  p.display("Simple randomize");

  p.length.rand_mode(0);          // Make length nonrandom,
  p.length = 42;                  // set it to a constant value
  `SV_RAND_CHECK(p.randomize());  // then randomize the payload
  p.display("Randomize with rand_mode");
end
```

In Sample 6.39, the packet size is stored in the random variable `length`. The first half of the test randomizes both the `length` variable and the contents of the `payload` dynamic array. The second half calls `rand_mode` to make `length` a nonrandom variable, sets it to 42, then calls `randomize`. The constraint sets the `payload` size at the constant 42, but the array is still filled with random values.

6.11.3   Checking Values Using Constraints

If you randomize an object and then modify some variables, you can check that the object is still valid by checking if all constraints are still obeyed. Call `handle.randomize(null)` and SystemVerilog treats all variables as nonrandom ("state variables") and just ensures that all constraints are satisfied, i.e all expressions are true. If any constraints are not satisfied, the `randomize` function returns 0.

6.11.4   Randomizing Individual Variables

Suppose you want to randomize a few variables inside a class. You can call `randomize` with the subset of variables. Only those variables passed in the argument list will be randomized; the rest will be treated as state variables and not randomized. All constraints remain in effect. In Sample 6.40, the first call to randomize only changes the values of two `rand` variables `med` and `hi`. The second call only changes the value of `med`, whereas `hi` retains its previous value. Surprisingly, you can pass a non-random variable, as shown in the last call, and `low` is given a random value, as long as it obeys the constraint.

Sample 6.40 Randomizing a subset of variables in a class

```
class Rising;
  bit [7:0] low;            // Not random
  rand bit [7:0] med, hi;   // Random variable
  constraint up
    { low < med; med < hi; } // See Section 6.4.2
endclass

initial begin
  Rising r;
  r = new();
  r.randomize();       // Randomize med, hi; low untouched
  r.randomize(med);    // Randomize only med
  r.randomize(low);    // Randomize only low, even though not rand
end
```

This trick of only randomizing a subset of the variables is not commonly used in real testbenches as you are restricting the randomness of your stimulus. You want your testbench to explore the full range of legal values, not just a few corners.

6.11.5   Turn Constraints Off and On

Sections 6.6 and 6.7 discuss valid constraints and `constraint_mode`. Turning off individual constraints is fine for error generation, but should be used in moderation.

6.11.6   Specifying a Constraint in a Test Using In-Line Constraints

If you keep adding constraints to a class, it becomes hard to manage and control. Soon, everyone is checking out the same file from your source control system. Many times a constraint is only used by a single test, so why have it visible to every test? One way to localize the effects of a constraint is to use in-line constraints, `random-ize with`, shown in Section 6.8. This works well if your new constraint is additive to the default constraints. If you follow the recommendations in Section 6.7 to create "valid constraints", you can quickly constrain valid sequences. For error injection, you can disable any constraint that conflicts with what you are trying to do. A test that injects a particular flavor of corrupted data would first turn off the particular validity constraint that checks for that error.

There are several tradeoffs with using in-line constraints. The first is that now your constraints are in multiple locations which can make it more difficult to understand all the active constraints. If you add a new constraint to the original class, it may conflict with the in-line constraint. The second is that it can be very hard for you to reuse an in-line constraint across multiple tests. By definition, an in-line constraint only exists in one piece of code. You could put it in a routine in a separate file and then call it as needed. At that point it has become nearly the same as an external constraint.

6.11.7   Specifying a Constraint in a Test with External Constraints

The body of a constraint does not have to be defined within the class, just as a routine body can be defined externally, as shown in Section 5.10 . Your data class could be defined in one file, with one empty constraint. Then each test could define its own version of this constraint to generate its own flavors of stimulus as shown in Samples 6.41 and 6.42.

Sample 6.41 Class with an external constraint

```
// packet.sv
class Packet;
  rand bit [7:0] length;
  rand bit [7:0] payload[];
  constraint c_valid {length > 0;
                      payload.size() == length;}
  constraint c_external;
endclass
```

Sample 6.42 Program defining an external constraint

```
// test.sv
program automatic test;
'include "packet.sv"
  constraint Packet::c_external {length == 1;}
  ...
endprogram
```

External constraints have several advantages over in-line constraints. They can be put in a file and thus reused between tests. An external constraint applies to all instances of the class, whereas an in-line constraint only affects the single call to `randomize`. Consequently, an external constraint provides a primitive way to change a class without having to learn advanced OOP techniques. Keep in mind that with this technique, you can only add constraints, not alter existing ones, and you need to define the external constraint prototype in the original class.

Like in-line constraints, external constraints can cause problems, as the constraints are spread across multiple files. The LRM requires external constraints to be defined in the same scope as the original class. A class defined in a package must have its external constraint also defined in the same package, limiting its usefulness. That is why Sample 6.42 includes the class definition rather than using a package.

A final consideration is what happens when the body for an external constraint is never defined. The SystemVerilog LRM does not currently specify what should happen in this case. Before you build a testbench with many external constraints, find out how your simulator handles missing definitions. Is this an error that prevents simulation, just a warning, or no message at all?

6.11.8 Extending a Class

In Chapter 8, you will learn how to extend a class. With this technique, you can take a testbench that uses a given class, and swap in an extended class that has additional or redefined constraints, routines, and variables. See Sample 8.10 for a typical testbench. Note that if you define a constraint in an extended class with the same name as one in the base class, the extended constraint replaces the base one.

Learning OOP techniques requires a little more study, but the flexibility of this new approach repays with great rewards.

6.12 Common Randomization Problems

You may be comfortable with procedural code, but writing constraints and understanding random distributions requires a new way of thinking. Here are some issues you may encounter when trying to create random stimulus.

6.12.1   *Use Signed Variables with Care*

When creating a testbench, you may be tempted to use the int, byte, or other
signed types for counters and other simple variables. Don't use them in random
constraints unless you really want signed values. What values are produced when
the class in Sample 6.43 is randomized? It has two random variables and wants to
make the sum of them 64.

Sample 6.43 Signed variables cause randomization problems

```
class SignedVars;
  rand byte pkt1_len, pkt2_len;
  constraint total_len {
    pkt1_len + pkt2_len == 64;
  }
endclass
```

Obviously, you could get pairs of values such as (32, 32) and (2, 62). Additionally,
you could see (−63, 127), as this is a legitimate solution of the equation, even though
it may not be what you wanted. To avoid meaningless values such as negative
lengths, use only unsigned random variables, as shown in Sample 6.44.

Sample 6.44 Randomizing unsigned 32-bit variables

```
class Vars32;
  rand bit [31:0] pkt1_len, pkt2_len;   // unsigned type
  constraint total_len {
    pkt1_len + pkt2_len == 64;
  }
endclass
```

Even this version causes problems, as large values of pkt1_len and pkt2_len,
such as 32'h80000040 and 32'h80000000, wrap around when added together
and give 32'd64 or 32'h40. You might think of adding another pair of constraints
to restrict the values of these two variables, but the best approach is to make them
only as wide as needed, and to avoid using 32-bit variables in constraints. In Sample
6.45, the sum of two 8-bit variables is compared to a 9-bit value.

Sample 6.45 Randomizing unsigned 8-bit variables

```
class Vars8;
  rand bit [7:0] pkt1_len, pkt2_len;   // 8-bits wide
  constraint total_len {
    pkt1_len + pkt2_len == 9'd64;      // 9-bit sum
  }
endclass
```

6.12.2   Solver Performance Tips

Each constraint solver has its strengths and weaknesses but there are some guidelines that you can follow to improve the speed of your simulations with constrained
random variables. Tools are always being improved, so check with your vendor for
more specific information.

If you just need to fill an array with raw data, don't use the solver as it has some
overhead choosing values, even for a variable that has no constraints. Don't declare
these arrays as `rand`, instead calculate the values in `pre_randomize` with `$urandom`
or `$urandom_range`. These functions calculate a value up to 100 times faster than the
solver, which is important when you need a 1000 values quickly. Generally, the larger
the array, the less important are the individual values, and the less likely that there
is a need to use a solver. Even if you need a non-uniform range of values, or there is
a simple relationship between values, you might be able to employ an `if` statement.

6.12.3   Choose the Right Arithmetic Operator to Boost Efficiency

 Simple arithmetic operators such as addition and subtraction, bit
extracts, and shifts are handled very efficiently by the solver in a
constraint. However, multiplication, division, and modulo are
very expensive with 32-bit values. Remember that any constant
without an explicit size, such as `42`, is treated as a 32-bit value,
`32'd42`.

If you want to generate random addresses that are near a page boundary, where a
page is 4096 bytes, you could write the following code, but the solver may take a
long time to find suitable values for `addr` if you use the constraint in Sample 6.46.

Sample 6.46 Expensive constraint with mod and unsized variable

```
rand bit [31:0] addr;
constraint slow_near_page_boundary {
  addr % 4096 inside {[0:20], [4075:4095]};
}
```

Many constants in hardware are powers of 2, so take advantage of this with a bit
extraction rather than division and modulo. Only constrain the bits that matter, not
the upper bits. Likewise, multiplication by a power of two can be replaced by a shift.
Note that some constraint solvers make these optimizations automatically Sample
6.47 replaces the MOD operator with a bit extract.

Sample 6.47 Efficient constraint with bit extract

```
rand bit [31:0] addr;
constraint near_page_boundry {
  addr[11:0] inside {[0:20], [4075:4095]};
}
```

6.13   Iterative and Array Constraints

The constraints presented so far allow you to specify limits on single variables. What if you want to randomize an array? The `foreach` constraint and several array functions let you shape the distribution of the values.

 Using the `foreach` constraint creates many constraints that can slow down simulation. A good solver can quickly solve hundreds of constraints but may slow down with thousands. Especially slow are nested `foreach` constraints, as they produce N^2 constraints for an array of size N. See Section 6.13.5 for an algorithm that used `randc` variables instead of nested `foreach`.

6.13.1   Array Size

The easiest array constraint to understand is the `size` function. In Sample 6.48, you are specifying the number of elements in a dynamic array or queue.

Sample 6.48 Constraining dynamic array size

```
class dyn_size;
  rand bit [31:0] d[];
  constraint d_size {d.size() inside {[1:10]}; }
endclass
```

Using the `inside` constraint lets you set a lower and upper boundary on the array size. In many cases you may not want an empty array, that is, `size==0`. Remember to specify an upper limit; otherwise, you can end up with thousands or millions of elements, which can cause the random solver to take an excessive amount of time.

6.13.2   Sum of Elements

You can send a random array of data into a design, but you can also use it to control the flow. Perhaps you have an interface that has to transfer four data words. The words can be sent consecutively or over many cycles. A strobe signal tells when the data signal is valid. Figure 6.2 shows some legal strobe patterns, sending four values over ten cycles.

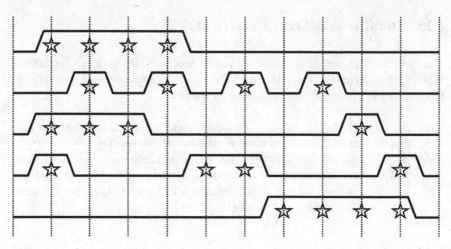

Fig. 6.2 Random strobe waveforms

You can create these patterns using a random array as shown in Sample 6.49. Constrain it to have four bits enabled out of the entire range using the sum function.

Sample 6.49 Random strobe pattern class

```
class StrobePat;
  rand bit strobe[10];
  constraint c_set_four { strobe.sum() == 4'h4; }
endclass

initial begin
  StrobePat sp;
  int count = 0;            // Index into data array

  sp = new();
  `SV_RAND_CHECK(sp.randomize());

  foreach (sp.strobe[i]) begin
    ##1 bus.cb.strobe <= sp.strobe[i];
    // If strobe is enabled, drive out next data word
    if (sp.strobe[i])
      bus.cb.data <= data[count++];
  end
end
```

As you remember from Chapter 2, the sum of an array of single-bit elements would normally be a single bit, e.g., 0 or 1. Sample 6.49 compares strobe.sum to a 4-bit value (4'h4), so the sum is calculated with 4-bit precision. The example uses 4-bit precision to store the maximum number of elements, which is 10.

6.13.3   Issues with Array Constraints

The sum function looks simple but can cause several problems because of Verilog's arithmetic rules. The following is a simple problem that one of the authors experienced creating constrained random stimulus. You want to generate from one to eight transactions, such that the total length of all of them is less than 1024 bytes. Sample 6.50 shows a first attempt, 6.51 has the test program, and 6.52 shows the output. The len field is a byte in the original transaction.

Sample 6.50   First attempt at sum constraint: bad_sum1

```
class bad_sum1;
  rand byte len[];
  constraint c_len {len.sum() < 1024;
                    len.size() inside {[1:8]};}

  function void display();
    $write("sum=%4d, val=", len.sum());
    foreach(len[i]) $write("%4d ", len[i]);
    $display;
  endfunction
endclass
```

Sample 6.51   Program to try constraint with array sum

```
program automatic test;
  bad_sum1 c;
  initial begin
    c = new();
    repeat (5) begin
      `SV_RAND_CHECK(c.randomize());
      c.display();
    end
  end
endprogram
```

Sample 6.52   Output from bad_sum1

```
sum=  81, val=  62 -20  39
sum=  39, val= -27  67   1  76 -97 -58  77
sum=  38, val=  60 -22
sum=  72, val=-120  29 123 102 -41 -21
sum= -53, val= -58 -85-115 112-101 -62
```

This generates some smaller lengths, but the sum is sometimes negative and is always less than 127 — definitely not what you wanted! Sample 6.53 shows another attempt, but this time replace the byte data type with an unsigned field. The display function is unchanged. Sample 6.54 shows the output.

Sample 6.53 Second attempt at sum constraint: bad_sum2

```
class bad_sum2;
  rand bit [7:0] len[];  // 8 bits unsigned, not byte
  constraint c_len {len.sum() < 1024;
                    len.size() inside {[1:8]};}
endclass
```

Sample 6.54 Output from bad_sum2

```
sum=  79, val=  88 100 246    2   14 228 169
sum= 120, val=  74  75 141   86
sum=  39, val=  39
sum= 193, val=  31 156 172   33   57
sum= 173, val=  59 150   25 101 138 212
```

Sample 6.53 has a subtle problem. The sum of all transaction lengths is always less than 256, even though you constrained the array sum to be less than 1024. The problem here is that in Verilog, the sum of many 8-bit values is computed using an 8-bit result. Sample 6.55 bumps the len field up to 32 bits using the uint type from Section 2.8 .

Sample 6.55 Third attempt at sum constraint: bad_sum3

```
class bad_sum3;
  rand uint len[];  // 32 bits
  constraint c_len {len.sum() < 1024;
                    len.size() inside {[1:8]};}
endclass
```

Sample 6.56 Output from bad_sum3

```
sum= 245, val=1348956995 3748256598 985546882 2507174362
sum= 600, val=2072193829 315191491 484497976 3050698208
 2300168220 3988671456 3998079060 970369544
sum=  17, val=1924767007 3550820640 4149215303 3260098955
sum= 440, val=3192781444 624830067 1300652226 4072252356
 3694386235
sum= 864, val=3561488468 733479692
```

Wow – what happened here in Sample 6.56? This is similar to the signed problem in Section 6.12.1, in that the sum of two very large numbers can wrap around to a small number. You need to limit the size based on the comparison in the constraint. Samples 6.57 and 6.58 show the next attempt and result.

Sample 6.57 Fourth attempt at sum constraint: `bad_sum4`

```
class bad_sum4;
  rand bit [9:0] len[];  // 10 bits, unsigned
  constraint c_len {len.sum() < 1024;
                     len.size() inside {[1:8]};}
endclass
```

Sample 6.58 Output from `bad_sum4`

```
sum= 989, val= 787 202
sum=1021, val= 564  76 132 235   0   8   6
sum= 872, val= 624 101 136  11
sum= 978, val= 890  88
sum= 905, val= 663 242
```

This does not work either as the individual `len` fields are more than 8 bits, so the `len` values are often greater than 255. You need to specify that each `len` field is between 1 and 255, but use a 10-bit field so they sum correctly. This requires constraining every element of the array, as shown in the following section.

6.13.4 Constraining Individual Array and Queue Elements

SystemVerilog lets you constrain individual elements of an array using `foreach`. While you might be able to write constraints for a fixed-size array by listing every element, the `foreach` style is more compact. The only practical way to constrain a dynamic array or queue is with `foreach` as shown in Samples 6.59 and 6.60.

Sample 6.59 Simple foreach constraint: `good_sum5`

```
class good_sum5;
  rand uint len[];
  constraint c_len {foreach (len[i])
                     len[i] inside {[1:255]};
                    len.sum() < 1024;
                    len.size() inside {[1:8]};}
endclass
```

Sample 6.60 Output from `good_sum5`

```
sum=1011, val=  83 249 197 187 152  95  40   8
sum=1012, val= 213 252 213  44 196  20  20  54
sum= 370, val= 118  76 176
sum= 976, val= 233 187  44 157 201  81  73
sum= 412, val= 172 167  73
```

The addition of the constraint for individual elements fixed the example. Note that the len array can be 10 or more bits wide, but must be unsigned.

You can specify constraints between array elements as long as you are careful about the endpoints. The class in Sample 6.61 creates an ascending list of values by comparing each element to the previous, except for the first.

Sample 6.61 Creating ascending array values with foreach

```
class Ascend;
  rand uint d[10];
  constraint c {
    foreach (d[i])        // For every element
      if (i>0)            // except the first
        d[i] > d[i-1];    // compare with previous element
  }
endclass
```

How complex can these constraints become? Constraints have been written to solve Einstein's problem (a logic puzzle with five people, each with five separate attributes), the Eight Queens problem (place eight queens on a chess board so that none can capture each other), and even Sudoku.

6.13.5 Generating an Array of Unique Values

How can you create an array of random unique values? If your array has N elements, and the element values range from 0..N-1, you can simply use the array shuffle function as described in Section 2.6.3 .

What if the range of values is greater than the number of array elements? If you try to make a randc array, each array element will be randomized independently, so you are almost certain to get repeated values.

You may be tempted to use a constraint solver to compare every element with every other with nested foreach loops as shown in Sample 6.62. This creates over 4000 individual constraints, which could slow down simulation.

Sample 6.62 Creating unique array values with foreach

```
class UniqueSlow;            // Bad code, do not use
  rand bit [7:0] ua[64];
  constraint c {
    foreach (ua[i])          // For every element,
      foreach (ua[j])
        if (i != j)          //    except the diagonals,
          ua[i] != ua[j];    //    compare to other elements
  }
endclass
```

Instead, you should use procedural code as shown in Sample 6.63 with a helper class containing a `randc` variable so that you can randomize the same variable over and over.

Sample 6.63   Creating unique array values with a randc helper class

```
class randc8;
  randc bit [7:0] val;
endclass

class LittleUniqueArray;
  bit [7:0] ua [64];      // Array of unique values

  function void pre_randomize();
    randc8 rc8;
    rc8 = new();
    foreach (ua[i]) begin
      `SV_RAND_CHECK(rc8.randomize());
      ua[i] = rc8.val;
    end
  endfunction
endclass
```

Samples 6.64 and 6.65 give a more general solution. For example, you may need to assign ID numbers to N bus drivers, which are in the range of 0 to MAX-1 where MAX >=N.

Sample 6.64   Unique value generator

```
// Create unique random values in a range 0:max-1
class RandcRange;
  randc bit [15:0] value;
  int max_value;   // Maximum possible value

  function new(input int max_value = 10);
    this.max_value = max_value;
  endfunction

  constraint c_max_value {value < max_value;}
endclass
```

Sample 6.65 Class to generate a random array of unique values

```
class UniqueArray;
  int max_array_size, max_value;
  rand bit [15:0] ua[];         // Array of unique values
  constraint c_size {ua.size() inside {[1:max_array_size]};}

  function new(input int max_array_size=2, max_value=2);
    this.max_array_size = max_array_size;
    // If max_value is smaller than max array size,
    // array could have duplicates, so adjust max_value
    if (max_value < max_array_size)
      this.max_value = max_array_size;
    else
      this.max_value = max_value;
  endfunction

  // Array a[] allocated in randomize(), fill w/unique vals
  function void post_randomize();
    RandcRange rr;
    rr = new(max_value);
    foreach (ua[i]) begin
      `SV_RAND_CHECK(rr.randomize());
      ua[i] = rr.value;
    end
  endfunction

  function void display();
    $write("Size: %3d:", ua.size());
    foreach (ua[i]) $write("%4d", ua[i]);
    $display;
  endfunction
endclass
```

Sample 6.66 has a program. Here is a program that uses the `UniqueArray` class.

Sample 6.66 Using the `UniqueArray` class

```
program automatic test;
  UniqueArray ua;
  initial begin
    ua = new(50);                         // Max array size = 50

    repeat (10) begin
      `SV_RAND_CHECK(ua.randomize());  // Create random array
      ua.display();                       // Display values
    end
  end
endprogram
```

6.13.6   *Randomizing an Array of Handles*

If you need to create multiple random objects, you might create a random array of handles. Unlike an array of integers, you need to allocate all the elements before randomization as the random solver never constructs objects. If you have a dynamic array, allocate the maximum number of elements you may need, and then use a constraint to resize the array as shown in Sample 6.67. A dynamic array of handles can remain the same size or shrink during randomization, but it can never increase in size.

Sample 6.67 Constructing elements in a random array class

```
parameter MAX_SIZE = 10;

class RandStuff;
  rand bit [31:0] value;
endclass

class RandArray;
  rand RandStuff array[];    // Don't forget rand!

  constraint c {array.size() inside {[1:MAX_SIZE]}; }

  function new();
    array = new[MAX_SIZE];    // Allocate maximum size
    foreach (array[i])
      array[i] = new();
  endfunction;
endclass

RandArray ra;
initial begin
  ra = new();                          // Construct array and all objects
  `SV_RAND_CHECK(ra.randomize()); // Randomize array
  foreach (ra.array[i])
    $display(ra.array[i].value);
end
```

The above code works well for a single array randomization. If you need to repeatedly randomize the same array over and over, allocate the array and construct the elements in pre_randomize. See Section 5.14.4 for more on arrays of handles.

6.14   Atomic Stimulus Generation vs. Scenario Generation

Up until now, you have seen atomic random transactions. You have learned how to make a single random bus transaction, a single network packet, or a single processor instruction. This is a good start, however your job is to verify that the design works with real-world stimuli. A bus may have long sequences of transactions such as DMA transfers or cache fills. Network traffic consists of extended sequences of packets as you simultaneously read e-mail, browse a web page, and download music from the net, all in parallel. Processors have deep pipelines that are filled with the code for routine calls, `for` loops, and interrupt handlers. Generating transactions one at a time is unlikely to mimic any of these scenarios.

6.14.1   An Atomic Generator with History

The easiest way to create a stream of related transactions is to have an atomic generator base some of its random values on ones from previous transactions. The class might constrain a bus transaction to repeat the previous command, such as a write, 80% of the time, and also use the previous destination address plus an increment. You can use the `post_randomize` function to make a copy of the generated transaction for use by the next call to `randomize`.

This scheme works well for smaller cases but gets into trouble when you need information about the entire sequence ahead of time. A DUT may need to know the length of a sequence of network transactions before it starts.

6.14.2   Random Array of Objects

If you want to generate stimulus for a complex, multi-level protocol, you could build up a combination of code and arrays of random objects. The UVM and VMM both allow you to generate random sequences through a sophisticated set of classes and macros. This section shows a simplified random sequence.

One way to generate random sequences is to randomize an entire array of objects. You can create constraints that refer to the previous and next objects in the array, and the SystemVerilog solver solves all constraints simultaneously. Since the entire sequence is generated at once, you can then extract information such as the total number of transactions or a checksum of all data values before the first transaction is sent. Alternatively, you can build a sequence for a DMA transfer that is constrained to be exactly 1024 bytes, and let the solver pick the right number of transactions to reach that goal.

Sample 6.68 shows a simple sequence of transactions, each one with a destination address that is greater than the one before. It builds on the array constraint shown in Sample 6.61.

Sample 6.68 Simple random sequence with ascending values

```
class Transaction;              // Simple transaction
  rand bit [3:0] src, dst;
endclass

class Transaction_seq;
  rand Transaction items[10]; // Array of transaction handles

  function new();               // Construct the sequence items
    foreach (items[i])
      items[i] = new();
  endfunction // new

  constraint c_ascend           // Each dst addr is greater than
    { foreach (items[i])        // the one before it
        if (i>0)
          items[i].dst > items[i-1].dst;
}
endclass // Transaction_seq

Transaction_seq seq;

initial begin
  seq = new();                  // Construct the sequence
  `SV_RAND_CHECK(seq.randomize());  // Randomize it
  foreach (seq.items[i])
    $display("item[%0d] = %0d", i, seq.items[i].dst);
end
```

6.14.3   Combining Sequences

You can combine multiple sequences together to make a more realistic flow of transactions. For example, for a network device, you could make one sequence that resembles downloading e-mail, a second that is viewing a web page, and a third that is entering single characters into web-based form.The techniques to combine these flows is beyond the scope of this book, but you can learn more from the VMM, as described in Bergeron, et al. (2005).

6.14.4   Randsequence

You may find it challenging to write random constraints as they don't execute sequentially like procedural statements. An alternative way to create random sequences is to describe the grammar of a protocol with a declarative style using a syntax similar to BNF (Backus-Naur Form) and random weighted case statements.

SystemVerilog's `randsequence` construct resembles the algorithmic code that you have traditionally used but can still be challenging.

Sample 6.69 generates a sequence called `stream`. A `stream` can be either `cfg_read`, `io_read`, or `mem_read`. The random sequence engine randomly picks one. The `cfg_read` label has a weight of 1, `io_read` has twice the weight and so is twice as likely to be chosen as `cfg_read`. The label `mem_read` is most likely to be chosen, with a weight of 5.

Sample 6.69 Command generator using `randsequence`

```
initial begin
  for (int i=0; i<15; i++) begin
    randsequence (stream)
      stream :  cfg_read := 1 |
                io_read  := 2 |
                mem_read := 5;
      cfg_read : { cfg_read_task(); } |
                 { cfg_read_task(); } cfg_read;
      mem_read : { mem_read_task(); } |
                 { mem_read_task(); } mem_read;
      io_read  : { io_read_task(); } |
                 { io_read_task(); } io_read;
    endsequence
  end // for
end

task cfg_read_task();
  ...
endtask
```

A `cfg_read` can be either a single call to `cfg_read_task`, or a call to the task followed by another `cfg_read`. As a result, the task is always called at least once, and possibly many times.

One big advantage of `randsequence` is that it is procedural code and you can debug it by stepping though the execution, or adding `$display` statements. When you call `randomize` for an object, it either all works or all fails, but you can't see the steps taken to get to a result.

There are several problems with using `randsequence`. The code to generate the sequence is separate and a very different style from the classes with data and constraints used by the sequence. So if you use both `randomize` and `randsequence`, you have to master two different forms of randomization. More seriously, if you want to modify a sequence, perhaps to add a new branch or action, you have to modify the original sequence code. You can't just make an extension. As you will see in Chapter 8, you can extend a class to add new code, data, and constraints without having to edit the original class.

6.15   Random Control

At this point you may be thinking that this process is a great way to create long streams of random input into your design. Or you may think that this is a lot of work if all you want to do is occasionally to make a random decision in your code. You may prefer a set of procedural statements that you can step through using a debugger.

6.15.1   Introduction to randcase

You can use `randcase` to make a weighted choice between several actions, without having to create a class and instance. Sample 6.70 chooses one of the three branches based on the weight. SystemVerilog adds up the weights (1+8+1 = 10), chooses a value in this range, and then picks the appropriate branch. The branches are not order dependent, the weights can be variables, and they do not have to add up to 100%. The function `$urandom_range` is described in Section 6.10.

Sample 6.70   Random control with `randcase` and `$urandom_range`

```
initial begin
  bit [15:0] len;
  randcase
    1: len = $urandom_range(0, 2);   // 10%: 0, 1, or 2
    8: len = $urandom_range(3, 5);   // 80%: 3, 4, or 5
    1: len = $urandom_range(6, 7);   // 10%: 6 or 7
  endcase
  $display("len=%0d", len);
end
```

You can write Sample 6.70 using a class and the `randomize` function. For this small case, the OOP version in Sample 6.71 is a little larger. However, if this were part of a larger class, the constraint would be more compact than the equivalent `randcase` statement.

Sample 6.71   Equivalent constrained class

```
class LenDist;
  rand bit [15:0] len;
  constraint c {len dist {[0:2] := 1, [3:5] := 8, [6:7] := 1}; }
endclass

initial begin
  LenDist lenD;
  lenD = new();
  `SV_RAND_CHECK(lenD.randomize());
  $display("len=%0d", lenD.len);
end
```

Code using `randcase` is more difficult to override and modify than random constraints. The only way to modify the random results is to rewrite the code or use variable weights.

Be careful using `randcase`, as it does not leave any tracks behind. For example, you could use it to decide whether or not to inject an error in a transaction. The problem is that the downstream transactors and scoreboard need to know of this choice. The best way to inform them would be to use a variable in the transaction or environment. However, if you are going to create a variable that is part of these classes, you could have made it a random variable and used constraints to change its behavior in different tests.

6.15.2   *Building a Decision Tree with* `randcase`

You can use the `randcase` statement to create a decision tree. Sample 6.72 has just two levels of procedural code, but you can see how it can be extended to use more.

Sample 6.72 Creating a decision tree with `randcase`

```
initial begin
  // Level 1
  randcase
    one_write_wt: do_one_write();
    one_read_wt:  do_one_read();
    seq_write_wt: do_seq_write();
    seq_read_wt:  do_seq_read();
  endcase
  end

// Level 2
task do_one_write();
  randcase
    mem_write_wt: do_mem_write();
    io_write_wt:  do_io_write();
    cfg_write_wt: do_cfg_write();
  endcase
endtask

task do_one_read();
  randcase
    mem_read_wt: do_mem_read();
    io_read_wt:  do_io_read();
    cfg_read_wt: do_cfg_read();
  endcase
endtask
```

6.16   Random Number Generators

How random is SystemVerilog? On the one hand, your testbench depends on an uncorrelated stream of random values to create stimulus patterns that go beyond any directed test. On the other hand, you need to repeat the patterns over and over during debug of a particular test, even if the design and testbench make minor changes.

6.16.1   Pseudorandom Number Generators

Verilog uses a simple PRNG that you could access with the $random function. The generator has an internal state that you can set by providing a seed to $random. All IEEE-1364-compliant Verilog simulators use the same algorithm to calculate values.

Sample 6.73 shows a simple PRNG, not the one used by SystemVerilog. The PRNG has a 32-bit state. To calculate the next random value, square the state to produce a 64-bit value, take the middle 32 bits, then add the original value.

Sample 6.73 Simple pseudorandom number generator

```
bit [31:0] state = 32'h12345678;
function bit [31:0] my_random();
  bit [63:0] s64;
  s64 = state * state;
  state = (s64 >> 16) + state;
  return state;
endfunction
```

You can see how this simple code produces a stream of values that seem random, but can be repeated by using the same seed value. SystemVerilog calls its own PRNG to generate a new value for randomize and randcase.

6.16.2   Random Stability — Multiple Generators

Verilog has a single PRNG that is used for the entire simulation. What would happen if SystemVerilog kept this approach? Testbenches often have several stimulus generators running in parallel, creating data for the design under test. If two streams share the same PRNG, they each get a subset of the random values.

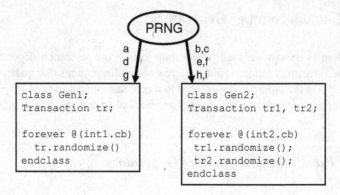

Fig. 6.3 Sharing a single random generator

In Fig. 6.3, there are two stimulus generators and a single PRNG producing values a, b, c, etc. Gen2 has two random objects, so during every cycle, it uses twice as many random values as Gen1.

A problem can occur when one of the classes changes as shown in Fig. 6.4. Gen1 gets an additional random variable, and so consumes two random values every time it is called. This approach changes the values used not only by Gen1, but also by Gen2.

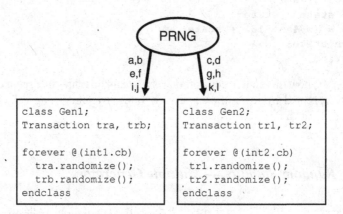

Fig. 6.4 First generator uses additional values

In SystemVerilog, there is a separate PRNG for every object and thread. Figure 6.5 shows how changes to one object don't affect the random values seen by others.

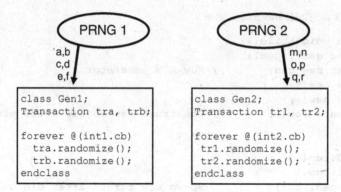

Fig. 6.5 Separate random generators per object

6.16.3   Random Stability and Hierarchical Seeding

In SystemVerilog, every object and thread has its own PRNG and unique seed. When a new object or thread is started, its PRNG is seeded from its parent's PRNG. Thus a single seed specified at the start of simulation can create many streams of random stimulus, each distinct.

When you are debugging a testbench, you add, delete, and move code. Even with random stability, your changes may cause the testbench to generate different random values. This can be very frustrating if you are in the middle of debugging a DUT failure, and the testbench no longer creates the same stimulus. You can minimize the effect of code modifications by adding any new objects or threads after existing ones. Sample 6.74 shows a routine from testbench that constructs objects, and runs them in parallel threads.

Sample 6.74   Test code before modification

```
function void build();
  pci_gen gen0, gen1;
  gen0 = new();
  gen1 = new();
  fork
    gen0.run();
    gen1.run();
  join
endfunction : build
```

Sample 6.75 adds a new generator, and runs it in a new thread. The new object is constructed after the existing ones, and the new thread is spawned after the old ones.

Sample 6.75 Test code after modification

```
function void build();
  pci_gen gen0, gen1;
  atm_gen new_gen;         // New ATM generator
  gen0 = new();
  gen1 = new();
  new_gen = new();         // Construct new object after old ones

  fork
    gen0.run();
    gen1.run();
    new_gen.run();         // Spawn new thread after old ones
  join
endfunction : build
```

As new code is added, you may not be able to keep the random streams the same as the old ones, but you might be able to postpone any side effects from these changes.

6.17   Random Device Configuration

 An important part of your DUT to test is the configuration of both the internal DUT settings and the system that surrounds it. As described in Section 6.2.1, your tests should randomize the environment so that you can be confident it has been tested in as many modes as possible.

Sample 6.76 shows a random testbench configuration that can be modified as needed at the test level. The EthCfg class describes the configuration for a 4-port Ethernet switch. It is instantiated in an environment class, which in turn is used in the test. The test overrides one of the configuration values, enabling all 4 ports.

Sample 6.76 Ethernet switch configuration class

```
class EthCfg;

  rand bit [ 3:0] in_use;        // Ports used in test:3,2,1,0
  rand bit [47:0] mac_addr[4];   // MAC addresses
  rand bit [ 3:0] is_100;        // 100mb mode for ports 3,2,1,0
  rand uint run_for_n_frames;    // # frames in test

  // Force some addr bits when running in unicast mode
  constraint local_unicast {
    foreach (mac_addr[i])
      mac_addr[i][41:40] == 2'b00;
  }

  constraint reasonable {        // Limit test length
    run_for_n_frames inside {[1:100]};
  }

endclass
```

The configuration class is used in the `Environment` class during several phases.
The configuration is constructed in the `Environment` constructor, but not randomized until the `gen_cfg` phase as shown in Sample 6.77. This allows you to turn
constraints on and off before `randomize` is called. Afterwards, you can override
the generated values before the `build` phase creates the virtual components around
the DUT. (The classes such as `EthGen` and `EthMii` are not shown).

Sample 6.77 Building environment with random configuration

```
class Environment;

  EthCfg cfg;
  EthGen gen[4];
  EthMii drv[4];

  function new();
    cfg = new();                              // Construct the cfg
  endfunction

  // Use random configuration to build the environment
  function void build();
    foreach (gen[i]) begin
      gen[i] = new();
      drv[i] = new();
      if (cfg.is_100[i])
          drv[i].set_speed(100);
    end
  endfunction
```

```
function void gen_cfg();
  `SV_RAND_CHECK(cfg.randomize());   // Randomize the cfg
endfunction

task run();
  foreach (gen[i])
    if (cfg.in_use[i]) begin
      // Only start the testbench transactors that are in-use
      gen[i].run();
      ...
    end
endtask

task wrap_up();
  // Not currently used
endtask
endclass : Environment
```

Now you have all the components to build a test, which is described in a program block. The test in Sample 6.78 instantiates the environment class and then runs each step.

Sample 6.78   Simple test using random configuration

```
program automatic test;

  Environment env;

  initial begin
    env = new();        // Construct environment
    env.gen_cfg();      // Create random configuration
    env.build();        // Build the testbench environment
    env.run();          // Run the test
    env.wrap_up();      // Clean up after test & report
  end

endprogram
```

You may want to override the random configuration, perhaps to reach a corner case. The test in Sample 6.79 randomizes the configuration class and then enables all the ports.

Sample 6.79   Simple test that overrides random configuration

```
program automatic test;

   Environment env;

   initial begin
      env = new();        // Construct environment
      env.gen_cfg();      // Create random configuration

      // Override random in_use - turn all 4 ports on
      env.cfg.in_use = '1;

      env.build();        // Build the testbench environment
      env.run();          // Run the test
      env.wrap_up();      // Clean up after test & report
   end

endprogram
```

Notice how in Sample 6.77 all generators were constructed, but only a few were run, depending on the random configuration. If you only constructed the generators that are in-use, you would have to surround any reference to gen [i] with a test of in_ use [i], otherwise your testbench would crash when it tried to refer to the non-existent generator. The extra memory taken up by these generators that are not used is a small price to pay for a more stable testbench.

6.18   Conclusion

Constrained-random tests are the only practical way to generate the stimulus needed to verify a complex design. SystemVerilog offers many ways to create a random stimulus and this chapter presents many of the alternatives.

A test needs to be flexible, allowing you either to use the values generated by default or to constrain or override the values so that you can reach your goals. Always plan ahead when creating your testbench by leaving sufficient "hooks" so that you can steer the testbench from the test without modifying existing code.

6.19   Exercises

1. Write the SystemVerilog code for the following items.

   a. Create a class `Exercise1` containing two random variables, 8-bit `data` and 4-bit `address`. Create a constraint block that keeps `address` to 3 or 4.
   b. In an `initial` block, construct an `Exercise1` object and randomize it. Check the status from randomization.

2. Modify the solution for Exercise 1 to create a new class `Exercise2` so that:

   a. `data` is always equal to 5
   b. The probability of `address==0` is 10%
   c. The probability of `address` being between [1:14] is 80%
   d. The probability of `address==15` is 10%

3. Using the solution to either Exercise 1 or 2, demonstrate its usage by generating 20 new `data` and `address` values and check for success from the constraint solver.

4. Create a testbench that randomizes the `Exercise2` class 1000 times.

   a. Count the number of times each `address` value occurs and print the results in a histogram. Do you see an exact 10% / 80% / 10% distribution? Why or why not?
   b. Run the simulation with 3 different random seeds, creating histograms, and then comment on the results. Here is how to run a simulation with the seed 42.

      VCS: > `simv +ntb_random_seed=42`
      IUS: > `irun exercise4.sv -svseed 42`
      Questa: > `vsim -sv_seed 42`

5. For the code in Sample 6.4, describe the constraints on the `len`, `dst`, and `src` variables.

6. Complete Table 6.9 below for the following constraints.

```
class MemTrans;
  rand bit x;
  rand bit [1:0] y;
  constraint c_xy {
    y inside{ [x:3] };
    solve x before y;
  }
endclass
```

Table 6.9 Solution probabilities

Solution	x	y	Probability
A	0	0	
B	0	1	
C	0	2	
D	0	3	
E	1	0	
F	1	1	
G	1	2	
H	1	3	

7. For the following class, create:

   a. A constraint that limits read transaction addresses to the range 0 to 7, inclusive.
   b. Write behavioral code to turn off the above constraint. Construct and randomize a MemTrans object with an in-line constraint that limits read transaction addresses to the range 0 to 8, inclusive. Test that the in-line constraint is working.

```
class MemTrans;
   rand bit rw; // read if rw=0, write if rw=1
   rand bit [7:0] data_in;
   rand bit [3:0] address;
endclass // MemTrans
```

8. Create a class for a graphics image that is 10x10 pixels. The value for each pixel can be randomized to black or white. Randomly generate an image that is, on average, 20% white. Print the image and report the number of pixels of each type.

9. Create a class, StimData, containing an array of integer samples. Randomize the size and contents of the array, constraining the size to be between 1 and 1000. Test the constraint by generating 20 transactions and reporting the size.

10. Expand the `Transaction` class below so back-to-back transactions of the same type do not have the same address. Test the constraint by generating 20 transactions.

```
package my_package;

  typedef enum {READ, WRITE} rw_e;

  class Transaction;
    rw_e old_rw;
    rand rw_e rw;
    rand bit [31:0] addr, data;
    constraint rw_c{if (old_rw == WRITE) rw != WRITE;};

    function void post_randomize;
      old_rw = rw;
    endfunction

    function void print_all;
      $display("addr = %d, data = %d, rw = %s",
               addr, data, rw);
    endfunction

  endclass

endpackage
```

11. Expand the `RandTransaction` class below so back-to-back transactions of the same type do not have the same address. Test the constraint by generating 20 transactions.

```
class Transaction;
  rand rw_e rw;
  rand bit [31:0] addr, data;
endclass


class RandTransaction;

  rand Transaction trans_array[];

  constraint  rw_c {foreach (trans_array[i])
    if ((i>0) && (trans_array[i-1].rw == WRITE))
      trans_array[i].rw != WRITE;}

  function new();
    trans_array = new[TESTS];
    foreach (trans_array[i])
      trans_array[i] = new();
  endfunction

endclass
```

Chapter 7
Threads and Interprocess Communication

In real hardware, the sequential logic is activated on clock edges, whereas combinational logic is constantly changing when any inputs change. All this parallel activity is simulated in Verilog RTL using `initial` and `always` blocks, plus the occasional gate and continuous assignment statement. To stimulate and check these blocks, your testbench uses many threads of execution, all running in parallel. Most blocks in your testbench environment are modeled with a transactor and run in their own thread.

The SystemVerilog scheduler is the traffic cop that chooses which thread runs next. You can use the techniques in this chapter to control the threads and thus your testbench.

Each of these threads communicates with its neighbors. In Fig. 7.1, the generator passes the stimulus to the agent. The environment class needs to know when the generator completes and then tell the rest of the testbench threads to terminate. This is done with interprocess communication (IPC) constructs such as the standard Verilog events, event control and `wait` statements, and the SystemVerilog mailboxes and semaphores.[1]

---

[1]The SystemVerilog LRM uses "thread" and "process" interchangeably. The term "process" is most commonly associated with Unix processes, in which each contains a program running in its own memory space. Threads are lightweight processes that may share common code and memory, and consume far fewer resources than a typical process. This book uses the term "thread." However, "interprocess communication" is such a common term that it is used in this book.

C. Spear and G. Tumbush, *SystemVerilog for Verification: A Guide to Learning the Testbench Language Features*, DOI 10.1007/978-1-4614-0715-7_7,
© Springer Science+Business Media, LLC 2012

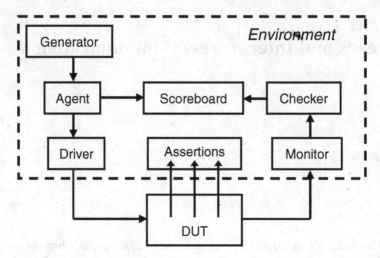

Fig. 7.1 Testbench environment blocks

7.1 Working with Threads

While all the thread constructs can be used in both modules and program blocks, your testbenches belong in program blocks. As a result, your code always starts with `initial` blocks that start executing at time 0. You cannot put an `always` block in a program. However, you can easily get around this by using a `forever` loop in an `initial` block.

Classic Verilog has two ways of grouping statements — with a `begin...end` or `fork...join`. Statements in a `begin...end` run sequentially, whereas those in a `fork...join` execute in parallel. The latter is very limited in that all statements inside the `fork...join` have to finish before the rest of the block can continue. As a result, it is rare for Verilog testbenches to use this feature.

SystemVerilog introduces two new ways to create threads — with the `fork...join_none` and `fork...join_any` statements, shown in Fig. 7.2.

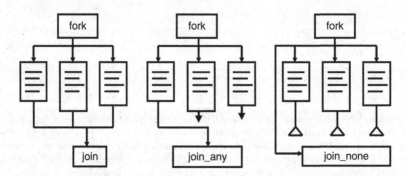

Fig. 7.2 Fork...join blocks

Your testbench communicates, synchronizes, and controls these threads with existing constructs such as events, @ event control, the `wait` and `disable` statements, plus new language elements such as semaphores and mailboxes.

7.1.1   Using *fork...join* and *begin...end*

Sample 7.1 has a `fork...join` parallel block with an enclosed `begin...end` sequential block, and shows the difference between the two.

Sample 7.1 Interaction of begin...end and fork...join

```
initial begin
      $display("@%0t: start fork...join example", $time);
  #10 $display("@%0t: sequential after #10", $time);
  fork
        $display("@%0t: parallel start", $time);
    #50 $display("@%0t: parallel after #50", $time);
    #10 $display("@%0t: parallel after #10", $time);
    begin
      #30 $display("@%0t: sequential after #30", $time);
      #10 $display("@%0t: sequential after #10", $time);
    end
  join
  $display("@%0t: after join", $time);
  #80 $display("@%0t: finish after #80", $time);
end
```

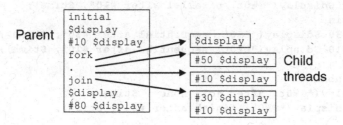

Fig. 7.3 Fork...join block

In the output below, the code in the fork…join executes in parallel, so statements with shorter delays execute before those with longer delays. As shown in Sample 7.2, the fork…join completes after the last statement, which starts with #50.

Sample 7.2 Output from begin…end and fork…join

```
@0: start fork...join example
@10: sequential after #10
@10: parallel start
@20: parallel after #10
@40: sequential after #30
@50: sequential after #10
@60: parallel after #50
@60: after join
@140: finish after #80
```

7.1.2 Spawning Threads with fork…join_none

A fork…join_none block schedules each statement in the block, but execution continues in the parent thread. Sample 7.3 is identical to Sample 7.1 except that the join has been converted to join_none.

Sample 7.3 Fork…join_none code

```
initial begin
  $display("@%0t: start fork...join_none example", $time);
  #10 $display("@%0t: sequential after #10", $time);
  fork
        $display("@%0t: parallel start", $time);
    #50 $display("@%0t: parallel after #50", $time);
    #10 $display("@%0t: parallel after #10", $time);
    begin
      #30 $display("@%0t: sequential after #30", $time);
      #10 $display("@%0t: sequential after #10", $time);
    end
  join_none
  $display("@%0t: after join_none", $time);
  #80 $display("@%0t: finish after #80", $time);
end
```

The diagram for this block is similar to Fig. 7.3. Note that the statement after the `join_none` block in Sample 7.4 executes before any statement inside the `fork...join_none`.

Sample 7.4 Fork...join_none output

```
@0: start fork...join_none example
@10: sequential after #10
@10: after join_none
@10: parallel start
@20: parallel after #10
@40: sequential after #30
@50: sequential after #10
@60: parallel after #50
@90: finish after #80
```

7.1.3 Synchronizing Threads with fork...join_any

A `fork...join_any` block schedules each statement in the block. Then, when the first statement completes, execution continues in the parent thread. All other remaining threads continue. Sample 7.5 is identical to the previous examples, except that the `join` has been converted to `join_any`.

Sample 7.5 Fork...join_any code

```
initial begin
  $display("@%0t: start fork...join_any example", $time);
  #10 $display("@%0t: sequential after #10", $time);
  fork
        $display("@%0t: parallel start", $time);
    #50 $display("@%0t: parallel after #50", $time);
    #10 $display("@%0t: parallel after #10", $time);
    begin
      #30 $display("@%0t: sequential after #30", $time);
      #10 $display("@%0t: sequential after #10", $time);
    end
  join_any
  $display("@%0t: after join_any", $time);
  #80 $display("@%0t: finish after #80", $time);
end
```

Note in Sample 7.6, the statement $display("after join_any") completes after the first statement in the parallel block.

Sample 7.6 Output from `fork...join_any`

```
@0: start fork...join_any example
@10: sequential after #10
@10: parallel start
@10: after join_any
@20: parallel after #10
@40: sequential after #30
@50: sequential after #10
@60: parallel after #50
@90: finish after #80
```

7.1.4   Creating Threads in a Class

You can use a `fork...join_none` to start a thread, such as the code for a random transactor generator. Sample 7.7 shows a generator / driver class with a `run` task that creates N packets. The full testbench has classes for the driver, monitor, checker, and more, all with transactors that need to run in parallel.

Sample 7.7 Generator / Driver class with a run task

```
class Gen_drive;

  // Transactor that creates N packets
  task run(input int n);
    Packet p;

    fork
      repeat (n) begin
        p = new();
        `SV_RAND_CHECK(p.randomize());
        transmit(p);
      end
    join_none        // Use fork-join_none so run() does not block
  endtask

  task transmit(input Packet p);
  ...
  endtask
endclass

Gen_drive gen;

initial begin
  gen = new();
  gen.run(10);
  // Start the checker, monitor, and other threads
  ...
end
```

There are several points to note with Sample 7.7. First, the transactor is not started in the new() function. The constructor should just initialize values, not start any threads. Separating the constructor from the code that does the real work allows you to change any variables before you start executing the code in the object. This allows you to inject errors, modify the defaults, and alter the behavior of the object. Next, the run task starts a thread in a fork...join_none block. The thread is a part of the transactor and should be spawned there, not in the parent class.

7.1.5   Dynamic Threads

Verilog's threads are very predictable. You can read the source code and count the initial, always, and fork...join blocks to know how many threads were in a module. On the other hand, SystemVerilog lets you create threads dynamically, and does not require you to wait for them to finish.

In Sample 7.8, the testbench generates random transactions and sends them to a DUT that stores them for some predetermined time, and then returns them. The testbench has to wait for the transaction to complete, but does not want to stop the generator.

Sample 7.8 Dynamic thread creation

```
program automatic test(bus_ifc.TB bus);
  // Code for interface not shown
  task check_trans(input Transaction tr);
    fork
      begin
      wait (bus.cb.data == tr.data);
      $display("@%0t: data match %d", $time, tr.data);
      end
    join_none    // Spawn thread, don't block
  endtask

  Transaction tr;

  initial begin
    repeat (10) begin
      tr = new();              // Create a random transaction
      `SV_RAND_CHECK(tr.randomize());
      transmit(tr);            // Send transaction into the DUT
      check_trans(tr);         // Wait for reply from DUT
    end
    #100; // Wait for final transaction to complete
  end
endprogram
```

When the `check_trans` task is called, it spawns off a thread to watch the bus for the matching transaction data. During a normal simulation, many of these threads run concurrently. In this simple example, the thread just prints a message, but you could add more elaborate controls.

7.1.6 *Automatic Variables in Threads*

 A common but subtle bug occurs when you have a loop that spawns threads and you don't save variable values before the next iteration. Sample 7.8 only works in a `program` or `module` with automatic storage. If `check_trans` used static storage, each thread would share the same variable `tr`, so later calls would overwrite the value set by earlier ones. Likewise, if the example had the `fork...join_none` inside the `repeat` loop, it would try to match incoming transactions using `tr`, but its value would change the next time through the loop. Always use automatic variables to hold values in concurrent threads.

Sample 7.9 has a `fork...join_none` inside a `for` loop. SystemVerilog schedules the threads inside a `fork...join_none` but they are not executed until after the original code blocks, here because of the #0 delay. So Sample 7.9 prints "3 3 3" which are the values of the index variable `j` when the loop terminates.

Sample 7.9 Bad `fork...join_none` inside a loop

```
program no_auto;
  initial begin
    for (int j=0; j<3; j++)
      fork
        $write(j);  // Bug - prints final value of index
      join_none
    #0 $display;
  end
endprogram
```

Sample 7.10 Execution of bad fork...join_none inside a loop

j	Statement	Comment
0	for (j=0; ...	
0	Spawn $write(j)	[thread 0]
1	j++	j=1
1	Spawn $write(j)	[thread 1]
2	j++	j=2
2	Spawn $write(j)	[thread 2]
3	j++	j=3
3	join_none	
3	#0	Delay before $display
3	$write(j)	[thread 0: j=3]
3	$write(j)	[thread 1: j=3]
3	$write(j)	[thread 2: j=3]
3	$display;	

The #0 delay blocks the current thread and reschedules it to start later during the current time slot. In Sample 7.10, the delay makes the current thread run after the threads spawned in the fork...join_none statement. This delay is useful for blocking a thread, but you should be careful, as excessive use causes race conditions and unexpected results.

You should use automatic variables inside a fork...join statement to save a copy of a variable as shown in Sample 7.11.

Sample 7.11 Automatic variables in a fork...join_none

```
initial begin
  for (int j=0; j<3; j++)
    fork
      automatic int k = j;    // Make copy of index
      begin
        $write(k);            // Print copy
      end
    join_none
  #0 $display;
end
```

The fork...join_none block is split into two parts, declarations and procedural code. The automatic variable declaration with initialization runs in the thread inside the for loop. During each loop, a copy of k is created and set to the current value of j. Then the body of the fork...join_none ($write) is scheduled, including a copy of k. After the loop finishes, #0 blocks the current thread, so the three threads run, printing the value of their copy of k. When the threads complete, and there is nothing else left during the current time-slot region, SystemVerilog advances to the next statement and the $display executes.

Sample 7.12 traces the code and variables from Sample 7.11. The three copies of the automatic variable k are called k0, k1, and k2 for this sample.

Sample 7.12 Steps in executing automatic variable code

j	k0	k1	k2	Statement
0				for (j=0; ...
0	0			Create k0, spawn $write(k) [thread 0]
1	0			j++
1	0	1		Create k1, spawn $write(k) [thread 1]
2	0	1		j++
2	0	1	2	Create k2, spawn $write(k) [thread 2]
3	0	1	2	j<3
3	0	1	2	join_none
3	0	1	2	#0
3	0	1	2	$write(k0) [thread 0]
3	0	1	2	$write(k1) [thread 1]
3	0	1	2	$write(k2) [thread 2]
3	0	1	2	$display;

Another way to write Sample 7.11 is to declare the automatic variable outside of the fork...join_none. Sample 7.13 works inside a program with automatic storage.

Sample 7.13 Automatic variables in a fork...join_none

```
program automatic bug_free;
  initial begin
    for (int j=0; j<3; j++) begin
      automatic int k = j;        // Make copy of index
      fork
        begin
          $write(k);              // Print copy
        end
      join_none
    end
    #0 $display;                  // New line after all threads end
  end
endprogram
```

7.1.7   Waiting for all Spawned Threads

In SystemVerilog, when all the initial blocks in the program are done, the simulator exits. Sample 7.14 shows how you can spawn many threads, which might still be running. Use the wait fork statement to wait for all child threads.

Sample 7.14 Using `wait fork` to wait for child threads

```
task run_threads();
  ...                      // Create some transactions
  fork
    check_trans(tr1);  // Spawn first thread
    check_trans(tr2);  // Spawn second thread
    check_trans(tr3);  // Spawn third thread
  join_none
  ...                      // Do some other work

  // Now wait for the above threads to complete
  wait fork;
endtask
```

7.1.8 Sharing Variables Across Threads

Inside a class's routines, you can use local variables, class variables, or variables defined in the program. If you forget to declare a variable, SystemVerilog looks up the higher scopes until it finds a match. This can cause subtle bugs if two parts of the code are unintentionally sharing the same variable, perhaps because you forgot to declare it in the innermost scope.

For example, if you like to use the index variable, i, be careful that two different threads of your testbench don't concurrently modify this variable by each using it in a `for` loop. Or you may forget to declare a local variable in a class, such as `Buggy`, shown below. If your program block declares a global i, the class just uses the global instead of the local that you intended. You might not even notice this unless two parts of the program try to modify the shared variable at the same time.

Sample 7.15 Bug using shared program variable

```
program automatic bug;

  class Buggy;
    int data[10];
    task transmit();
      fork
        for (i=0; i<10; i++)  // i is not declared here
          send(data[i]);
      join_none
    endtask
  endclass

  int i;                       // Program-level i, shared
  Buggy b;
  event receive;

  initial begin
    b = new();
    for (i=0; i<10; i++)       // i is not declared here
      b.data[i] = i;
    b.transmit();

    for (i=0; i<10; i++)       // i is not declared here
      @(receive) $display(b.data[i]);
    end
endprogram
```

The solution is to declare all your variables in the smallest scope that encloses all uses of the variable. In Sample 7.15, declare index variables inside the for loops, not at the program or class level. Better yet, use the foreach statement whenever possible.

7.2 Disabling Threads

Just as you need to create threads in the testbench, you also need to stop them. The Verilog disable statement works on SystemVerilog threads. The following sections show how you can asynchronously disable threads. This can cause unexpected behavior, so you should watch out for side effects when a thread is stopped midstream. You may, instead, want to design your algorithm to check for interrupts at stable points, then gracefully give up its resources.

7.2.1   Disabling a Single Thread

Here is the check_trans task, this time using a fork...join_any plus a disable to create a watch with a time-out. In this case, you are disabling a labelled block, to precisely specify what to stop.

   The outermost fork...join_none is identical to Sample 7.8. This version implements a time-out with two threads inside a fork...join_any so that the simple wait statement is executed in parallel with a delayed $display. If the correct bus data comes back quickly enough, the wait construct completes, the join_any executes, and then the disable kills off the remaining thread. However, if the bus data does not get the right value before the TIME_OUT delay completes, the error message is printed, the join_any executes, and the disable kills the thread with the wait.

Sample 7.16   Disabling a thread

```
parameter TIME_OUT = 1000ns;

task check_trans(input Transaction tr);
  fork

    begin
    // Wait for response, or some maximum delay
    fork : timeout_block
      begin
        wait (bus.cb.data == tr.data);
        $display("@%0t: data match %d", $time, tr.data);
      end
      #TIME_OUT $display("@%0t: Error: timeout", $time);
    join_any
    disable timeout_block;
    end

  join_none     // Spawn thread, don't block
endtask
```

Watch out, as you might unintentionally stop too many threads with disable label. This statement stops every process executing the labeled block, as might occur if you have multiple driver or monitor objects running in parallel. If your code only has one instance, disable label is a safe way to stop a thread.

7.2.2   Disabling Multiple Threads

Sample 7.16 used the classic Verilog disable statement to stop the threads in a named block. SystemVerilog introduces the disable fork statement so you can stop all child threads that have been spawned from the current thread.

Watch out, as you might unintentionally stop too many threads with `disable fork`, such as those created from surrounding task calls. You should always surround the target code with a `fork…join` to limit the scope of a `disable fork` statement.

The next few samples use the `check_trans` task from Sample 7.16. You can just think of this task as doing a `#TIME_OUT`. Sample 7.17 has an additional `begin…end` block inside the `fork…join` to make the statements sequential.

Sample 7.17 Limiting the scope of a disable fork

```
initial begin
  check_trans(tr0);           // thread 0

  // Create a thread to limit scope of disable
  fork                        // thread 1
    begin
      check_trans(tr1);       // thread 2
      fork                    // thread 3
        check_trans(tr2);     // thread 4
      join

      // Stop threads 2-4, but leave 0 alone
      #(TIME_OUT/10) disable fork;
    end
  join
end
```

Fig. 7.4 shows a diagram of the spawned threads.

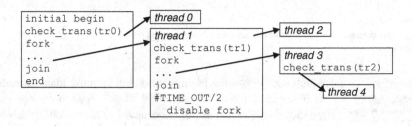

Fig. 7.4 Fork…join block diagram

The code calls `check_trans` that starts thread 0. Next a `fork…join` creates thread 1. Inside this thread, one is spawned by the `check_trans` task and one by the innermost `fork…join`, which spawns thread 4 by calling the task. After a delay, a `disable fork` stops and all the child threads, 2–4. Thread 0 is outside the `fork… join` block that has the `disable`, so it is unaffected.

Sample 7.18 is the more robust version of Sample 7.17, with `disable` with a label that explicitly names the threads that you want to stop.

Sample 7.18 Using `disable` label to stop threads

```
initial begin
  check_trans(tr0);        // thread 0
  fork                     // thread 1
    begin : threads_inner
      check_trans(tr1);    // thread 2
      check_trans(tr2);    // thread 3
    end

    // Stop threads 2 & 3, but leave 0 alone
    #(TIME_OUT/10) disable threads_inner;
  join
end
```

7.2.3   Disable a Task that was Called Multiple Times

Be careful when you disable a block from inside that block - you might end up stopping more than you expected. As expected, if you disable a task from inside the task, it is like a return statement, but it also kills all threads started by the task. Additionally, a single `disable` label terminates all threads using that code, not just the current one.

In Sample 7.19, the `wait_for_time_out` task is called three times, spawning three threads. Then, thread 0 also disables the task after #2ns. When you run this code, you will see the three threads starting, but none finishes, because of the `disable` in thread 0 stops all three threads, not just one. If this task was inside a driver class that was instantiated multiple times, a `disable` label in one could stop all the blocks.

Sample 7.19 Using `disable` label to stop a task

```
task wait_for_time_out(input int id);
  if (id == 0)
    fork
      begin
        #2ns;
        $display("@%0t: disable wait_for_time_out", $time);
        disable wait_for_time_out;
      end
    join_none

  fork : just_a_little
    begin
      $display("@%0t: %m: %0d entering thread", $time, id);
      #TIME_OUT;
      $display("@%0t: %m: %0d done", $time, id);
    end
  join_none
endtask
```

```
initial begin
  wait_for_time_out(0); // Spawn thread 0
  wait_for_time_out(1); // Spawn thread 1
  wait_for_time_out(2); // Spawn thread 2
  #(TIME_OUT*2) $display("@%0t: All done", $time);
end
```

7.3 Interprocess Communication

All these threads in your testbench need to synchronize and exchange data. At the most basic level, one thread waits for another, such as the environment object waiting for the generator to complete. Multiple threads might try to access a single resource such as bus in the DUT, so the testbench needs to ensure that one and only one thread is granted access. At the highest level, threads need to exchange data such as transaction objects that are passed from the generator to the agent. All of this data exchange and control synchronization is called interprocess communication (IPC), which is implemented in SystemVerilog with events, semaphores, and mailboxes. These are described in the remainder of this chapter.

There are generally three parts to IPC: a producer that creates the information, a consumer that accepts the information, and the channel that carries the information. The producer and consumer are in separate threads.

7.4 Events

A Verilog event synchronizes threads. It is similar to a phone, where one person waits for a call from another person. In Verilog a thread waits for an event with the @ operator. This operator is edge sensitive, so it always blocks, waiting for the event to change. Another thread triggers the event with the -> operator, unblocking the first thread.

System Verilog enhances the Verilog event in several ways. An event is now a handle to a synchronization object that can be passed around to routines. This feature allows you to share events across objects without having to make the events global. The most common way is to pass the event into the constructor for an object.

There is always the possibility of a race condition in Verilog where one thread blocks on an event at the same time another triggers it. If the triggering thread executes before the blocking thread, the trigger is missed. SystemVerilog introduces the `triggered` status that lets you check whether an event has been triggered, including during the current time-slot. A thread can wait on this function instead of blocking with the @ operator.

7.4.1 Blocking on the Edge of an Event

When you run Sample 7.20, one initial block starts, triggers its event, and then blocks on the other event, as shown in the output in Sample 7.21. The second block starts, triggers its event (waking up the first), and then blocks on the first event. However, the second thread locks up because it missed the first event, as it is a zero-width pulse.

Sample 7.20 Blocking on an event in Verilog

```verilog
event e1, e2;
initial begin
  $display("@%0t: 1: before trigger", $time);
  -> e1;
  @e2;
  $display("@%0t: 1: after trigger", $time);
end

initial begin
  $display("@%0t: 2: before trigger", $time);
  -> e2;
  @e1;
  $display("@%0t: 2: after trigger", $time);
 end
```

Sample 7.21 Output from blocking on an event

```
@0: 1: before trigger
@0: 2: before trigger
@0: 1: after trigger
```

7.4.2 Waiting for an Event Trigger

Instead of the edge-sensitive block @e1, use the level-sensitive wait(e1.triggered). This does not block if the event has been triggered during this time step. Otherwise, it waits until the event is triggered.

Sample 7.22 Waiting for an event

```
event e1, e2;

initial begin
  $display("@%0t: 1: before trigger", $time);
  -> e1;
  wait (e2.triggered);
  $display("@%0t: 1: after trigger", $time);
end

initial begin
  $display("@%0t: 2: before trigger", $time);
  -> e2;
  wait (e1.triggered);
  $display("@%0t: 2: after trigger", $time);
end
```

When you run Sample 7.22, one initial block starts, triggers its event, and then blocks on the other event. The second block starts, triggers its event (waking up the first) and then blocks on the first event, producing the output in Sample 7.23.

Sample 7.23 Output from waiting for an event

```
@0: 1: before trigger
@0: 2: before trigger
@0: 1: after trigger
@0: 2: after trigger
```

Several of these samples have race conditions and may not execute exactly the same on every simulator. For example, the output in Sample 7.23 assumes that when the second block triggers e2, execution jumps back to the first block. It would also be legal for the second block to trigger e2, wait on e1, and display a message before control is returned back to the first block.

7.4.3   Using Events in a Loop

You can synchronize two threads with an event, but use caution.

If you use wait (handshake.triggered) in a loop, be sure to advance the time before waiting again. Otherwise your code will go into a zero delay loop as the wait continues over and over again on a single event trigger. Sample 7.24 incorrectly uses a level-sensitive blocking statement for notification that a transaction is ready.

Sample 7.24 Waiting on event causes a zero delay loop

```
forever begin
  // This is a zero delay loop!
  wait(handshake.triggered);
  $display("Received next event");
  process_in_zero_time();
end
```

Just as you learned to always put a delay inside an `always` block you need to put a delay in a transaction process loop. The edge-sensitive delay statement in Sample 7.25 continues once and only once per event trigger.

Sample 7.25 Waiting for an edge on an event

```
forever begin
  // This prevents a zero delay loop!
  @handshake;
  $display("Received next event");
  process_in_zero_time();
end
```

You should avoid events if you need to send multiple notifications in a single time slot, and look at other IPC methods with built-in queuing such as semaphores and mailboxes, discussed later in this chapter.

7.4.4 Passing Events

As described above, an event in SystemVerilog can be passed as an argument to a routine. In Sample 7.26, an event is used by a transactor to signal when it has completed.

Sample 7.26 Passing an event into a constructor

```
program automatic test;

class Generator;
  event done;
  function new (input event done); // Pass event from TB
    this.done = done;
  endfunction

  task run();
    fork
      begin
        ...                          // Create transactions
        -> done;                     // Tell the test we are done
      end
    join_none
  endtask
endclass

  event gen_done;
  Generator gen;

  initial begin
    gen = new(gen_done);             // Instantiate testbench
    gen.run();                       // Run transactor
    wait(gen_done.triggered);        // Wait for finish
  end
endprogram
```

7.4.5   Waiting for Multiple Events

In Sample 7.26, you had a single generator that fired a single event. What if your testbench environment class must wait for multiple child processes to finish, such as N generators? The easiest way is to use `wait fork`, that waits for all child processes to end. The problem is that this also waits for all the transactors, drivers, and any other threads that were spawned by the environment. You need to be more selective. You still want to use events to synchronize between the parent and child threads.

You could use a `for` loop in the parent to wait for each event, but that would only work if thread 0 finished before thread 1, which finished before thread 2, etc. If the threads finish out of order, you could be waiting for an event that triggered many cycles ago.

The solution is to make a new thread and then spawn children from there that each block on an event for each generator, as shown in Sample 7.27. Now you can do a `wait fork` because you are being more selective.

Sample 7.27 Waiting for multiple threads with wait fork

```
event done[N_GENERATORS];

initial begin
  foreach (gen[i]) begin
    gen[i] = new(done[i]); // Create N generators
    gen[i].run();          // Start them running
  end

  // Wait for all gen to finish by waiting for each event
  foreach (gen[i])
    fork
      automatic int k = i;
      wait (done[k].triggered);
    join_none

  wait fork;  // Wait for all those triggers to finish
end
```

Another way to solve this problem is to keep track of the number of events that have triggered, as shown in Sample 7.28.

Sample 7.28 Waiting for multiple threads by counting triggers

```
event done[N_GENERATORS];
int done_count;

initial begin
  foreach (gen[i]) begin
    gen[i] = new(done[i]); // Create N generators
    gen[i].run();          // Start them running
  end

  // Wait for all generators to finish
  foreach (gen[i])
    fork
      automatic int k = i;
      begin
        wait (done[k].triggered);
        done_count++;
      end
    join_none
  wait (done_count==N_GENERATORS); // Wait for triggers

end
```

That was slightly less complicated. Why not get rid of all the events and just wait on a count of the number of running generators? This count can be a static variable in the `Generator` class. Note that most of the thread manipulation code has been replaced with a single `wait` construct. The last block in Sample 7.29 waits for the count using the class scope resolution operator, `::`. You could have used any handle, such as `gen[0]`, but that would be less direct.

Sample 7.29 Waiting for multiple threads using a thread count

```
class Generator;
  static int thread_count = 0;

  task run();
    thread_count++;          // Start another thread
    fork
      begin
      // Do the real work in here
      // And when done, decrement the thread count
      thread_count--;
      end
    join_none
  endtask
endclass

Generator gen[N_GENERATORS];

initial begin
  // Create N generators
  foreach (gen[i])
    gen[i] = new();

  // Start them running
  foreach (gen[i])
    gen[i].run();

  // Wait for all the generators to complete
  wait (Generator::thread_count == 0);
end
```

7.5 Semaphores

A semaphore allows you to control access to a resource. Imagine that you and your spouse share a car. Obviously, only one person can drive it at a time. You can manage this situation by agreeing that whoever has the key can drive it. When you are done with the car, you give up the car so that the other person can use it. The key is the semaphore that makes sure only one person has access to the car. In operating system terminology, this is known as "mutually exclusive access," so a semaphore is known as a "mutex" and is used to control access to a resource.

Semaphores can be used in a testbench when you have a resource, such as a bus, that may have multiple requestors from inside the testbench but, as part of the physical design, can only have one driver. In SystemVerilog, a thread that requests a key when one is not available always blocks. Multiple blocking threads are queued in FIFO order.

7.5.1 Semaphore Operations

There are three basic operations for a semaphore. You create a semaphore with one or more keys using the new method, get one or more keys with the blocking task get(), and return one or more keys with put(). If you want to try to get a semaphore, but not block, use the try_get() function. If keys are available, try_get() obtains them and returns 1. If there are not sufficient keys, it just returns a 0. Sample 7.30 shows how to control access to a resource with a semaphore.

Sample 7.30 Semaphores controlling access to hardware resource

```
program automatic test(bus_ifc.TB bus);
  semaphore sem;                 // Create a semaphore
  initial begin
    sem = new(1);                // Allocate with 1 key
    fork
      sequencer();               // Spawn two threads that both
      sequencer();               //    do bus transactions
    join
  end

  task sequencer();
    repeat($urandom()%10)        // Random wait, 0-9 cycles
      @bus.cb;
    sendTrans();                 // Execute the transaction
  endtask

  task sendTrans();
    sem.get(1);                  // Get the key to the bus
    @bus.cb;                     // Drive signals onto bus
    bus.cb.addr <= tr.addr;
    ...
    sem.put(1);                  // Put it back when done
  endtask
endprogram
```

7.5.2   Semaphores with Multiple Keys

There are two things you should watch out for with semaphores. First, you can put
more keys back than you took out. Suddenly you may have two keys but only one car!
Secondly, be careful if your testbench needs to get and put multiple keys. Perhaps you
have one key left, and a thread requests two, causing it to block. Now a second thread
requests a single semaphore – what should happen? In SystemVerilog the second
request, get(1), sneaks ahead of the earlier get(2), bypassing the FIFO ordering.

　　If you are mixing different sized requests, you can always write your own class.
That way you can be very clear on who gets priority.

7.6   Mailboxes

How do you pass information between two threads? Perhaps your generator needs to
create many transactions and pass them to a driver. You might be tempted to just have
the generator thread call a task in the driver. If you do that, the generator needs to know
the hierarchical path to the driver task, making your code less reusable. Additionally,
this style forces the generator to run at the same speed as the driver, that can cause
synchronization problems if one generator needs to control multiple drivers.

Think of your generator and driver as transactors that are autonomous
objects that communicate through a channel. Each object gets a transaction from an upstream object (or creates it, as in the case of a generator), does some processing, and then passes it to a downstream
object. The channel must allow its driver and receiver to operate asynchronously. You may be tempted to just use a shared array or queue, but it can be
difficult to create threads that read, write, and blocks safely.

　　The solution is a SystemVerilog mailbox. From a hardware point of view, the easiest
way to think about a mailbox is that it is just a FIFO, with a source and sink. The
source puts data into the mailbox, and the sink gets values from the mailbox.
Mailboxes can have a maximum size or can be unlimited. When the source thread
tries to put a value into a sized mailbox that is full, that thread blocks until the value
is removed. Likewise, if a sink threads tries to remove a value from a mailbox that
is empty, that thread blocks until a value is put into the mailbox.

　　Figure 7.5 shows a mailbox connecting a generator and driver.

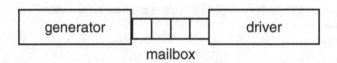

Fig. 7.5　A mailbox connecting two transactors

A mailbox is an object and thus has to be instantiated by calling the `new` function. This takes an optional `size` argument to limit the number of entries in the mailbox. If the size is 0 or not specified, the mailbox is unbounded and can hold an unlimited number of entries.

You put data into a mailbox with the `put()` task, and remove it with the blocking `get()` task. A `put()` blocks if the mailbox is full, and `get()` blocks if the mailbox is empty. Use `try_put()` if you want to see if the mailbox is full. and `try_get()` to see if it is empty. The `peek()` task gets a copy of the data in the mailbox but does not remove it.

The data is a single value, such as an integer, or logic of any size or a handle. A mailbox never contains objects, only references to them. By default, a mailbox does not have a type, so you could put any mix of data into it. Don't do it! Enforce one data type per mailbox by sticking with parameterized mailboxes as shown in Sample 7.31 to catch type mismatches at compile time.

Sample 7.31   Mailbox declarations

```
mailbox #(Transaction) mbx_tr;   // Parameterized: recommended
mailbox mbx_untyped;             // Unspecialized: avoid
```

A classic mailbox bug, shown in Sample 7.32, is a loop that randomizes objects and puts them in a mailbox, but the object is only constructed once, outside the loop. Since there is only one object, it is randomized over and over.

Sample 7.32   Bad generator creates only one object

```
task generator_bad(input int n,
                   input mailbox #(Transaction) mbx);
  Transaction tr;
  tr = new();                    // Create just one transaction
  repeat (n) begin
    `SV_RAND_CHECK(tr.randomize());
    $display("GEN: Sending addr=%h", tr.addr);
    mbx.put(tr);                 // Send transaction to driver
  end
endtask
```

Figure 7.6 shows all the handles pointing to a single object. A mailbox only holds handles, not objects, so you end up with a mailbox containing multiple handles that all point to the single object. The code that gets the handles from the mailbox just sees the last set of random values.

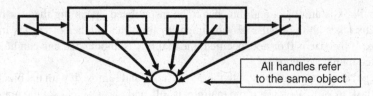

Fig. 7.6 A mailbox with multiple handles to one object

The solution, shown in Sample 7.33, is to make sure your loop has all three steps of constructing the object, randomizing it, and putting it in the mailbox. This bug is so common that it is also mentioned in Section 5.14.3.

Sample 7.33 Good generator creates many objects

```
task generator_good(input int n,
                    input mailbox #(Transaction) mbx);
  Transaction tr;
  repeat (n) begin
    tr = new();                // Create a new transaction
    `SV_RAND_CHECK(tr.randomize());
    $display("GEN: Sending addr=%h", tr.addr);
    mbx.put(tr);               // Send transaction to driver
  end
endtask
```

The result, shown in Fig. 7.7, is that every handle points to a unique object. This type of generator is known as the Blueprint Pattern and described in Section 8.2.

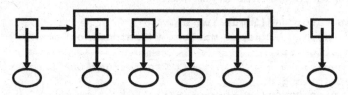

Fig. 7.7 A mailbox with multiple handles to multiple objects

Sample 7.34 shows the driver that waits for transactions from the generator.

Sample 7.34 Good driver receives transactions from mailbox

```
task driver(input mailbox #(Transaction) mbx);
  Transaction tr;
  forever begin
    mbx.get(tr);               // Get transacton from mailbox
    $display("DRV: Received addr=%h", tr.addr);
    // Drive transaction into DUT
  end
endtask
```

If you don't want your code to block when accessing the mailbox, use the `try_get()` and `try_peek()` functions. If they are successful, they return a nonzero value; otherwise, they return 0. These are more reliable than the `num()` function, as the number of entries can change between when you measure it and when you next access the mailbox.

7.6.1 Mailbox in a Testbench

Sample 7.35 shows a program with a Generator and Driver exchanging transactions using a mailbox.

Sample 7.35 Exchanging objects using a mailbox: the Generator class

```
program automatic mailbox_example(bus_ifc.TB bus);

class Generator;
  Transaction tr;
  mailbox #(Transaction) mbx;

  function new(input mailbox #(Transaction) mbx);
    this.mbx = mbx;
  endfunction

  task run(input int count);
    repeat (count) begin
      tr = new();
      `SV_RAND_CHECK(tr.randomize);
      mbx.put(tr);     // Send out transaction
    end
  endtask
endclass


class Driver;
  Transaction tr;
  mailbox #(Transaction) mbx;

  function new(input mailbox #(Transaction) mbx);
    this.mbx = mbx;
  endfunction

  task run(input int count);
    repeat (count) begin
      mbx.get(tr);     // Fetch next transaction
      // Drive transaction here
    end
  endtask
endclass
```

```
   mailbox #(Transaction) mbx;     // Mailbox connecting gen & drv
   Generator gen;
   Driver drv;
   int count;

   initial begin
      mbx = new();                  // Construct the mailbox
      gen = new(mbx);               // Construct the generator
      drv = new(mbx);               // Construct the driver
      count = $urandom_range(50);  // Run up to 50 transactions

      fork
         gen.run(count);            // Spawn the generator
         drv.run(count);            // Spawn the driver
      join                          // Wait for both to finish

   end
endprogram
```

7.6.2 Bounded Mailboxes

By default, mailboxes are similar to an unlimited FIFO — a producer can put any
number of objects into a mailbox before the consumer gets the objects out. However,
you may want the two threads to operate in lockstep so that the producer blocks
until the consumer is done with the object.

   You can specify a maximum size for the mailbox when you construct it. The
default mailbox size is 0 which creates an unbounded mailbox. Any size greater
than 0 creates a bounded mailbox. If you attempt to put more objects than this limit,
put() blocks until you get an object from the mailbox, creating a vacancy.

Sample 7.36 Bounded mailbox

```
program automatic bounded;
   mailbox #(int) mbx;

   initial begin
      mbx = new(1);   // Mailbox size = 1
      fork

      // Producer thread
      for (int i=1; i<4; i++) begin
         $display("Producer: before put(%0d)", i);
         mbx.put(i);
         $display("Producer: after  put(%0d)", i);
      end
```

```
    // Consumer thread
    repeat(4) begin
      int j;
      #1ns mbx.get(j);
      $display("Consumer: after  get(%0d)", j);
    end
  join
end
endprogram
```

Sample 7.36 creates the smallest possible mailbox, which can hold a single message. The Producer thread tries to put three messages (integers) in the mailbox, and the Consumer thread slowly gets messages every 1ns. As Sample 7.37 shows, the first put() succeeds, then the Producer tries put(2) which blocks. The Consumer wakes up, gets a message 1 from the mailbox, so now the Producer can finish putting the message 2.

Sample 7.37 Output from bounded mailbox

```
Producer: before put(1)
Producer: after  put(1)
Producer: before put(2)
Consumer: after  get(1)
Producer: after  put(2)
Producer: before put(3)
Consumer: after  get(2)
Producer: after  put(3)
Consumer: after  get(3)
```

The bounded mailbox acts as a buffer between the two processes. You can see how the Producer generates the next value before the Consumer reads the current value.

7.6.3   Unsynchronized Threads Communicating with a Mailbox

 In many cases, two threads that are connected by a mailbox should run in lockstep, so that the producer does not get ahead of the consumer. The benefit of this approach is that your entire chain of stimulus generation now runs in lock step. The highest level generator only completes when the last low level transaction completes transmission. Now your testbench can tell precisely when all stimulus has been sent. In another example, if your generator gets ahead of the driver, and you are gathering functional coverage on the generator, you might record that some transactions were tested, even if the test stopped prematurely. So even though a mailbox allows you to decouple the two sides, you may still want to keep them synchronized.

If you want two threads to run in lockstep, you need a handshake in addition to the mailbox. In Sample 7.38 the Producer and Consumer are now classes that exchange integers using a mailbox, with no explicit synchronization between the two objects. As a result, as shown in Sample 7.39, the producer runs to completion before the consumer even starts.

Sample 7.38 Producer–consumer without synchronization

```
program automatic unsynchronized;

  mailbox #(int) mbx;

  class Producer;
    task run();
      for (int i=1; i<4; i++) begin
        $display("Producer: before put(%0d)", i);
        mbx.put(i);
      end
    endtask
  endclass

  class Consumer;
    task run();
      int i;
      repeat (3) begin
        mbx.get(i);        // Get integer from mbx
        $display("Consumer: after  get(%0d)", i);
      end
    endtask
  endclass

  Producer p;
  Consumer c;

  initial begin
    // Construct mailbox, producer, consumer
    mbx = new();        // Unbounded
    p = new();
    c = new();

    // Run the producer and consumer in parallel
    fork
      p.run();
      c.run();
    join
  end
endprogram
```

The above sample holds the mailbox in a global variable to make the code more compact. In real code, you should pass the mailbox into the class through the constructor and save a reference to it in a class-level variable.

Sample 7.38 has no synchronization so the Producer puts all three integers into the mailbox before the Consumer can get the first one. This is because a thread continues running until there is a blocking statement, and the Producer has none. The Consumer thread blocks on the first call to `mbx.get`.

Sample 7.39 Producer–consumer without synchronization output

```
Producer: before put(1)
Producer: before put(2)
Producer: before put(3)
Consumer: after  get(1)
Consumer: after  get(2)
Consumer: after  get(3)
```

This example has a race condition, so on some simulators the consumer could activate earlier. The result is still the same as the values are determined by the producer, not by how quickly the consumer sees them.

7.6.4   Synchronized Threads Using a Bounded Mailbox and a Peek

In a synchronized testbench, the Producer and Consumer operate in lock step. This way, you can tell when the input stimuli is complete by waiting for any of the threads. If the threads operate unsynchronized, you need to add extra code to detect when the last transaction is applied to the DUT.

To synchronize two threads, the Producer creates and puts a transaction into a mailbox, then blocks until the Consumer finishes with it. This is done by having the Consumer remove the transaction from the mailbox only when it is finally done with it, not when the transaction is first detected.

Sample 7.40 show the first attempt to synchronize two threads, this time with a bounded mailbox. The Consumer uses the built-in mailbox method `peek()` to look at the data in the mailbox without removing. When the Consumer is done processing the data, it removes the data with `get()`. This frees up the Producer to generate a new value. If the Consumer loop started with a `get()` instead of the `peek()`, the transaction would be immediately removed from the mailbox, so the Producer could wake up before the Consumer finished with the transaction. Sample 7.41 has the output from this code.

Sample 7.40   Producer–consumer synchronized with bounded mailbox

```
program automatic synch_peek;
// Uses Producer from Sample 7-38

  mailbox #(int) mbx;

  class Consumer;
    task run();
      int i;
      repeat (3) begin
        mbx.peek(i);        // Peek integer from mbx
        $display("Consumer: after peek(%0d)", i);
        mbx.get(i);         // Remove from mbx
      end
    endtask
  endclass : Consumer

  Producer p;
  Consumer c;

  initial begin
    // Construct mailbox, producer, consumer
    mbx = new(1);    // Bounded mailbox - limit 1!
    p = new();
    c = new();

    // Run the producer and consumer in parallel
    fork
      p.run();
      c.run();
    join
  end
endprogram
```

Sample 7.41   Output from producer–consumer with bounded mailbox

```
Producer: before put(1)
Producer: before put(2)
Consumer: after   peek(1)
Consumer: after   peek(2)
Producer: before put(3)
Consumer: after   peek(3)
```

You can see that the Producer and Consumer are in lockstep, but the Producer is still one transaction ahead of the Consumer. This is because a bounded mailbox with size=1 only blocks when you try to do a put of the second transaction.[2]

---

[2] This behavior is different from the VMM channel. If you set a channel's full level to 1, the very first call to put() places the transaction in the channel, but does not return until the transaction is removed.

7.6.5   *Synchronized Threads Using a Mailbox and Event*

You may want the two threads to use a handshake so that the Producer never gets ahead of the Consumer. The Consumer already blocks, waiting for the Producer using a mailbox. The Producer needs to block, waiting for the Consumer to finish the transaction. Do this by adding a blocking statement to the Producer such as an event, a semaphore, or a second mailbox. Sample 7.42 uses an event to block the Producer after it puts data in the mailbox. The Consumer triggers the event after it consumes the data.

If you use wait (handshake.triggered) in a loop, be sure to advance the time before waiting again, as previously shown in Section 7.4.3. This wait blocks only once in a given time slot, so you need move into another. Sample 7.42 uses the edge-sensitive blocking statement @handshake instead to ensure that the Producer stops after sending the transaction. The edge-sensitive statement works multiple times in a time slot but may have ordering problems if the trigger and block happen in the same time slot.

Sample 7.42 Producer–consumer synchronized with an event

```
program automatic mbx_evt;
  mailbox #(int) mbx;
  event handshake;

  class Producer;
    task run();
      for (int i=1; i<4; i++) begin
        $display("Producer: before put(%0d)", i);
        mbx.put(i);
        @handshake;
        $display("Producer: after  put(%0d)", i);
      end
    endtask
  endclass : Producer

  class Consumer;
    task run();
      int i;
      repeat (3) begin
        mbx.get(i);
        $display("Consumer: after  get(%0d)", i);
        ->handshake;
      end
    endtask
  endclass : Consumer
```

```
   Producer p;
   Consumer c;

   initial begin
     p = new();
     c = new();
     mbx = new();

     // Run the producer and consumer in parallel
     fork
       p.run();
       c.run();
     join
   end
endprogram
```

Now the Producer does not advance until the Consumer triggers the event, as shown in Sample 7.43.

Sample 7.43 Output from producer–consumer with event

```
Producer: before put(1)
Consumer: after  get(1)
Producer: after  put(1)
Producer: before put(2)
Consumer: after  get(2)
Producer: after  put(2)
Producer: before put(3)
Consumer: after  get(3)
Producer: after  put(3)
```

You can see that the Producer and Consumer are successfully running in lockstep by the fact that the Producer never produces a new value until after the old one is read.

7.6.6 Synchronized Threads Using Two Mailboxes

Another way to synchronize the two threads is to use a second mailbox that sends a completion message back to the Producer, as shown in Sample 7.44.

Sample 7.44 Producer–consumer synchronized with a mailbox

```
program automatic mbx_mbx2;
  mailbox #(int) mbx, rtn;
  class Producer;
    task run();
      int k;
      for (int i=1; i<4; i++) begin
        $display("Producer: before put(%0d)", i);
        mbx.put(i);
        rtn.get(k);
        $display("Producer: after  get(%0d)", k);
      end
    endtask
  endclass : Producer

  class Consumer;
    task run();
      int i;
      repeat (3) begin
        $display("Consumer: before get");
        mbx.get(i);
        $display("Consumer: after  get(%0d)", i);
        rtn.put(-i);
      end
    endtask
  endclass : Consumer

  Producer p;
  Consumer c;
  initial begin
    p = new();
    c = new();
    mbx = new();
    rtn = new();

    // Run the producer and consumer in parallel
    fork
      p.run();
      c.run();
    join
  end
endprogram
```

The return message in the rtn mailbox is just a negative version of the original integer. You could use any value, but this one can be checked against the original for debugging purposes.

Sample 7.45 Output from producer–consumer with mailbox

```
Producer: before put(1)
Consumer: before get
Consumer: after   get(1)
Consumer: before get
Producer: after   get(-1)
Producer: before put(2)
Consumer: after   get(2)
Consumer: before get
Producer: after   get(-2)
Producer: before put(3)
Consumer: after   get(3)
Producer: after   get(-3)
```

You can see from Sample 7.45 that the Producer and Consumer are successfully running in lockstep.

7.6.7   Other Synchronization Techniques

You can also complete the handshake by blocking on a variable or a semaphore. An event is the simplest construct, followed by blocking on a variable. A semaphore is comparable to using a second mailbox, but no information is exchanged. System Verilog's bounded mailbox just does not work as well as these other techniques as there is no way to block the producer when it puts the first transaction in. Sample 7.41 shows that the Producer is always one transaction ahead of the Consumer.

7.7   Building a Testbench with Threads and IPC

Way back in Section 1.10 you learned about layered testbenches. Figure 7.8 shows the relationship between the different parts. Now that you know how to use threads and IPC, you can construct a basic testbench with transactors.

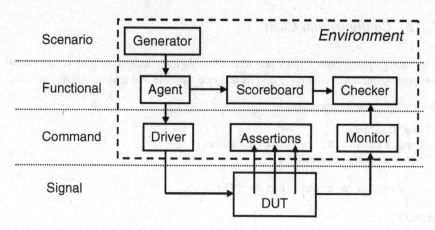

Fig. 7.8 Layered testbench with environment

7.7.1 Basic Transactor

Sample 7.46 is the Agent class that sits between the Generator and the Driver.

Sample 7.46 Basic Transactor

```
class Agent;

  mailbox #(Transaction) gen2agt, agt2drv;
  Transaction tr;

  function new(input mailbox #(Transaction) gen2agt, agt2drv);
    this.gen2agt = gen2agt;
    this.agt2drv = agt2drv;
  endfunction

  task run();
    forever begin
      gen2agt.get(tr);      // Get transaction from upstream block
      ...                   // Do some processing
      agt2drv.put(tr);      // Send it to downstream block
    end
  endtask

  task wrap_up();    // Empty for now
  endtask

endclass
```

7.7.2 Configuration Class

The configuration class allows you to randomize the configuration of your system
for every simulation. Sample 7.47 has just one variable and a basic constraint.

Sample 7.47 Configuration class

```
class Config;
  rand bit [31:0] run_for_n_trans;
  constraint reasonable
    {
      run_for_n_trans inside {[1:1000]};
    }
endclass
```

7.7.3 Environment Class

The Environment class, shown as a dashed line in Fig. 7.8, holds the Generator,
Agent, Driver, Monitor, Checker, Scoreboard, and Config objects, and the mailboxes between them. Sample 7.48 shows a basic Environment class.

Sample 7.48 Environment class

```
class Environment;

  Generator   gen;
  Agent       agt;
  Driver      drv;
  Monitor     mon;
  Checker     chk;
  Scoreboard  scb;
  Config      cfg;
  mailbox #(Transaction) gen2agt, agt2drv, mon2chk;

  extern function new();
  extern function void gen_cfg();
  extern function void build();
  extern task run();
  extern task wrap_up();
endclass


function Environment::new();
  cfg = new();
endfunction

function void Environment::gen_cfg();
  `SV_RAND_CHECK(cfg.randomize);
endfunction
```

```
function void Environment::build();
  // Initialize mailboxes
  gen2agt = new();
  agt2drv = new();
  mon2chk = new();

  // Initialize transactors
  gen = new(gen2agt);
  agt = new(gen2agt, agt2drv);
  drv = new(agt2drv);
  mon = new(mon2chk);
  chk = new(mon2chk);
  scb = new(cfg);
endfunction

task Environment::run();
  fork
    gen.run(cfg.run_for_n_trans);
    agt.run();
    drv.run();
    mon.run();
    chk.run();
    scb.run(cfg.run_for_n_trans);
  join
endtask

task Environment::wrap_up();
  fork
    gen.wrap_up();
    agt.wrap_up();
    drv.wrap_up();
    mon.wrap_up();
    chk.wrap_up();
    scb.wrap_up();
  join
endtask
```

Chapter 8 shows more details on how to build these classes.

7.7.4   Test Program

Sample 7.49 shows the main test, which is in a program block. As discussed in Section 4.3.4, you can also put a test in a module, but at a slight increase in the chances of race conditions.

Sample 7.49 Basic test program

```
program automatic test;

  Environment env;

  initial begin
    env = new();
    env.gen_cfg();
    env.build();
    env.run();
    env.wrap_up();
  end

endprogram
```

7.8   Conclusion

Your design is modeled as many independent blocks running in parallel, so your testbench must also generate multiple stimulus streams and check the responses using parallel threads. These are organized into a layered testbench, orchestrated by the toplevel environment. SystemVerilog introduces powerful constructs such as fork...join_none and fork...join_any for dynamically creating new threads, in addition to the standard fork...join. These threads communicate and synchronize using events, semaphores, mailboxes, and the classic @ event control and wait statements. Lastly, the disable command is used to terminate threads.

These threads and the related control constructs complement the dynamic nature of OOP. As objects are created and destroyed, they can run in independent threads, allowing you to build a powerful and flexible testbench environment.

7.9 Exercises

1. For the following code determine the order and time of execution for each statement if a `join` or `join_none` or `join_any` is used. Hint: the order and time of execution between the `fork` and `join`/`join_none`/`join_any` is the same, only the order and execution time of the statements after the `join` are different.

```
initial begin
  $display("@%0t: start fork...join example", $time);
  fork
    begin
      #20 $display("@%0t: sequential A after #20", $time);
      #20 $display("@%0t: sequential B after #20", $time);
    end
    $display("@%0t: parallel start", $time);
    #50 $display("@%0t: parallel after #50", $time);
    begin
      #30 $display("@%0t: sequential after #30", $time);
      #10 $display("@%0t: sequential after #10", $time);
    end
  join // or join_any or join_none
  $display("@%0t: after join", $time);
  #80 $display("@%0t: finish after #80", $time);
end
```

2. For the following code what would the output be with and without a `wait fork` inserted in the indicated location?

```
initial begin

  fork
    transmit(1);
    transmit(2);
  join_none

  fork: receive_fork
    receive(1);
    receive(2);
  join_none

  // What is the output with/without a wait fork here?

  #15ns disable receive_fork;
  $display("%0t: Done", $time);
end

task transmit(int index);
  #10ns;
  $display("%0t: Transmit is done for index = %0d",
            $time, index);
endtask

task receive(int index);
  #(index * 10ns);
  $display("%0t: Receive is done for index = %0d",
            $time, index);
endtask
```

3. What would be displayed with the following code? Assume that the events and task `trigger` is declared inside a program declared as automatic.

```
event e1, e2;
task trigger(event local_event, input time wait_time);
  #wait_time;
  ->local_event;
endtask

initial begin
  fork
    trigger(e1, 10ns);
    begin
      wait(e1.triggered());
      $display("%0t: e1 triggered", $time);
    end
  join
end

initial begin
  fork
    trigger(e2, 20ns);
    begin
      wait(e2.triggered());
      $display("%0t: e2 triggered", $time);
    end
  join
end
```

4. Create a task called `wait10` that for 10 tries will wait for 10ns and then check for 1 semaphore key to be available. When the key is available, quit the loop and print out the time.

5. What would be displayed with the following code that calls the task from Exercise 4?

```
initial begin
  fork
    begin
      sem = new(1);
      sem.get(1);
      #45ns;
      sem.put(2);
    end
    wait10();
  join
end
```

6. What would be displayed with the following code?

```
program automatic test;
  mailbox #(int) mbx;
  int value;
  initial begin
    mbx = new(1);
    $display("mbx.num()=%0d", mbx.num());
    $display("mbx.try_get= %0d", mbx.try_get(value));
    mbx.put(2);
    $display("mbx.try_put= %0d", mbx.try_put(value));
    $display("mbx.num()=%0d", mbx.num());
    mbx.peek(value);
    $display("value=%0d", value);
  end
endprogram
```

7. Look at Fig. 7.8 "Layered testbench with environment" on page 265 and create the Monitor class. You can make the following assumptions.

   a. The Monitor class has knowledge of class OutputTrans with member variables out1 and out2.
   b. The DUT and Monitor are connected with an interface called my_bus, with signals out1 and out2.
   c. The interface my_bus has a clocking block, cb.
   d. On every active clock edge, the Monitor class will sample the DUT outputs, out1 and out2, assign them to an object of type OutputTrans, and place the object in a mailbox.

Chapter 8
Advanced OOP and Testbench Guidelines

How would you create a complex class for a bus transaction that also performs error injection and has random delays? The first approach is to put everything in a large, flat class. This approach is simple to build, easy to understand (all the code is right there in one class) but can be slow to develop and debug. Additionally, such a large class is a maintenance burden, as anyone who wants to make a new transaction behavior has to edit the same file. Just as you would never create a complex RTL design using just one Verilog module, you should break classes down into smaller, reusable blocks.

Another approach is composition. As you learned in Chapter 5, you can instantiate one class inside another, just as you instantiate modules inside another, building up a hierarchical testbench. You write and debug your classes from the top down or bottom up, always looking for natural partitions when deciding what variables and method go into the various classes. A pixel could be partitioned into its color and coordinate. A packet might be divided into header and payload. You might break an instruction into opcode and operands. See Section 8.4 for guidelines on partitioning.

Sometimes it is difficult to divide the functionality into separate parts. Consider injecting errors during a bus transaction. When you write the original class for the transaction, you may not think of all the possible error cases. Ideally, you would like to make a class for a good transaction, and later add different error injectors. The transaction has data fields and an error-checking checksum field generated from the data. One form of error injection is corruption of the checksum field. If you use composition, you need separate classes for good transactions and error transactions. Testbench code that used good objects would have to be rewritten to process the new error objects. What you need is a class that resembles the original class but adds a few new variables and methods. This result is accomplished through inheritance.

C. Spear and G. Tumbush, *SystemVerilog for Verification: A Guide to Learning the Testbench Language Features*, DOI 10.1007/978-1-4614-0715-7_8, © Springer Science+Business Media, LLC 2012

Inheritance allows a new class to be extended from an existing one by adding new variables and methods. The original class is known as the base class. Since the new class extends the capability of the base class, it is called the extended class. Inheritance provides reusability by overlaying features, such as error injection, on an existing class, without modifying that class.

A real power of OOP is that it gives you the ability to take an existing class, such as a transaction, and selectively update parts of its behavior by replacing methods, but without having to change the surrounding infrastructure. All your original tests that depend on the base class keep working, and you can now create new tests with the extended class. With some planning, you can create a testbench solid enough to send basic transactions, but able to accommodate any extensions needed by the test.

Note that this chapter goes into a wide range of advanced OOP topics, many of which you won't need when learning SystemVerilog. Feel free to skip the later sections for now, and save them for when you are digging into the internals of UVM and VMM.

8.1   Introduction to Inheritance

Figure 8.1 shows a simple testbench. The test controls the generator. The generator creates transactions, randomizes them, and sends them to the driver along the dotted line. The driver breaks down the transaction into pin wiggles and sends it into the DUT along the dashed line. The rest of the testbench is left out.

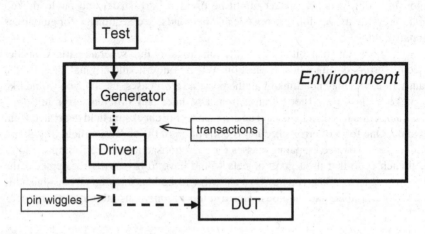

Fig. 8.1 Simplified layered testbench

8.1.1 Basic Transaction

The basic transaction class in Sample 8.1 has variables for the source and destination addresses, eight data words, and a checksum for error checking, plus methods for displaying the contents and calculating the checksum. The calc_csm function is tagged as virtual so that it can be redefined if needed, as shown in the next section. Virtual methods are explained in more detail later in this chapter in Section 8.3.2. The class is simple enough that it uses the default SystemVerilog constructor that allocates memory and initializes variables to their default value.

Sample 8.1 Base Transaction class

```
class Transaction;
  rand bit [31:0] src, dst, data[8];  // Random variables
  bit [31:0] csm;                      // Calculated variable

  virtual function void calc_csm();
    csm = src ^ dst ^ data.xor;
  endfunction

  virtual function void display(input string prefix="");
    $display("%sTr: src=%h, dst=%h, csm=%h, data=%p",
              prefix, src, dst, csm, data);
  endfunction
endclass
```

Normally calculating the checksum would be done in post_randomize(), but in this example it has been separated from the randomization to show how to inject errors.

Figure 8.2 shows a diagram for the class with both the variables and methods.

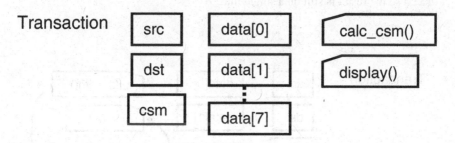

Fig. 8.2 Base Transaction class diagram

8.1.2 Extending the Transaction Class

Suppose you have a testbench that sends good transactions through the DUT and now you want to inject errors. If you follow the guidelines from Chapter 1, you would want to make as few code changes as possible to your existing testbench. So how can you reuse the existing Transaction class? Take the existing class and extend it to create a new class. This is done by declaring a new class, BadTr, as an extension of the current class. Transaction is the base class, and BadTr is the extended class. The code is shown in Sample 8.2 and in a diagram in Fig. 8.3.

Sample 8.2 Extended Transaction class

```
class BadTr extends Transaction;
  rand bit bad_csm;

  virtual function void calc_csm();
    super.calc_csm();           // Compute good csm
    if (bad_csm) csm = ~csm;    // Corrupt the csm bits
  endfunction

  virtual function void display(input string prefix="");
    $write("%sBadTr: bad_csm=%b, ", prefix, bad_csm);
    super.display();
  endfunction
```

Note that in Sample 8.2, the variable csm is does not need a hierarchical identifier. The BadTr class can see all the variables from the original Transaction plus its own variables such as bad_csm, as shown in Fig. 8.3. The calc_csm function in the extended class calls calc_csm in the base class using the super prefix. You can call a single level up, but going across multiple levels such as super.super. new is not allowed in SystemVerilog. This style, that reaches across multiple levels, would violate the rules of encapsulation by reaching across multiple boundaries.

The original display method printed a single line, starting with the prefix. So the extended display method prints the prefix, class name, and bad_csm with $write so the result is still on a single line.

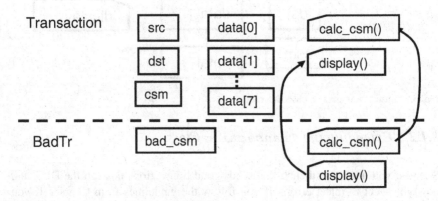

Fig. 8.3 Extended Transaction class diagram

Always declare methods inside a class as virtual so that they can be redefined in an extended class. This applies to all tasks and functions except the `new` function, which is called when the object is constructed, so there is no way to extend it. SystemVerilog always calls the `new` function based on the handle's type. Virtual methods are described fully in Section 8.3.2.

8.1.3 More OOP Terminology

Here is a quick glossary of terms. As explained in Chapter 5, the OOP term for a variable in a class is "property," and a task or function is called a "method." A base class is one that is not derived from any other class. When you extend a class, the original class (such as `Transaction`) is called the parent class or superclass. The extended class (`BadTr`) is also known as the derived or subclass. The "prototype" for a method is just the first line that shows the argument list and return type, if any. The prototype is used when you move the body of the method outside the class, but is needed to describe how the method communicates, as shown in Section 5.10.

8.1.4 Constructors in Extended Classes

When you start extending classes, there is one rule about constructors (`new` functions) to keep in mind. If your base class constructor has any arguments, the extended class must have a constructor and must call the base's constructor on its first line. In Sample 8.3, since `Base::new` has an argument, `Extended::new` must call it.

Sample 8.3 Constructor with arguments in an extended class

```
class Base;
  int val;
  function new(input int val); // Has an argument
    this.val = val;
  endfunction
endclass

class Extended extends Base;
  function new(input int val);
    super.new(val);  // Must be first line of new
    // Other constructor actions
  endfunction
endclass
```

8.1.5   Driver Class

The driver class in Sample 8.4 receives transactions from the generator and drives them into the DUT.

Sample 8.4 Driver class

```
class Driver;
 mailbox #(Transaction) gen2drv; // Mbx between Generator and here

   function new(input mailbox #(Transaction) gen2drv);
     this.gen2drv = gen2drv;
   endfunction

   virtual task run();
     Transaction tr;          // Handle to a Transaction object or
                              // a class extended from Transaction
     forever begin
       gen2drv.get(tr);       // Get transaction from generator
       tr.calc_csm();         // Process the transaction
       @ifc.cb;
       ifc.cb.src <= tr.src;  // Send transaction
       ...
     end
   endtask
endclass
```

This class receives `Transaction` objects from the generator though the mailbox `gen2drv`, breaks them down into signal changes in the interface to stimulate the DUT. What happens if your generator instead sends a `BadTr` object into the class? OOP rules say that if you have a handle of the base type (`Transaction`), it can also point to an object of an extended type (`BadTr`). The handle `tr` can only reference things in the base class such as the variables `src`, `dst`, `csm`, and `data`, and the method `calc_csm`. So you can send `BadTr` objects into the driver without changing the `Driver` class.

See Chapter 10 and 11 for examples of fully functional drivers with advanced features such as virtual interfaces and callbacks.

When the driver calls `tr.calc_csm`, which one will be called, the one in `Transaction` or `BadTr`? Since `calc_csm` was declared as a virtual method in the base class in Sample 8.1, SystemVerilog chooses the proper method based on the type of object stored in `tr`. If the object is of type `Transaction`, SystemVerilog calls the task `Transaction::calc_csm`. If it is of type `BadTr`, SystemVerilog calls the function `BadTr::calc_csm`.

8.1.6   Simple Generator Class

The generator in Sample 8.5 for this testbench creates a random transaction and puts it in the mailbox to the driver. The following (bad) example shows how you might create the class from what you have learned so far. Note that this avoids a very common testbench bug by constructing a new transaction object every pass through the loop instead of just once outside. This bug is discussed in more detail in Section 7.6 on mailboxes.

Sample 8.5   Bad generator class

```
// Generator class that uses Transaction objects
// First attempt... too limited
class Generator;
  mailbox #(Transaction) gen2drv; // Carries transactions to driver
  Transaction tr;

  function new(input mailbox #(Transaction) gen2drv);
    this.gen2drv = gen2drv;      // this-> class-level var
  endfunction

  virtual task run(input int num_tr = 10);
    repeat (num_tr) begin
      tr = new();                      // Construct transaction
      'SV_RAND_CHECK(tr.randomize()); // Randomize it
      gen2drv.put(tr.copy());          // Send copy to driver
    end
  endtask
endclass
```

There is a big limitation with this generator. The `run` task constructs a transaction and immediately randomizes it. This means that the transaction uses whatever constraints are turned on by default. The only way you can change this would be to edit the `Transaction` class, which goes against the verification guidelines presented in this book. Worse yet, the generator only uses `Transaction` objects — there is no way to use an extended object such as `BadTr`. The fix is to separate the construction of `tr` from its randomization as shown below in Section 8.2.

As you build data-oriented classes such as network and bus transactions, you will see that they have common properties (`id`) and methods (`display`). Control-oriented classes such as the `Generator` and `Driver` classes also have a common structure. You can enforce this by making both of these classes extensions of a base `Transactor` class, with virtual methods for `run`, and `wrap_up`. Both the UVM and VMM has an extensive set of base classes for transactors, data, and much more.

8.2   Blueprint Pattern

A useful OOP technique is the "blueprint pattern." If you have a machine to make signs, you don't need to know the shape of every possible sign in advance. You just need a stamping machine and then change the die to cut different shapes. Likewise, when you want to build a transactor generator, you don't have to know how to build every type of transaction; you just need to be able to stamp new ones that are similar to a given transaction.

Instead of constructing and then immediately using an object, as in Sample 8.5, construct a blueprint object (the cutting die), and then modify its constraints with `constraint_mode`, or even replace it with an extended object, as shown in Fig. 8.4. Now when you randomize this blueprint, it will have the random values that you want. Make a copy of this object and send the copy to the downstream transactor.

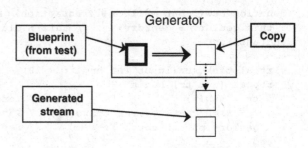

Fig. 8.4 Blueprint pattern generator

The beauty of this technique is that if you change the blueprint object, your generator creates an object of a different type. Using the sign analogy, you change the cutting die from a square to a triangle to make Yield signs, as shown in Fig. 8.5.

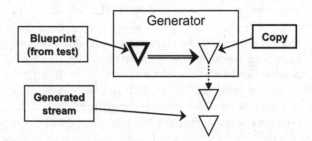

Fig. 8.5 Blueprint generator with new pattern

The blueprint is the "hook" that allows you to change the behavior of the generator class without having to change its code. You need to make a copy method that can make a copy of the blueprint to transmit, so that the original blueprint object is kept around for the next pass through the loop.

Sample 8.6 shows the generator class using the blueprint pattern. The important thing to notice is that the blueprint object is constructed in one place (the `new` function)

Sample 8.6 Generator class using blueprint pattern

```
class Generator;
  mailbox #(Transaction) gen2drv;
  Transaction blueprint;

  function new(input mailbox #(Transaction) gen2drv);
    this.gen2drv = gen2drv;
    blueprint = new();
  endfunction

  virtual task run(input int num_tr = 10);
    repeat(num_tr) begin
      `SV_RAND_CHECK(blueprint.randomize);
      gen2drv.put(blueprint.copy()); // Send copy to the driver
    end
  endtask
endclass
```

and used in another (the run task). Previous coding guidelines in this book said to separate the declaration and construction; similarly, you need to separate the construction and randomization of the blueprint object.

The copy method, which makes a duplicate of an object by copying its variables into a new object, is discussed in Sections 5.15 and 8.5. For now, remember that you must add it to the Transaction and BadTr classes. Sample 8.34 on page 304 shows an advanced generator using templates.

This generator constructs a new transaction every time the blueprint is randomized. This coding style prevents the classic OOP mailbox bug, as the mailbox will store handles to multiple unique objects, not that same single object.

Another advantage of randomizing the blueprint object over and over is that randc variables work correctly. The bad generator in Sample 8.5 constructed new objects every pass through the loop. Every object with a randc variable maintains a history of previous values generated for the variable. Every time you construct a new object, that history is lost, and the bad generator creates objects with separate randc variables. In Sample 8.6, only the blueprint object is randomized, so the randc history is maintained.

Section 8.2.3 shows how to change the blueprint.

8.2.1 The Environment Class

Chapter 1 discussed the three phases of execution: Build, Run, and Wrap-up. Sample 8.7 shows the environment class that instantiates all the testbench components, and runs these three phases. Also notice how the mailbox gen2drv carries transactions from the generator to the driver, and so is passed into the constructor for each.

Sample 8.7 Environment class

```
// Testbench environment class
class Environment;
  Generator gen;
  Driver drv;
  mailbox #(Transaction) gen2drv;

  virtual function void build();   // Build the environment by
    gen2drv = new();               // constructing the mailbox,
    gen = new(gen2drv);            // the generator,
    drv = new(gen2drv);            // and driver
  endfunction

  virtual task run();
    fork
      gen.run();
      drv.run();
    join
  endtask

  virtual task wrap_up();
    // Empty for now - call scoreboard for report
  endtask

endclass
```

8.2.2   A Simple Testbench

The test is contained in the top-level program shown in Sample 8.8. The basic test
just lets the environment run with all the defaults.

Sample 8.8 Simple test program using environment defaults

```
program automatic test;

  Environment env;
  initial begin
    env = new();            // Construct the environment
    env.build();            // Build testbench objects
    env.run();              // Run the test
    env.wrap_up();          // Clean up afterwards
  end
endprogram
```

8.2.3   Using the Extended *Transaction* Class

To inject an error, you need to change the blueprint object from a `Transaction` object to a `BadTr`. You do this between the build and run phases in the environment. The top-level testbench in Sample 8.9 runs each phase of the environment and changes the blueprint. Note how all the references to `BadTr` are in this one file, so you don't have to change the `Environment` or `Generator`

classes. You want to restrict the scope of where `BadTr` can be used, so a standalone begin...end block is used in the middle of the `initial` block. This makes a visually distinctive block of code. You can take a shortcut and construct the extended class in the declaration.

Sample 8.9 Injecting an extended transaction into testbench

```
program automatic test;

  Environment env;
  initial begin
    env = new();
    env.build();                  // Construct generator, etc.

    begin
      BadTr bad = new();          // Replace blueprint with
      env.gen.blueprint = bad;    // the "bad" one
    end

    env.run();                    // Run the test with BadTr
    env.wrap_up();                // Clean up afterwards
  end
endprogram
```

8.2.4   Changing Random Constraints with an Extended Class

In Chapter 6 you learned how to generate constrained random data. Most of your tests are going to need to further constrain the data, which is best done with inheritance. In Sample 8.10, the original Transaction class is extended to include a new constraint that keeps the destination address in the range of +/−100 of the source address.

Sample 8.10 replaces the generator's blueprint with an extended object that has an additional constraint. As you will learn later in this chapter, the `Nearby` class should have a `copy` method, but hold on for a few sections.

Sample 8.10 Adding a constraint with inheritance

```
class Nearby extends Transaction;
  constraint c_nearby {
    dst inside {[src-100:src+100]};
  }
  // copy method not shown
endclass


program automatic test;
  Environment env;
  initial begin
    env = new();
    env.build();                       // Construct generator, etc.

    begin
      Nearby nb = new();               // Create a new blueprint
      env.gen.blueprint = nb;          // Replace the blueprint
    end

    env.run();                         // Run the test with Nearby
    env.wrap_up();                     // Clean up afterwards
  end
endprogram
```

Note that if you define a constraint in an extended class with the same name as one in the base class, the extended constraint replaces the base one. This allows you to change the behavior of existing constraints.

8.3 Downcasting and Virtual Methods

As you start to use inheritance to extend the functionality of classes, you need a few OOP techniques to control the objects and their functionality. In particular, a handle can refer to an object for a certain class, or any extended class. So what happens when a base handle points to an extended object? What happens when you call a method that exists in both the base and extended classes? This section explains what happens using several examples.

8.3.1 Downcasting with $cast

Downcasting or conversion is the act of casting a base class handle to point to an object that is a class extended from that base type. Consider the base and extended classes in Sample 8.11 and Fig. 8.6.

Sample 8.11 Base and extended class

```
class Transaction;
  rand bit [31:0] src;
  virtual function void display(input string prefix="");
    $display("%sTransaction: src=%0d", prefix, src);
  endfunction
endclass

class BadTr extends Transaction;
  bit bad_csm;
  virtual function void display(input string prefix="");
    $display("%sBadTr: bad_csm=%b", prefix, bad_csm);
    super.display(prefix);
  endfunction
endclass
```

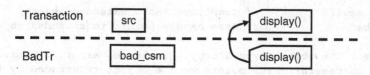

Fig. 8.6 Simplified extended transaction

You can assign an extended handle to a base handle, and no special code is needed, as shown in Sample 8.12. When a class is extended, all the base class variables and methods are included, so src is in the extended object. The assignment to tr is permitted, as any reference using the base handle tr is valid, such as tr.src and tr.display.

Sample 8.12 Copying extended handle to base handle

```
Transaction tr;
BadTr bad;
bad = new();        // Construct BadTr extended object
tr = bad;           // Base handle points to extended obj
                    // tr is downcast to point to BadTr type
$display(tr.src);   // Display variable in base class
tr.display;         // Calls BadTr::display
```

What if you try going in the opposite direction, copying a handle to a base object into an extended handle, as shown in Sample 8.13? This fails because the base object is missing properties that only exist in the extended class, such as bad_csm. The SystemVerilog compiler does a static check of the handle types and will not compile the second line.

Sample 8.13 Copying a base handle to an extended handle

```
tr = new();            // Construct base object
bad = tr;              // ERROR: WILL NOT COMPILE
$display(bad.bad_csm); // bad_csm is only in extended object
```

It is not always illegal to assign a base handle to an extended handle, but you must always use $cast. The assignment is allowed when the base handle points to an extended object. The $cast method checks the type of object referenced by the handles, not just the handle. If the source object is the same type as the destination, or a class extended from the destination's class, you can copy the address of the extended object from the base handle, tr, into the extended handle, bad2.

Sample 8.14 Using $cast to copy handles

```
Transaction tr;
BadTr bad, bad2;

bad = new();           // Construct BadTr extended object
tr = bad;              // Base handle points to extended object

// Check the object type & copy. Simulation error if mismatch
// If successful, bad2 points to the object referenced by tr
$cast(bad2, tr);

// Check for type mismatch, no simulation error
if($cast(bad2, tr))
  $display(bad2.bad_csm); // bad_csm exists in original object
else
  $display("ERROR: cannot assign tr to bad2");
```

When you use $cast as a task, SystemVerilog checks the type of the source object at run time and gives an error if it is not compatible with the destination. When you use $cast as a function, SystemVerilog still checks the type, but no longer prints an error if there is a mismatch. The $cast function returns zero when the types are incompatible, and one for compatible types.

As an alternative to the if statement in Sample 8.14, you could use something like the **SV_RAND_CHECK** macro from Section 6.3.2. You should not use an immediate assert statement as the assertion expression is not evaluated if you disable assertions, which means the $cast and bad2 assignment will never execute.

8.3.2 Virtual Methods

By now you should be comfortable using handles with extended classes. What happens if you try to call a method using one of these handles? Sample 8.15 and 8.16 show base and extended classes and code that calls methods inside these classes.

Sample 8.15 Transaction and BadTr classes

```
class Transaction;
  rand bit [31:0] src, dst, data[8];   // Variables
  bit [31:0] csm;

  virtual function void calc_csm();    // XOR all fields
    csm = src ^ dst ^ data.xor;
  endfunction
endclass : Transaction

class BadTr extends Transaction;
  rand bit bad_csm;
  virtual function void calc_csm();
    super.calc_csm();                  // Compute good csm
    if (bad_csm) csm = ~csm;           // Corrupt the csm bits
  endfunction
endclass : BadTr
```

Sample 8.16 contains a block of code that uses handles of different types.

Sample 8.16 Calling class methods

```
Transaction tr;
BadTr bad;

initial begin
  tr = new();
  tr.calc_csm();      // Calls Transaction::calc_csm

  bad = new();
  bad.calc_csm();     // Calls BadTr::calc_csm

  tr = bad;           // Base handle points to ext obj
  tr.calc_csm();      // Calls BadTr::calc_csm
end
```

To decide which virtual method to call, SystemVerilog uses the object's type, not the handle's type. In the last statement of Sample 8.16, tr points to an extended object (BadTr) and so BadTr::calc_csm is called.

If you leave out the virtual modifier on Transaction::calc_csm, SystemVerilog checks the type of the handle tr (Transaction), not the object. That last statement in Sample 8.16 calls Transaction::calc_csm – probably not what you wanted.

The OOP term for multiple methods sharing a common name is "polymorphism." It solves a problem similar to what computer architects faced when trying to make a processor that could address a large address space but had only a small amount of physical memory. They created the concept of virtual memory, where the code and data for a program could reside in memory or on a disk. At compile time, the program didn't know where its parts resided — that was all taken care of by the hardware plus operating system at run time. A virtual address could be mapped to some RAM chips, or the swap file on the disk. Programmers no longer needed to worry about this virtual memory mapping when they wrote code — they just knew that the processor would find the code and data at run time. See also Denning (2005).

8.3.3 Signatures and Polymorphism

There is a downside to using virtual methods: once you define one, all extended classes that define the same method must use the same "signature," i.e., the same number and type of arguments, plus return value, if any. You cannot add or remove an argument in an extended virtual method. This means you need to plan ahead.

There is a good reason that SystemVerilog and other OOP languages require that a virtual method must have the same signature as the one in the parent (or grandparent). If you were able to add an additional argument, or turn a task into a function, polymorphism would no longer work. Your code needs to be able to call a virtual method with the assurance that a method in a extended class will have the same interface.

8.3.4 Constructors are Never Virtual

When you call a virtual method, SystemVerilog checks the type of the object to decide if it should call the method in the base class or the extended. Now you can see why a constructor can not be virtual. When you call it, there is no object whose type can be checked. The object only exists after the constructor call starts.

8.4 Composition, Inheritance, and Alternatives

As you build up your testbench, you have to decide how to group related variables and methods together into classes. In Chapter 5 you learned how to build basic classes and include one class inside another. Previously in this chapter, you saw the basics of inheritance. This section shows you how to decide between the two styles, and also shows an alternative.

8.4.1 Deciding Between Composition and Inheritance

How should you tie together two related classes? Composition uses a "has-a" relationship. A packet has a header and a body. Inheritance uses an "is-a" relationship.

A BadTr is a Transaction, just with more information. Table 8.1 is a quick guide, with more detail below.

Table 8.1 Comparing inheritance to composition

Question	Inheritance (is-a relationship)	Composition (has-a relationship)
1. Do you need to group multiple extended classes together? (SystemVerilog does not support multiple inheritance)	No	Yes
2. Does the higher-level class represent objects at a similar level of abstraction?	Yes	No
3. Is the lower-level information always present or required?	Yes	No
4. Does the additional data need to remain attached to the original class while it is being processed by pre-existing code?	Yes	No

1. Are there several small classes that you want to combine into a larger class? For example, you may have a data class and header class and now want to make a packet class. SystemVerilog does not support multiple inheritance, where one class extends from several classes at once. Instead you have to use composition. Alternatively, you could extend one of the classes to be the new class, and manually add the information from the others.
2. In Sample 8.15, the Transaction and BadTr classes are both bus transactions created in a generator and driven into the DUT, so inheritance makes sense.
3. The lower-level information such as src, dst, and data must always be present for the Driver to send a transaction.
4. In Sample 8.15, the new BadTr class has a new field bad_csm and the extended calc_csm function. The Generator class just transmits a transaction and does not care about the additional information. If you use composition to create the error bus transaction, the Generator class would have to be rewritten to handle the new type.

If two objects seem to be related by both "is-a" and "has-a," you may need to break them down into smaller components.

8.4.2   Problems with Composition

The classical OOP approach to building a class hierarchy partitions functionality into small blocks that are easy to understand. However, testbenches are not standard software development projects, as was discussed in Section 5.16 on public vs. local attributes. Concepts such as information hiding (using local variables) conflict with building a testbench that needs maximum visibility and controllability. Similarly, dividing a transaction into smaller pieces may cause more problems than it solves.

When you are creating a class to represent a transaction, you may want to partition it to keep the code more manageable. For example, you may have an Ethernet MAC frame and your testbench uses two flavors, normal (II) and Virtual LAN (VLAN). Using composition, you could create a basic cell EthMacFrame with all the common fields such as da and sa and a discriminant variable, kind, to indicate the type as shown in Sample 8.17. There is a second class to hold the VLAN information, which is included in EthMacFrame.

Sample 8.17 Building an Ethernet frame with composition

```
// Not recommended
class EthMacFrame;
  typedef enum {II, IEEE} kind_e;
  rand kind_e kind;
  rand bit [47:0] da, sa;
  rand bit [15:0] len;

  ...
  rand Vlan vlan_h;
endclass

class Vlan;
  rand bit [15:0] vlan;
endclass
```

There are several problems with composition. First, it adds an extra layer of hierarchy, so you are constantly having to add an extra name to every reference. The VLAN information is called eth_h.vlan_h.vlan. If you start adding more layers, the hierarchical names become a burden.

A more subtle issue occurs when you want to instantiate and randomize the hierarchy of classes. What does the EthMacFrame constructor create? Since kind is random, you don't know whether to construct a Vlan object when new is called. When you randomize the class, the constraints set variables in both the EthMacFrame and Vlan objects based on the random kind field. You have a circular dependency in that randomization only works on objects that have been instantiated, but you can't instantiate these objects until kind has been chosen.

The only solution to the construction and randomization problems is to always instantiate all objects in EthMacFrame::new. However, if you are always using all alternatives, why divide the Ethernet cell into two different classes?

8.4.3   *Problems with Inheritance*

Inheritance can solve some of these issues. Variables in the extended classes can be referenced without the extra hierarchy as in `eth_h.vlan`. You don't need a discriminant, but you may find it easier to have one variable to test rather than doing type-checking as shown in Sample 8.18.

Sample 8.18   Building an Ethernet frame with inheritance

```
// Not recommended
class EthMacFrame;
   typedef enum {II, IEEE} kind_e;
   rand kind_e kind;
   rand bit [47:0] da, sa;
   rand bit [15:0] len;
   ...
endclass

class Vlan extends EthMacFrame;
   rand bit [15:0] vlan;
endclass
```

On the downside, a set of classes that use inheritance always requires more effort to design, build, and debug than a set of classes without inheritance. Your code must use `$cast` whenever you have an assignment from a base handle to an extended handle. Building a set of virtual methods can be challenging, as they all have to have the same signature. If you need an extra argument, you need to go back and edit the entire set, and possibly the method calls too.

There are also problems with randomization. How do you make a constraint that randomly chooses between the two kinds of frame and sets the proper variables? You can't put a constraint in `EthMacFrame` that references the `vlan` field.

The final issue is with multiple inheritance. In Fig. 8.7, you can see how the VLAN frame is extended from a normal MAC frame. The problem is that these different standards reconverged. SystemVerilog does not support multiple inheritance, so you could not create the VLAN / Snap / Control frame through inheritance.

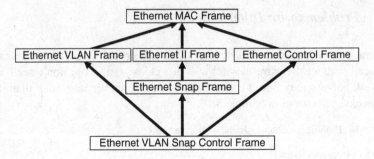

Fig. 8.7 Multiple inheritance problem

8.4.4 A Real-World Alternative

If composition leads to large hierarchies, but inheritance requires extra code and planning to deal with all the different classes, and both have difficult construction and randomization, what can you do? You can instead make a single, flat class that has all the variables and methods. This approach leads to a very large class, but it handles all the variants cleanly. You have to use the discriminant variable often to tell which variables are valid, as shown in Sample 8.19. It contains several conditional constraints, which apply in different cases, depending on the value of `kind`.

Sample 8.19 Building a flat Ethernet frame

```
class eth_mac_frame;
  typedef enum {II, IEEE} kind_e;
  rand kind_e kind;
  rand bit [47:0] da, sa;
  rand bit [15:0] len, vlan;
  rand bit [ 7:0] data[];
  ...
  constraint eth_mac_frame_II {
    if (kind == II) {
      data.size() inside {[46:1500]};
      len == data.size();
  }}
  constraint eth_mac_frame_ieee {
    if (kind == IEEE) {
      data.size() inside {[46:1500]};
      len < 1522;
  }}
endclass
```

Regardless of how you build your classes, define the typical behavior and constraints in the class, and then use inheritance to inject new behavior at the test level.

8.5 Copying an Object

In Sample 8.6, the generator first randomized, and then copied the blueprint to make a new transaction. Take a closer look at the copy function in Sample 8.20. Also see Section 5.15 for more examples of copy functions.

Sample 8.20 Base transaction class with a virtual copy function

```
class Transaction;
  rand bit [31:0] src, dst, data[8];  // Variables
  bit [31:0] csm;

  virtual function Transaction copy();
    copy = new();               // Construct destination object
    copy.src  = this.src;       // Copy data fields
    copy.dst  = this.dst;       // The prefix "this." is
    copy.data = this.data;      //    not needed, but makes code
    copy.csm  = this.csm;       //    more explicit
    return copy;                // Return handle to copy
  endfunction
endclass
```

When you extend the Transaction class to make the class BadTr, the copy function still has to return a Transaction object. This is because the extended virtual function must match the base Transaction::copy, including all arguments and return type, as shown in Sample 8.21

Sample 8.21 Extended transaction class with virtual copy method

```
class BadTr extends Transaction;
  rand bit bad_csm;

  virtual function Transaction copy();
    BadTr bad;
    bad = new();                    // Construct extended object
    bad.src      = this.src;   // Copy data fields
    bad.dst      = this.dst;
    bad.data     = this.data;
    bad.csm      = this.csm;
    bad.bad_csm  = this.bad_csm;
    return bad;                     // Return handle to copy
  endfunction
endclass : BadTr
```

8.5.1   Specifying a Destination for Copy

The previous copy methods always constructed a new object. An improvement for copy is to specify the location where the copy should be put. This technique is useful when you want to reuse an existing object, and not allocate a new one.

Sample 8.22   Base transaction class with copy function

```
class Transaction;

  virtual function Transaction copy(input Transaction to=null);
    if (to == null)
      copy = new();              // Construct new object
    else
      copy = to;                 // or use existing
    copy.src  = this.src;        // Copy data fields
    copy.dst  = this.dst;
    copy.data = this.data;
    copy.csm  = this.csm;
    return copy;
  endfunction
```

The only difference is the additional argument to specify the destination, and the code to test that a destination object was passed to this method. If nothing was passed (the default), construct a new object, or else use the existing one.

Since you have added a new argument to a virtual method in the base class, you will have to add it to the same method in the extended classes, such as BadTr.

Sample 8.23   Extended transaction class with new copy function

```
class BadTr extends Transaction;

  virtual function Transaction copy(input Transaction to=null);
    BadTr bad;
    if (to == null)
      bad = new();                    // Create a new object
    else
      $cast(bad, to);                 // Reuse existing one
    super.copy(bad);                  // Copy base data fields
    bad.bad_csm = this.bad_csm;       // Copy extended fields
    return bad;
  endfunction
endclass : BadTr
```

Notice how BadTr::copy only needs to copy the fields in the extended class and can use the base class method, Transaction::copy to copy its own fields.

8.6 Abstract Classes and Pure Virtual Methods

By now you have seen classes with methods to perform common operations such as copying and displaying. One goal of verification is to create code that can be shared across multiple projects. If your company standardizes on a common set of classes and methods, it is easier to reuse code between projects.

OOP languages such as SystemVerilog have two constructs to allow you to build a shareable base class. The first is an abstract class, which is a class that can be extended, but not instantiated directly. It is defined with the `virtual` keyword. The second is a pure virtual method, which is a prototype without a body. A class extended from an abstract class can only be instantiated if all pure virtual methods have bodies. The `pure` keyword specifies that a method declaration is a prototype, and not just an empty virtual method. A pure method has no `endfunction` or `endtask`. Lastly, pure virtual methods can only be declared in an abstract class. An abstract class can contain pure virtual methods, virtual methods with and without a body, and non-virtual methods. Note that if you define a virtual method without a body, i.e. no code inside, you can call it but it just immediately returns.

Sample 8.24 shows an abstract class, `BaseTr`, which is a base class for transactions. It starts with a some useful properties such as `id` and `count`. The constructor makes sure every instance has a unique ID. Next are pure virtual methods to compare, copy, and display the object.

Sample 8.24 Abstract class with pure virtual methods

```
virtual class BaseTr;
   static int count;      // Number of instance created
   int id;                // Unique transaction id

   function new();
      id = count++;       // Give each object a unique ID
   endfunction

   pure virtual function bit compare(input BaseTr to);
   pure virtual function BaseTr copy(input BaseTr to=null);
   pure virtual function void display(input string prefix="");

endclass : BaseTr
```

You can declare handles of type `BaseTr`, but you cannot construct objects of this type. You need to extend the class and provide implementations for all the pure virtual methods.

Sample 8.25 shows the definition of the `Transaction` class, which has been extended from `BaseTr`. Since `Transaction` has bodies for all the pure virtual methods extended from `BaseTr`, you can construct objects of this type in your testbench.

Sample 8.25 Transaction class extends abstract class

```
class Transaction extends BaseTr;
  rand bit [31:0] src, dst, csm, data[8];

  extern function new();
  extern virtual function bit compare(input BaseTr to);
  extern virtual function BaseTr copy(input BaseTr to=null);
  extern virtual function void display(input string prefix="");

endclass

function Transaction::new();
  super.new();
endfunction : new

function bit Transaction::compare(input BaseTr to);
  Transaction tr;
  if(!$cast(tr, to))  // Check if 'to' is correct type
    $finish;
  return ((this.src  == tr.src) &&
          (this.dst  == tr.dst) &&
          (this.csm  == tr.csm) &&
          (this.data == tr.data));
endfunction : compare

function BaseTr Transaction::copy(input BaseTr to=null);
  Transaction cp;
  if (to == null) cp = new();
  else               $cast(cp, to);
  cp.src  = this.src;    // Copy the data fields
  cp.dst  = this.dst;
  cp.data = this.data;
  cp.csm  = this.csm;
  return cp;
endfunction : copy

function void Transaction::display(input string prefix="");
  $display("%sTransaction %0d src=%h, dst=%x, csm=%x",
           prefix, id, src, dst, csm);
endfunction : display
```

Abstract classes and pure virtual methods let you build testbenches that have a common look and feel. This allows any engineer to read your code and quickly understand the structure.

8.7 Callbacks

One of the main guidelines of this book is to create a single verification environment that you can use for all tests with no changes. The key requirement is that this testbench must provide a "hook" where the test program can inject new code without modifying the original classes. Your driver may want to do the following.

- Inject errors
- Drop the transaction
- Delay the transaction
- Synchronize this transaction with others
- Put the transaction in the scoreboard
- Gather functional coverage data

Rather than try to anticipate every possible error, delay, or disturbance in the flow of transactions, the driver just needs to "call back" a method that is defined in the top-level test. The beauty of this technique is that the callback method can be defined differently in every test. As a result, the test can add new functionality to the driver using callbacks, without editing the `Driver` class. For some drastic behaviors such as dropping a transaction, you need to code this in the class ahead of time, but this is a known pattern. The reason why the transaction is dropped is left to the callback.

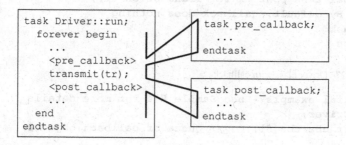

Fig. 8.8 Callback flow

In Fig. 8.8, the `Driver::run` task loops forever with a call to a `transmit` task. Before sending the transaction, `run` calls the pre-transmit callback, if any. After sending the transaction, it calls the post-callback task, if any. By default, there are no callbacks, so `run` just calls `transmit`.

You could make `Driver::run` a virtual method and then override its behavior in an extended class, perhaps `MyDriver::run`. The drawback to this is that you might have to duplicate all the original method's code in the new method if you are

injecting new behavior. Now if you made a change in the base class, you would
have to remember to propagate it to all the extended classes. Additionally, you can
inject a callback without modifying the code that constructed the original object.

8.7.1 Creating a Callback

A callback task is created in the top-level test and called from the driver, the lowest
level of the environment. However, the driver does not have to have any knowledge
of the test – it just has to use a generic class that the test can extend. The driver in
Sample 8.27 uses a queue to hold the callback objects, which allows you to add
multiple objects. The base callback class in Sample 8.26 is an abstract class that
must be extended before being used. Your callback is a task so it can have delays.

Sample 8.26 Base callback class

```
virtual class Driver_cbs;  // Driver callbacks

  virtual task pre_tx(ref Transaction tr, ref bit drop);
    // By default, callback does nothing
  endtask

  virtual task post_tx(ref Transaction tr);
    // By default, callback does nothing
  endtask
endclass
```

Sample 8.27 Driver class with callbacks

```
// Partial example - see Sample 8-4 for more details
class Driver;
  Driver_cbs cbs[$];     // Queue of callback objects

  task run();
    bit drop;
    Transaction tr;

    forever begin
      drop = 0;
      agt2drv.get(tr);  // Agent to driver mailbox
      foreach (cbs[i]) cbs[i].pre_tx(tr, drop);
      if (drop) continue;
      transmit(tr);       // Actual work
      foreach (cbs[i]) cbs[i].post_tx(tr);
    end
  endtask
endclass
```

Note that while `Driver_cbs` is an abstract class, `pre_tx` and `post_tx` are not pure virtual methods. This is because a typical callback uses only one of them. If a class has even one pure virtual method without an implementation, OOP rules won't allow you to instantiate it.

Callbacks are part of both VMM and UVM. This callback technique is not related to Verilog PLI callbacks or SVA callbacks.

8.7.2 Using a Callback to Inject Disturbances

A common use for a callback is to inject some disturbance such as causing an error or delay. The testbench in Sample 8.28 randomly drops packets using a callback object. Callbacks can also be used to send data to the scoreboard or to gather functional coverage values. Note that you can put callback objects in the queue with the `push_back()` or `push_front()` depending on the order in which you want these to be called. For example, you probably want the scoreboard called after any tasks that may delay, corrupt, or drop a transaction. You should only gather coverage after a transaction has been successfully transmitted.

Sample 8.28 Test using a callback for error injection

```
class Driver_cbs_drop extends Driver_cbs;
  virtual task pre_tx(ref Transaction tr, ref bit drop);
    // Randomly drop 1 out of every 100 transactions
    drop = ($urandom_range(0,99) == 0);
  endtask
endclass


program automatic test;
  Environment env;

  initial begin
    env = new();
    env.gen_cfg();
    env.build();

    begin                   // Create error injection callback
      Driver_cbs_drop dcd = new();
      env.drv.cbs.push_back(dcd); // Put into driver's Q
    end

    env.run();
    env.wrap_up();
  end

endprogram
```

8.7.3 A Quick Introduction to Scoreboards

The design of your scoreboard depends on the design under test. A DUT that processes atomic transactions such as packets may have a scoreboard that contains a transform function to turn the input transactions into expected values, a memory to hold these values, and a compare method. A processor design needs a reference model to predict the expected output, and the comparison between expected and actual values may happen at the end of simulation.

Sample 8.29 shows a simple scoreboard that stores transactions in a queue of expected values. The first method saves an expected transaction, and the second tries to find an expected transaction that matches an actual one that was received by the testbench. Note that when you search through a queue, you can get 0 matches (transaction not found), 1 match (ideal case) or multiple matches (you need to do a more sophisticated match).

Sample 8.29 Simple scoreboard for atomic transactions

```
class Scoreboard;
  Transaction scb[$];  // Store expected tr's in queue

  function void save_expected(input Transaction tr);
    scb.push_back(tr);
  endfunction

  function void compare_actual(input Transaction tr);
    int q[$];

    q = scb.find_index(x) with (x.src == tr.src);
    case (q.size())
      0: $display("No match found");
      1: scb.delete(q[0]);
      default:
        $display("Error, multiple matches found!");
    endcase
  endfunction : compare_actual
endclass : Scoreboard
```

8.7.4 Connecting to the Scoreboard with a Callback

The testbench in Sample 8.30 creates its own extension of the driver's callback class and adds a reference to the driver's callback queue. Note that the scoreboard callback needs a handle to the scoreboard so it can call the method to save the expected transaction. This example does not show the monitor side, which will need its own callback to send the actual transaction to the scoreboard for comparison.

Sample 8.30   Test using callback for scoreboard

```
class Driver_cbs_scoreboard extends Driver_cbs;
  Scoreboard scb;

  virtual task pre_tx(ref Transaction tr, ref bit drop);
    // Put transaction in the scoreboard
    scb.save_expected(tr);
  endtask

  function new(input Scoreboard scb);
    this.scb = scb;
  endfunction
endclass


program automatic test;
  Environment env;

  initial begin
    env = new();
    env.gen_cfg();
    env.build();

    begin                            // Create scoreboard callback
      Driver_cbs_scoreboard dcs = new(env.scb);
      env.drv.cbs.push_back(dcs);  // Put into driver's Q
    end

    env.run();
    env.wrap_up();
  end

endprogram
```

The VMM recommends that you use callbacks for scoreboards and functional coverage. The monitor transactor can use a callback to compare received transactions with expected ones. The monitor callback is also the perfect place to gather functional coverage on transactions that are actually sent by the DUT.

You may have thought of putting the scoreboard or functional coverage group in a transactor, and connect it to the testbench using a mailbox. This is a poor solution for several reasons. These testbench components are almost always passive and asynchronous, so they only wake up when the testbench has data for them, plus they never pass information to a downstream transactor. Thus a transactor that has to monitor multiple mailboxes concurrently is an overly complex solution. Additionally, you may sample data from several points in your testbench, but a transactor is designed for a single source. Instead, put methods in your scoreboard and coverage classes to gather data, and connect them to the testbench with callbacks.

The UVM recommends a TLM analysis port for connecting monitors / drivers to scoreboards and functional coverage. A description of this construct is beyond the scope of this book, but you can think of it as a mailbox with an optional consumer.

8.7.5   Using a Callback to Debug a Transactor

If a transactor with callbacks is not working as expected, you can add a debug callback. You can start by adding a callback to display the transaction. If there are multiple instances of the transactor, create a unique identifier for each. Put debug code before and after the other callbacks to locate the one that is causing the problem. Even for debug, you want to avoid making changes to the testbench environment.

8.8   Parameterized Classes

As you become more comfortable with classes, you may notice that a class, such as a stack or generator, only works on a single data type. This section shows how you can define a single parameterized class that works with multiple data types.

8.8.1   A Simple Stack

A common data structure is a stack, which has push and pop methods to store and retrieve data. Sample 8.31 shows a simple stack that works with the int data type.

Sample 8.31 Stack using the int type

```
parameter int SIZE = 100;
class IntStack;
  local int stack[SIZE];                  // Holds data values
  local int top;

  function void push(input int i);  // Push value on top
    stack[top++] = i;
  endfunction : push

  function int pop();                      // Remove value from top
    return stack[--top];
  endfunction

endclass : IntStack
```

The problem with this class is that it only works with integers. If you want to make a stack for real numbers, you would have to copy the class, and change the data type from int to real. This quickly leads to a proliferation of classes, which can become a maintenance problem if you ever want to add new operations such as traversing or printing the stack contents.

In SystemVerilog you can add a data type parameter to a class and then specify a type when you declare handles to that class. This is similar to, but more powerful than, a parameterized module, where you can specify a value such as bus width when it is instantiated. SystemVerilog's parameterized classes are similar to templates in C++.

Sample 8.32 is a parameterized class for a stack. Notice how the type T is defined on the first line with a default type of int.

Sample 8.32 Parameterized class for a stack

```
parameter int SIZE = 100;
class Stack #(type T=int);
  local T stack[SIZE];                 // Holds data values
  local int top;

  function void push(input T i);   // Push new value on top
    stack[top++] = i;
  endfunction : push

  function T pop();                    // Remove value from top
    return stack[--top];
  endfunction

endclass : Stack
```

The step of specifying values to a parameterized class is called specialization. Sample 8.33 declares a handle to the stack class with a real data type.

Sample 8.33 Creating the parameterized stack class

```
initial begin
  Stack #(real) rStack;                // Specialize the stack class

  rStack = new();                      // Construct a stack of reals
  for(int i=0; i<SIZE; i++)
    rStack.push(i*2.0);                // Push values onto stack

  for(int i=0; i<SIZE; i++)
    $display("%f", rStack.pop()); // Pop values off stack
end
```

Generators are a great example of a class that can be parameterized. Once you have defined the class for one, the same structure works for any data type. Sample 8.34 takes the atomic generator from Sample 8.6 and adds a parameter so you can generate any random object. The generator should be part of a package of verification classes. It needs to specify a the default type, so it uses `BaseTr` from Sample 8.24 as this abstract class should also be part of the verification package.

Sample 8.34 Parameterized generator class using blueprint pattern

```
class Generator #(type T=BaseTr);
  mailbox #(Transaction) gen2drv;
  T blueprint;                              // Blueprint object

  function new(input mailbox #(Transaction) gen2drv);
    this.gen2drv = gen2drv;
    blueprint = new();                      // Create default
  endfunction

  task run(input int num_tr = 10);
    T tr;
    repeat (num_tr) begin
      `SV_RAND_CHECK(blueprint.randomize);
      $cast(tr, blueprint.copy()); // Make a copy
      gen2drv.put(tr);                      // Send to driver
    end
  endtask
endclass
```

Using the `Transaction` class from Sample 8.25 and the generator in Sample 8.34, you can build a simple testbench like in Sample 8.35. It starts the generator and prints the first five transactions, using the mailbox synchronization shown in Sample 7.40.

Sample 8.35 Simple testbench using parameterized generator class

```
program automatic test;

  initial begin
    Generator #(Transaction) gen;
    mailbox #(Transaction) gen2drv;
    gen2drv = new(1);
    gen = new(gen2drv);

    fork
      gen.run();

      repeat (5) begin
        Transaction tr;
        gen2drv.peek(tr);   // Get next transaction
        tr.display();
        gen2drv.get(tr);    // Remove transaction
      end

    join_any
  end
endprogram // test
```

8.8.2   Sharing Parameterized Classes

When you specialize a parameterized class, as in the `real` stack in Sample 8.33, you are creating a new data type, with no OOP relationship to any other specialization. For example, you can not use `$cast()` to convert between a stack of real variables and one of integers. For that, you need a common base class as shown in Sample 8.36.

Sample 8.36   Common base class for parameterized generator class

```
class GenBase;
endclass

class Generator #(type T=BaseTr) extends GenBase;
  // See Generator class in Sample 8-34
endclass

  GenBase gen_queue[$];
  Generator #(Transaction) gen_good;
  Generator #(BadTr)       gen_bad;

  initial begin
    gen_good = new();                    // Construct good generator
    gen_queue.push_back(gen_good);       // Save it in the queue
    gen_bad = new();                     // Construct bad generator
    gen_queue.push_back(gen_bad);        // Save it in the same queue
  end
```

Upcoming sections show more examples of parameterized classes.

8.8.3   Parameterized Class Suggestions

When creating parameterized classes, you should start with a non-parameterized class, debug it thoroughly, and then add parameters. This separation reduces your debug effort.

A common set of virtual methods in your transaction class help you when creating parameterized classes. The `Generator` class uses the `copy` method, knowing that it always has the same signature. Likewise, the `display` method allows you to easily debug transactions as they flow through your testbench components.

The system functions `$typename()` and `$bits()` are helpful when your class needs to know the name and width of the parameter. The `$typename(T)` function returns the name of the parameter type such as `int`, `real`, or the class name for a handle. The `$bits()` function returns the width of the parameter. For complex types such as structures and arrays, it returns the number of bits required to hold an expression as a bit stream. The UVM transaction print methods use this function to get the fields to line up correctly.

Macros are an alternative to parameterized classes. For example, you could define a macro for the generator and pass it the transaction data type. Macros are harder to debug than parameterized classes, unless your compiler outputs the expanded code.

If you need to define several related classes that all share the same transaction type, you could use parameterized classes or a single large macro. In the end, how you define your classes is not as important as what goes into them.

8.9 Static and Singleton Classes

This section and the next show advanced OOP concepts that are used extensively in the UVM and VMM. You could try to understand UVM's factory mechanism by reading the source code with its many methods, but this section should save you several days of experimentation with a greatly simplified example. This chapter shows several alternatives so you can understand why the UVM did not pick a more simple alternative.

One of the goals of OOP is to eliminate global variables and methods as the resulting code is hard to maintain and reuse. Their names exist in the global name space, potentially causing name space collisions. Does `packet_count` refer to TCP/IP packets or some other protocol? Instead, put a variable called `count` in the `Packet` class to avoid any ambiguity.

8.9.1 Dynamic Class to Print Messages

Sometimes, however, you really need globals. For example, all verification methodologies provide a print service so you can filter messages and count errors. If you try to build such a class with what you have learned so far, it might look something like Sample 8.37.

Sample 8.37 Dynamic print class with static variables

```
class Print;
  static bit [31:0] error_count = 0, error_limit = -1;
  string class_name, instance_name;

  function new(input string class_name, instance_name);
    this.class_name    = class_name;
    this.instance_name = instance_name;
  endfunction

  function void error(input string ID, input string message);
    $display("@%0t %m [%s-%s] [%s] %s",
             $realtime, class_name, instance_name, ID, message);
    if (++error_count >= error_limit) begin
      $display("FATAL: Maximum error limit reached");

      $finish;
    end
  endfunction

endclass
```

This is a greatly simplified version of the VMM log class. The VMM code allows you to filter messages by the class and instance names, and many other features.

Sample 8.38 has a class that prints an error message with the `Print` class from Sample 8.37.

Sample 8.38   Transactor class with dynamic print object

```
class Xactor;
  Print p;
  function new();
    p = new("Xactor", "solo");
  endfunction // new

  task run();
    p.error("NYI", "This Xactor is not yet implemented");
  endtask
endclass // Xactor
```

The biggest limitation for the `Print` class is that every component in your test-bench needs to instantiate it. The simple `Print` class above has a small footprint, but a realistic one, like VMM's, could have many strings and arrays, consuming a significant amount of memory. This overhead, when added to a transactor class might not be significant, but could overwhelm a small transaction class, such as an ATM cell, which only has 53 bytes.

8.9.2   Singleton Class to Print Messages

An alternative to constructing all these print objects is to not construct any. As you saw in section 5.11.4, you could declare the methods in the `Print` class to be static. These methods can only reference static variables, as shown in Sample 8.39.

Sample 8.39   Static print class

```
class Print;
  static bit [31:0] error_count = 0, error_limit = -1;
  static function void error(input string ID,
                             input string message);
    $display("@%0t %m [%s] %s", $realtime, ID, message);
    error_count++;
    if (error_count >= error_limit) begin
      $display("Maximum error limit reached");
      $finish;
    end
  endfunction
endclass
```

Now that the class is static, you can no longer have per-instance information such as the parent class's name and instance. Any filtering has to be based on other criteria.

Sample 8.40 Transactor class with static print class

```
class Xactor;
  task run();
    Print::error("NYI", "This Xactor is not yet implemented");
  endtask
endclass
```

Sample 8.40 shows the call to the `error()` method using the `Print` class name.

This style of class is known as a singleton class, as there is only one copy, the one allocated at elaboration time with the static variables.

As your static classes, such as the one in Sample 8.39, grow larger, you have to label everything with the `static` keyword, a small annoyance. Next, the class is allocated before simulation time, even if you never use it. Additionally, there is no handle to this class, so you can not pass it around your testbench. The alternative to a static class is a singleton class (or singleton pattern) with a single instance, which is a non-static class that is only constructed once. They are more difficult to create initially, but they can simplify your program's architecture. Many of the UVM's classes are singletons.

The singleton pattern is implemented by creating a class with a method that creates a new instance of the class if one does not exist. If an instance already exists, it simply returns a handle to that object. To make sure that the object cannot be instantiated any other way, you must make the constructor `protected`. Don't make it `local`, because an extended class might need to access the constructor.

8.9.3 Configuration Database with Static Parameterized Class

Another good use for static classes in verification is a database of configuration parameters. At the start of simulation you randomize the configuration of your system. In a small system, you can simply store these in a single class or hierarchy of classes and pass them around the testbench as needed. At some point though, this becomes too complicated as handles are passed up and down the hierarchy. Instead, create a global database of parameters, indexed by a name, that you can access anywhere in the testbench. UVM 1.0 introduced this concept, which is the basis for the following set of examples. This code has a single string index into the database, while a real database such as UVM's could have a property name, instance name, and other values. You could concatenate these to create a more complex index string.

One issue with a database is that you need to store values of different types, such as bit vectors, integers, real numbers, enumerated values, string, class handles, virtual interfaces, and more in a single database. While you could find a few common types such as bit vectors and a common base class, there are some type such as virtual interfaces that are unique, so there is no easy way to store them in a common database. Earlier versions of OVM and UVM recommended creating a class wrapper around virtual interfaces, but this required extra coding and was a common source of bugs.

What if you made a different database for each data type? You could use an associative array indexed by the parameter name. A real database might also have an instance name, but for this simple example, you can just concatenate all the names together to make a single index. Sample 8.41 shows the code for an integer database made from global methods.

Sample 8.41 Configuration database with global methods

```
int db_int[string];
function void db_int_set(input string name, input int value);
  db_int[name] = value;
endfunction

function void db_int_get(input string name, ref int value);
  value = db_int[name];
endfunction

function void db_int_print();
  foreach (db_int[i])
    $display("db_int[%s] = %0d", i, db_int[i]);
endfunction
```

You can generalize this into a parameterized class with the concepts from Section 8.8, as shown in Sample 8.42.

Sample 8.42 Configuration database with parameterized class

```
class config_db #(type T=int);
  T db[string];
  function void set(input string name, input T value);
    db[name] = value;
  endfunction

  function void get(input string name, ref T value);
    value = db[name];
  endfunction

  function void print();
    $display("Configuration database %s", $typename(T));
    foreach (db[i])
      $display("db[%s] = %p", i, db[i]);
  endfunction
endclass
```

You can now construct objects for an integer database, a real database, etc. The final problem is that each instance of the database is local to the scope where this class is instantiated. The solution shown in Sample 8.43 is to go global and make this a static class, that is a class with static properties and methods.

Sample 8.43 Configuration database with static parameterized class

```
class config_db #(type T=int);
  static T db[string];
  static function void set(input string name, input T value);
    db[name] = value;
  endfunction

  static function void get(input string name, ref T value);
    value = db[name];
  endfunction

  static function void print();
    $display("\nConfiguration database %s", $typename(T));
    foreach (db[i])
      $display("db[%s] = %0p", i, db[i]);
  endfunction
endclass
```

You can test the above code with Sample 8.44 and see how the parameterized class creates a new database for each type.

Sample 8.44 Testbench for configuration database

```
class Tiny;
   int i;
endclass // Tiny

int i = 42, j = 43, k;                 // Integers for database
real pi = 22.0/7.0, r;                 // Reals for database
Tiny t;                                // Handle for database

initial begin
  config_db#(int)::set("i", i);        // Save an int in db
  config_db#(int)::set("j", j);        // Save another
  config_db#(real)::set("pi", pi);     // Save a real in db

  t = new();
  t.i = 8;
  config_db#(Tiny)::set("t", t);       // Save a handle in db
  config_db#(Tiny)::set("null", null); // Test null handles

  config_db#(int)::get("i", k);        // Fetch an int
  $display("Fetched value (%0d) of i (%0d) ", i, k);

  config_db#(int)::print();            // Print int db
  config_db#(real)::print();           // Print real db
  config_db#(Tiny)::print();           // Print Tiny db
end
```

With singletons implemented as single instances instead of static class members, you can initialize the singleton lazily, creating it only when it is needed.

The UVM database allows wildcards and other regular expressions, which requires a more complex lookup scheme than associative arrays.

8.10   Creating a Test Registry

In a real design, compiling your test and DUT takes a significant amount of time. If you want to run 100 tests, each in a separate program block, you need to recompile before each test, 100 times in all. This is a waste of CPU time as most of the code has not changed. If you make 100 program blocks, each with a single test, and connect all these programs in the model, you then need a way to disable all but one program block. The best solution is to include all tests and testbenches inside one program block, compile this once with the DUT. This section shows how you can select one test per run with a Verilog command line switch.

8.10.1   Test registry with Static Methods

Earlier examples in this book have a program that contains one test. For this new approach, each test is a separate class, all which are in a single program block, either imported from a package or included at compile time. The test classes are constructed, registered in a test registry, and then, at run time, you can choose the desired test at runtime. This follows an early VMM style.

First you need a base test class that your tests can extend from. Sample 8.45 shows an abstract class that contains a handle for the Environment class and a pure virtual task that is a placeholder for the method that contains your test code.

Sample 8.45   Base test class

```
virtual class TestBase;
  Environment env;
  pure virtual task run_test();
  function new();
    env = new();
  endfunction
endclass
```

The core of the test registry class is an associative array of handles to all the tests, indexed by the test name. The TestRegistry class, shown in Sample 8.46, is a static class with only static variables and methods, and is never constructed. The get_test() method reads the Verilog command line argument to determine which test to execute.

Sample 8.46   Test registry class

```
class TestRegistry;
  static TestBase registry[string];

  static function void register(string name, TestBase t);
    registry[name] = t;
  endfunction // register

  static function TestBase get_test();
    string name;
    if (!$value$plusargs("TESTNAME=%s", name))
      $display("ERROR: No +TESTNAME switch found");
    return registry[name];
  endfunction
endclass // TestRegistry
```

Sample 8.47 show how you can extend `TestBase` to create a simple test that
runs all the environment phases. The last line of the example is a declaration that
calls the constructor, which also registers the test. All the test objects are constructed,
but only one is run.

Sample 8.47   Simple test in a class

```
// Repeat for each test
class TestSimple extends TestBase;

  function new();
    env = new();
    TestRegistry::register("TestSimple", this);
  endfunction

  virtual task run_test();
    $display("%m");
    env.gen_config();
    env.build();
    env.run();
    env.wrap_up();
  endtask
endclass

TestSimple TestSimple_handle = new();  // Needed for each class
```

The program in Sample 8.48 now just asks the test registry for a test object and
runs it. The test classes can be declared in a package and imported, or declared
inside or outside the program block.

Sample 8.48 Program block for test classes

```
program automatic test;
  TestBase tb;
  initial begin
    tb = TestRegistry::get_test();
    tb.run_test();
  end
endprogram
```

Sample 8.49 shows how you can create a test class that injects new behavior by changing the generator's blueprint to create bad transactions.

Sample 8.49 Test class that puts a bad transaction in the generator

```
class TestBad extends TestBase;
  function new();
    env = new();
    TestRegistry::register("TestBad", this);
  endfunction // new

  virtual task run_test();
    $display("%m");
    env.gen_config();
    env.build();
    begin
      BadTr bad = new();
      env.gen.blueprint = bad;
    end
    env.run();
    env.wrap_up();
  endtask
endclass

TestBad TestBad_handle = new(); // Declaration & constructing
```

This short example allows you to compile many tests into a single simulation executable and choose your test at runtime, saving many recompiles. This pattern is fine when you are starting out with a handful of tests, but the next section shows more powerful approach.

8.10.2   Test Registry with a Proxy Class

The previous section's test registry works well for smaller test environments, but has several limitations for real projects. First, you need to remember to construct every test class, otherwise the registry can not locate it. Second, every test gets constructed at the start of simulation, even though only one is actually run.

When verifying a large design, there could be hundreds of tests, so constructing all of them wastes valuable simulation time and memory.

Consider this analogy. When you are looking to buy a car, you can go to a dealer to see the choices. If there are only a few variants, white or black, with or without sunroof, the dealer can stock one of each model with little overhead. This is what you saw in the previous section, where the test registry had an object of each test type.

What if there are many different models, each in one of a dozen colors, with variants such as radios, sunroofs, air conditioning, sports packages, and engines? The dealer could never have one of each type on his lot as there are hundreds of combinations. Instead he would show you a catalog with all the choices. You pick the options that you want, and the factory builds one to your specification. Likewise, the test registry can have a lot of small classes, each which knows how to build a complete test. The small class has low overhead, so even a thousand objects would not consume much memory. Now when you want to run test N, imagine flipping through the catalog (test registry) until you find a picture of your test, and you then tell the factory to build an object of that type.

The test registry needs a table (analogous to the above catalog) that goes from test names to objects. In section 8.10.1, this table is an associative array of `TestBase` handles, indexed by a string, shown in Sample 8.46. What if instead, you had a parameterized class whose only job is to construct a test? The UVM uses a design pattern called a proxy class whose only role is to build the actual desired class. The proxy class is lightweight in that it only contains a few properties and methods, and thus consumes little memory or CPU time. It acts like the picture in the car dealer's catalog, holding a representation of what you can build.

The next few code samples show how the UVM class factory works. Because the code in this book is a simplified version of the real UVM classes, the name has been changed to SVM, SystemVerilog Methodology, so that you won't confuse it with the real thing. Hopefully you will find this explanation of a simple factory easier to understand than trying to read the UVM source code.

First is Sample 8.50 which has the common base class from which everything else is built. It is a abstract class because you should never construct an object of this type, only classes extended from this one.

Sample 8.50 Common SVM base class

```
virtual class svm_object;
  // Empty class
endclass
```

Next is the component class in Sample 8.51. In the UVM, a component is a time-consuming object that forms the testbench hierarchy, similar to a VMM transactor. In this simplified example, the hierarchical parent handle has been removed.

Sample 8.51 Component class

```
virtual class svm_component extends svm_object;
  protected svm_component m_children[string];
  string name;

  function new(string name);
    this.name = name;
    $display("%m name='%s'", name);
  endfunction

  pure virtual task run_test();
endclass
```

Now define `svm_object_wrapper`, the abstract common base class for the proxy class as shown in Sample 8.52. It has pure virtual methods to return the name of the class type, and create an object of this type.

Sample 8.52 Common base class for proxy class

```
virtual class svm_object_wrapper;
  pure virtual function string get_type_name();
  pure virtual function svm_object create_object(string name);
endclass
```

Now for the crucial class, **svm_component_registry** shown in Sample 8.53. This is a lightweight class that can be constructed with little overhead. It is parameterized with the test class type and name. Once you have an instance of this class, your testbench can construct the actual test class at any time, using the **create_object** method. This is a singleton class as you only need one copy to create an instance of the test class. At the start of simulation, the static handle **me** is initialized by calling the **get()** method that constructs the first instance if needed.

Sample 8.53 Parameterized proxy class

```
class svm_component_registry #(type T=svm_component,
                               string Tname="<unknown>")
      extends svm_object_wrapper;

  typedef svm_component_registry #(T,Tname) this_type;

  virtual function string get_type_name();
    return Tname;
  endfunction

  local static this_type me = get();    // Handle to singleton

  static function this_type get();
    if (me == null) begin                 // Is there an instance?
      svm_factory f = svm_factory::get(); // Build factory
      me = new();                         // Build the singleton
      f.register(me);                     // Register class
    end
    return me;
  endfunction

  virtual function svm_object create_object (string name="");
    T obj;
    obj = new(name);
    return obj;
  endfunction

  static function T create(string name);
    create = new(name);
  endfunction

endclass : svm_component_registry
```

The last major class is **svm_factory**, which, at its core, is just a singleton class that holds the array, **m_type_names**, to go from test case name to the proxy class that creates an instance of the test class. Also in this class in Sample 8.54 is the get_test method that reads the test name from the simulation run command line and constructs an instance of the test class. Unlike Sample 8.46, you even get a little self checking.

Sample 8.54 Factory class

```
class svm_factory;
  // Assoc array from string to svm_object_wrapper handle
  static svm_object_wrapper m_type_names[string];

  static svm_factory m_inst;   // Handle to this singleton

  static function svm_factory get();
    if (m_inst == null) m_inst = new();
    return m_inst;
  endfunction

  static function void register(svm_object_wrapper c);
    m_type_names[c.get_type_name()] = c;
  endfunction

  static function svm_component get_test();
    string name;
    svm_object_wrapper test_wrapper;
    svm_component test_comp;

    if (!$value$plusargs("SVM_TESTNAME=%s", name)) begin
      $display("FATAL +SVM_TESTNAME not found");
      $finish;
    end
    $display("%m found +SVM_TESTNAME=%s", name);
    test_wrapper = svm_factory::m_type_names[name];
    $cast (test_comp, test_wrapper.create_object(name));
    return test_comp;
  endfunction
endclass : svm_factory
```

Lastly is a base test class, extended from svm_component shown in Sample 8.55. It uses the macro **svm_component_utils** to define a new data type, type_id, that points to the proxy class. The macro stringifies the token T that holds the class name, and turns it into a string containing the value of T with the syntax: `` `"T`" ``.

Sample 8.55 Base test class and registration macro

```
`define svm_component_utils(T) \
  typedef svm_component_registry #(T,`"T`") type_id; \
  virtual function string get_type_name (); \
    return `"T`"; \
  endfunction

class TestBase extends svm_component;
  Environment env;
  `svm_component_utils(TestBase)

  function new(string name);
    super.new(name);
    $display("%m");
    env = new();
  endfunction

  virtual task run_test();
  endtask
endclass : TestBase
```

Sample 8.56 Test program

```
program automatic test;
  initial begin
    svm_component test_obj;
    test_obj = svm_factory::get_test();
    test_obj.run_test();
  end
endprogram
```

Here are the steps that happen when you start a simulation with the command line switch +SVM_TESTNAME=TestBase.

- With the macro svm_component_utils, the class TestBase defines the type type_id based on the class svm_component_registry, with the parameters TestBase and "TestBase". Because this is a new type, the simulator initializes the static variable svm_component_registry::me by calling the get method that instantiates the class. This instance is registered in the factory. What does all this mean? There is now an object that can construct the TestBase class, and you can get to it through the factory.
- Simulation now starts and the factory's get_test method reads the test name from the command line. This string is used an index into the registry to get a handle to the proxy object. This object's create_object method constructs an instance of the TestBase object.

- The program calls the test object's run_test method, which calls the steps for the specific class. Now the TestBase class in Sample 8.55 does not do anything interesting, but add a call to svm_component_utils macro to the test classes in Sample 8.47 and Sample 8.49 and you can run tests.

Now you can see the basic UVM flow to start tests. The registry contains a list of proxy classes that can construct test objects.

8.10.3   UVM Factory Build

The UVM factory can also construct objects for any class in the testbench with the create method in Sample 8.53. Sample 8.57 show how to build a driver.

Sample 8.57 UVM factory build example

```
driver drv;
drv = driver::type_id::create("drv", this);
```

The above code calls the static method create to construct an object of type driver. In UVM, the second argument points to the parent of the component being created.

The UVM factory allows you to override the component so that when you build a component, you get an extended one instead.

You may have noticed a change in terminology. In classic OOP, you "construct" a class by calling the new method, based on the handle type and assigning the address to the handle on the left side of the assignment statement. With the UVM factory pattern, you "build" an object by calling the static create method. This could make an object of the same type as the handle, or an extended type.

8.11   Conclusion

The software concept of inheritance, where new functionality is added to an existing class, parallels the hardware practice of extending the design's features for each generation, while still maintaining backwards compatibility.

For example, you can upgrade your PC by adding a larger capacity disk. As long as it uses the same interface as the old one, you do not have to replace any other part of the system, yet the overall functionality is improved.

Likewise, you can create a new test by "upgrading" the existing driver class to inject errors. If you use an existing callback in the driver, you do not have to change any of the testbench infrastructure.

You need to plan ahead if you want use these OOP techniques. By using virtual methods and providing sufficient callback points, your test can modify the behavior of the testbench without changing its code. The result is a robust testbench that does not need to anticipate every type of disturbance (error-injection, delays, synchronization) that you may want as long as you leave a hook where the test can inject its own behavior.

The testbench is more complex than what you have previously constructed, but there is a payback in that the tests become smaller and easier to write. The testbench does the hard work of sending stimulus and checking responses, so the test only has to make small tweaks to cause specialized behavior. An extra few lines of testbench code might replace code that would have to be repeated in every single test.

Lastly, OOP techniques improve your productivity by allowing you to reuse classes. For example, a parameterized class for a stack that operates on any other class, rather than a single type, saves you from having to create duplicate code.

8.12   Exercises

1. Given the following class, create a method in an extended class ExtBinary that multiplies val1 and val2 and returns an integer.

```
class Binary;
  rand bit [3:0] val1, val2;

  function new(input bit [3:0] val1, val2);
    this.val1 = val1;
    this.val2 = val2;
  endfunction

  virtual function void print_int(input int val);
    $display("val=0d%0d", val);
  endfunction

endclass
```

2. Starting with the solution to Exercise 1, use the ExtBinary class to initialize val1=15, val2=8, and print out the multiplied value.

3. Starting with the solution to Exercise 1, create an extended class Exercise3 that constrains val1 and val2 to be less than 10.

4. Starting with the solution to Exercise 3, use the Exercise3 class to randomize val1 and val2, and print out the multiplied value.

5. Given the class in Exercise 1, and the following declarations, and an extended class `ExtBinary`, what will handles mc, mc2, and b point to after executing each code snippet a-d, or will a compile error occur?

```
Binary b;
ExtBinary mc, mc2;
```

a. `mc = new(15,8);`
   `b = mc;`
b. `b = new(15, 8);`
   `mc = b;`
c. `mc = new(15, 8);`
   `b = mc;`
   `mc2 = b;`
d. `mc = new(15, 8);`
   `b = mc;`
   `if($cast(mc2, b))`
      `$display("Success");`
   `else`
      `$display("Error: cannot assign");`

6. Given the classes `Binary` and `ExtBinary` in Exercise 1 and the following copy function for class `Binary`, create the function `ExtBinary::copy`.

```
virtual function Binary Binary::copy();
   copy = new(15,8);
   copy.val1 = val1;
   copy.val2 = val2;
endfunction
```

7. From the solution to Exercise 6, use the copy function to copy the object pointed to by the extended class handle mc to the extended class handle mc2.

8. Using code Sample 8.26 to Sample 8.28 in Section 8.7.1 and 8.7.2 of the text, add the ability to randomly delay a transaction between 0 and 100ns.

9. Create a class that can compare any data type using the case equality operators, `===` and `!==`. It contains a compare function that returns a 1 if the two values match, 0 otherwise. By default it compares two 4-bit data types.

10. Using the solution from Exercise 9, use the comparator class to compare two 4-bit values, expected_4bit and actual_4bit. Next, compare two values of type color_t, expected_color and actual_color. Increment an error counter if an error occurs.

Chapter 9
Functional Coverage

As designs become more complex, the only effective way to verify them effectively is with constrained-random testing (CRT). This approach elevates you above the tedium of writing individual directed tests, one for each feature in the design. However, if your testbench is taking a random walk through the space of all design states, how do you know if you have reached your destination? Even directed tests should be double checked with functional coverage. Over the life of a project, small changes in the DUT's timing or functionality can subtly alter the results from a directed test, so it no longer verifies the same features. Whether you are using random or directed stimulus, you can gauge progress using coverage.

Functional coverage is a measure of which design features have been exercised by the tests. Start with the design specification and create a verification plan with a detailed list of what to test and how. For example, if your design connects to a bus, your tests need to exercise all the possible interactions between the design and bus, including relevant design states, delays, and error modes. The verification plan is a map to show you where to go. For more information on creating a verification plan, see Bergeron (2006).

In many complex systems, you may never achieve 100% coverage as schedules don't allow you to reach every possible corner case. After all, you didn't have time to write directed tests to get sufficient coverage, and even CRT is limited by the time it takes you to create and debug test cases, and analyze the results.

Figure 9.1 shows the feedback loop to analyze the coverage results and decide on which actions to take in order to converge on 100% coverage. Your first choice is to run existing tests with more seeds; the second is to build new constraints. Only resort to creating directed tests if absolutely necessary.

Back when you exclusively wrote directed tests, the verification planning was limited. If the design specification listed 100 features, all you had to do was write 100 tests. Coverage was implicit in the tests — the "register move" test moved all combinations of registers back and forth. Measuring progress was easy: if you had completed 50 tests, you were halfway done. This chapter uses "explicit" and "implicit" to describe how coverage is specified. Explicit coverage is described

C. Spear and G. Tumbush, *SystemVerilog for Verification: A Guide to Learning the Testbench Language Features*, DOI 10.1007/978-1-4614-0715-7_9, © Springer Science+Business Media, LLC 2012

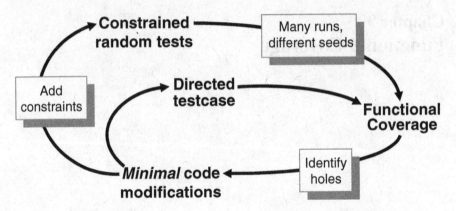

Fig. 9.1 Coverage convergence

directly in the test environment using SystemVerilog features. Implicit coverage is implied by a test — when the "register move" directed test passes, you have hopefully covered all register transactions.

With CRT, you are freed from hand crafting every line of input stimulus, but now you need to write code that tracks the effectiveness of the test with respect to the verification plan. You are still more productive, as you are working at a higher level of abstraction. You have moved from tweaking individual bits to describing the interesting design states. Reaching for 100% functional coverage forces you to think more about what you want to observe and how you can direct the design into those states.

9.1 Gathering Coverage Data

You can run the same random testbench over and over, simply by changing the random seed to generate new stimulus. Each individual simulation generates a database of functional coverage information, the trail of footprints from the random walk. You can then merge all this information together to measure your overall progress using functional coverage as shown in Figure 9.2.

You then analyze the coverage data to decide how to modify your tests. If the coverage levels are steadily growing, you may just need to run existing tests with new random seeds, or even just run longer tests. If the coverage growth has started to slow, you can add additional constraints to generate more "interesting" stimuli. When you reach a plateau, some parts of the design are not being exercised, so you need to create more tests. Lastly, when your functional coverage values near 100%, check the bug rate. If bugs are still being found, you may not be measuring true coverage for some areas of your design. Don't be in too big of a rush to reach 100% coverage, which just shows that you looked for bugs in all the usual places. While you are trying to verify your design, take many random walks through the stimulus space; this can create many unanticipated combinations, as shown in van der Schoot (2007).

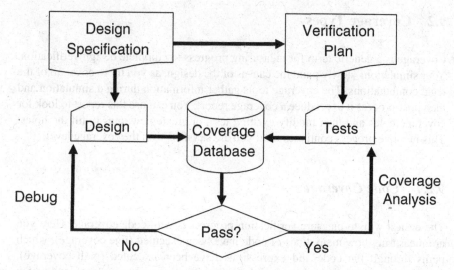

Fig. 9.2 Coverage flow

Each simulation vendor has its own format for storing coverage data and as well as its own analysis tools. You need to perform the following actions with those tools.

- **Run a test with multiple seeds**. For a given set of constraints and coverage groups, compile the testbench and design into a single executable. Now you need to run this constraint set over and over with different random seeds. You can use the Unix system clock as a seed, but be careful, as your batch system may start multiple jobs simultaneously. These jobs may run on different servers or may start on a single server with multiple processors. So combine all these values to make a truly unique seeds. The seed must be saved with the simulation and coverage results for repeatability.
- **Check for pass/fail**. Functional coverage information is only valid for a successful simulation. When a simulation fails because there is a design bug, the coverage information must be discarded. The coverage data measures how many items in the verification plan are complete, and this plan is based on the design specification. If the design does not match the specification, the coverage values are useless. Some verification teams periodically measure all functional coverage from scratch so that it reflects the current state of the design.
- **Analyze coverage across multiple runs**. You need to measure how successful each constraint set is, over time. If you are not yet getting 100% coverage for the areas that are targeted by the constraints, but the amount is still growing, run more seeds. If the coverage level has plateaued, with no recent progress, it is time to modify the constraints. Only if you think that reaching the last few test cases for one particular section may take too long for constrained-random simulation should you consider writing a directed test. Even then, continue to use random stimulus for the other sections of the design, in case this "background noise" finds a bug.

9.2   Coverage Types

Coverage is a generic term for measuring progress to complete design verification. Your simulations slowly paint the canvas of the design, as you try to cover all of the legal combinations. The coverage tools gather information during a simulation and then post-process it to produce a coverage report. You can use this report to look for coverage holes and then modify existing tests or create new ones to fill the holes. This iterative process continues until you are satisfied with the coverage level.

9.2.1   Code Coverage

The easiest way to measure verification progress is with code coverage. Here you are measuring how many lines of code have been executed (line coverage), which paths through the code and expressions have been executed (path coverage), which single-bit variables have had the values 0 or 1 (toggle coverage), and which states and transitions in a state machine have been visited (FSM coverage). You don't have to write any extra HDL code. The tool instruments your design automatically by analyzing the source code and adding hidden code to gather statistics. You then run all your tests, and the code coverage tool creates a database.

Most simulators include a code coverage tool. A post-processing tool converts the database into a readable form. The end result is a measure of how much your tests exercise the design code. Note that you are primarily concerned with analyzing the design code, not the testbench. Untested design code could conceal a hardware bug, or may be just redundant code.

Code coverage measures how thoroughly your tests exercised the "implementation" of the design specification, but not the verification plan. Just because your tests have reached 100% code coverage, your job is not done. What if you made a mistake that your test didn't catch? Worse yet, what if your implementation is missing a feature? The module in Sample 9.1 is for a D-flip flop. Can you see the mistake?

Sample 9.1 Incomplete D-flip flop model missing a path

```
module dff(output logic q, q_1,
           input  logic clk, d, reset_1);

  always @(posedge clk or negedge reset_1) begin
    q <= d;
    q_1 <= !d;
  end
endmodule
```

The reset logic was accidently left out. A code coverage tool would report that every line had been exercised, yet the model was not implemented correctly. Go back to the functional specification that describes reset behavior and make sure your verification plan includes a requirement to verify this. Then gather functional coverage information on the design during reset.

9.2.2 Functional Coverage

The goal of verification is to ensure that a design behaves correctly in its real environment, be that an MP3 player, network router, or cell phone. The design specification details how the device should operate, whereas the verification plan lists how that functionality is to be stimulated, verified, and measured. When you gather measurements on what functions were covered, you are performing "design" coverage. For example, the verification plan for a D-flip flop would mention not only its data storage but also how it resets to a known state. Until your test checks both these design features, you will not have 100% functional coverage.

Functional coverage is tied to the design intent and is sometimes called "specification coverage," while code coverage measures how well you have tested the RTL code and is known as, "implementation coverage." These are two very different metrics. Consider what happens if a block of code is missing from the design. Code coverage cannot catch this mistake and could report that you have executed 100% of the lines, but functional coverage will show that the functionality does not exist.

9.2.3 Bug Rate

An indirect way to measure coverage is to look at the rate at which fresh bugs are found, show in the graph in Fig. 9.3. You should keep track of how many bugs you found each week, over the life of a project. At the start, you may find many bugs through inspection as you create the testbench. As you read the design spec, you may find inconsistencies, which hopefully are fixed before the RTL is written. Once the testbench is up and running, a torrent of bugs is found as you check each module in the system. The bug rate drops, hopefully to zero, as the design nears tape-out. However, you are not yet done. Every time the rate sags, it is time to find different ways to create corner cases.

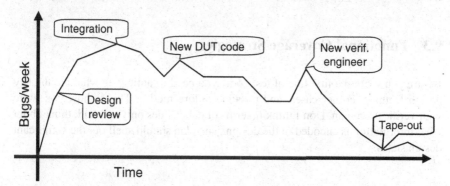

Fig. 9.3 Bug rate during a project

The bug rate can vary per week based on many factors such as project phases, recent design changes, blocks being integrated, personnel changes, and even vacation schedules. Unexpected changes in the rate could signal a potential problem.

As shown in Fig. 9.3, it is not uncommon to keep finding bugs even after tape-out, and even after the design ships to customers.

9.2.4 Assertion Coverage

Assertions are pieces of declarative code that check the relationships between design signals, either once or over a period of time. These can be simulated along with the design and testbench, or proven by formal tools. Sometimes you can write the equivalent check using SystemVerilog procedural code, but many assertions are more easily expressed using SystemVerilog Assertions (SVA).

Assertions can have local variables and perform simple data checking. If you need to check a more complex protocol, such as determining whether a packet successfully went through a router, procedural code is often better suited for the job. There is a large overlap between sequences that are coded procedurally or using SVA. See Vijayaraghavan and Ramanadhan (2005), Cohen et al. (2005), and Chapters 3 and 7 in the VMM book, Bergeron et al. (2006) for more information on SVA.

The most familiar assertions look for errors such as two signals that should be mutually exclusive or a request that was never followed by a grant. These error checks should stop the simulation as soon as they detect a problem. Assertions can also check arbitration algorithms, FIFOs, and other hardware. These are coded with the `assert property` statement.

Some assertions might look for interesting signal values or design states, such as a successful bus transaction. These are coded with the `cover property` statement. You can measure how often these assertions are triggered during a test by using assertion coverage. A cover property observes sequences of signals, whereas a cover group (described below) samples data values and transactions during the simulation. These two constructs overlap in that a cover group can trigger when a sequence completes. Additionally, a sequence can collect information that can be used by a cover group.

9.3 Functional Coverage Strategies

Before you write the first line of test code, you need to anticipate what are the key design features, corner cases, and possible failure modes. This is how you write your verification plan. Don't think in terms of data values only; instead, think about what information is encoded in the design. The plan should spell out the significant design states.

9.3.1 Gather Information, not Data

A classic example is a FIFO. How can you be sure you have thoroughly tested a 1K FIFO memory? You could measure the values in the read and write indices, but there are over a million possible combinations. Even if you were able to simulate that many cycles, you would not want to read the coverage report.

At a more abstract level, a FIFO can hold from 0 to N–1 possible values. So what if you just compare the read and write indices to measure how full or empty the FIFO is? You would still have 1K coverage values. If your testbench pushed 100 entries into the FIFO, then pushed 100 more, do you really need to know if the FIFO ever had 150 values? Not as long as you can successfully read out all values.

The corner cases for a FIFO are Full and Empty. If you can make the FIFO go from Empty (the state after reset) through Full and back down to Empty, you have covered all the levels in between. Other interesting states involve the indices as they pass between all 1's and all 0's. A coverage report for these cases is easy to understand.

You may have noticed that the interesting states are independent of the FIFO size. Once again, look at the information, not the data values.

Design signals with a large range (more than a few dozen possible values) should be broken down into smaller ranges, plus corner cases. For example, your DUT may have a 32-bit address bus, but you certainly don't need to collect 4 billion samples. Check for natural divisions such as memory and IO space. For a counter, pick a few interesting values, and always try to rollover counter values from all 1's back to 0.

9.3.2 Only Measure What you are Going to Use

Gathering functional coverage data can be expensive, so only measure what you will analyze and use to improve your tests. Your simulations may run slower as the simulator monitors signals for functional coverage, but this approach has lower overhead than gathering waveform traces and measuring code coverage. Once a simulation completes, the database is saved to disk. With multiple testcases and multiple seeds, you can fill disk drives with functional coverage data and reports. But if you never look at the final coverage reports, don't perform the initial measurements.

There are several ways to control cover data: at compilation, instantiation, or triggering. You could use switches provided by the simulation vendor, conditional compilation, or suppression of the gathering of coverage data. The last of these is less desirable because the post-processing report is filled with sections with 0% coverage, making it harder to find the few enabled ones.

9.3.3 Measuring Completeness

Like your kids in the backseat on a family vacation, your manager constantly asks you, "Are we there yet?" How can you tell if you have fully tested a design? You need to look at all coverage measurements and consider the bug rate to see if you have reached your destination.

At the start of a project, both code and functional coverage are low. As you develop tests, run them over and over with different random seeds until you no longer see increasing values of functional coverage. Create additional constraints and tests to explore new areas. Save test/seed combinations that give high coverage, so that you can use them in regression testing.

Fig. 9.4 Coverage comparison

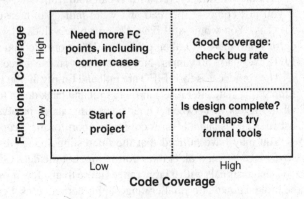

What if the functional coverage is high but the code coverage is low as shown in the upper left of Figure 9.4? Your tests are not exercising the full design, perhaps from an inadequate verification plan. It may be time to go back to the hardware specifications and update your verification plan. Then you need to add more functional coverage points to locate untested functionality.

A more difficult situation is high code coverage but low functional coverage. Even though your testbench is giving the design a good workout, you are unable to put it in all the interesting states. First, see if the design implements all the specified functionality. If it is there, but your tests can't reach it, you might need a formal verification tool that can extract the design's states and create appropriate stimulus.

The goal is both high code and functional coverage. However, don't plan your vacation yet. What is the trend of the bug rate? Are significant bugs still popping up?

Worse yet, are they being found deliberately, or did your testbench happen to stumble across a particular combination of states that no one had anticipated? On the other hand, a low bug rate may mean that your existing strategies have run out of steam, and you should look into different approaches. Try different approaches such as new combinations of design blocks and error generators.

9.4   Simple Functional Coverage Example

To measure functional coverage, you begin with the verification plan and write an executable version of it for simulation. In your System Verilog testbench, sample the values of variables and expressions. These sample locations are known as cover points. Multiple cover points that are sampled at the same time (such as when a transaction completes) are placed together in a cover group.

Sample 9.2 has a transaction that comes in eight flavors. The testbench generates the dst variable randomly, and the verification plan requires that every value be tried.

Sample 9.2   Functional coverage of a simple object

```
program automatic test(busifc.TB ifc);

  class Transaction;
    rand bit [31:0] data;
    rand bit [ 2:0] dst;             // Eight dst port numbers
  endclass

  Transaction tr;                    // Transaction to be sampled

  covergroup CovDst2;
    coverpoint tr.dst;               // Measure coverage
  endgroup

  initial begin
    CovDst2 ck;
    ck = new();                      // Instantiate group
    repeat (32) begin                // Run a few cycles
      @ifc.cb;                       // Wait a cycle
      tr = new();
      `SV_RAND_CHECK(tr.randomize);  // Create a transaction
      ifc.cb.dst  <= tr.dst;         //    and transmit
      ifc.cb.data <= tr.data;        //    onto interface
      ck.sample();                   // Gather coverage
    end
  end
endprogram
```

Sample 9.2 creates a random transaction and drives it out to an interface. The testbench samples the value of the dst field using the CovDst2 cover group. Eight possible values, 32 random transactions — did your testbench generate them all? Samples 9.3 and 9.4 have part of a coverage report from VCS. Because of randomization, every simulator will give different results.

As you can see, the testbench generated dst values of 1, 2, 3, 4, 5, 6, and 7, but never generated a 0. The at least column specifies how many hits are needed before a bin is considered covered. See Section 9.10.3 for the at_least option.

To improve your functional coverage, the easiest strategies are to run more simulation cycles, or to try new random seeds. For Sample 9.2, the very next random transaction (#33) has a dst value of 0, giving 100% coverage. Or, if you started simulation with a different seed, you may reach 100% in fewer transactions, for this trivial case. On a real design, you may see a plateau in coverage, with most coverage points getting hit more and more, but a few stubborn points that are never hit, no matter how long you run, regardless of seed values. In this case, you probably have to try a new strategy, as the testbench is not creating the proper stimulus. The most important part of any coverage report are the points with 0 hits.

Sample 9.3 Coverage report for a simple object

```
Coverpoint Coverage report
CoverageGroup: CovDst2
    Coverpoint: tr.dst
Summary
    Coverage: 87.50
    Goal: 100
    Number of Expected auto-bins: 8
    Number of User Defined Bins: 0
    Number of Automatically Generated Bins: 7
    Number of User Defined Transitions: 0

    Automatically Generated Bins

    Bin             # hits     at least
    =================================
    auto[1]           7           1
    auto[2]           7           1
    auto[3]           1           1
    auto[4]           5           1
    auto[5]           4           1
    auto[6]           2           1
    auto[7]           6           1
    =================================
```

Sample 9.4 Coverage report for a simple object, 100% coverage

```
Coverpoint Coverage report
CoverageGroup: CovDst2
    Coverpoint: tr.dst
Summary
    Coverage: 100
    Goal: 100
    Number of Expected auto-bins: 8
    Number of User Defined Bins: 0
    Number of Automatically Generated Bins: 8
    Number of User Defined Transitions: 0

    Automatically Generated Bins

    Bin             # hits     at least
    =================================
    auto[0]           1           1
    auto[1]           7           1
    auto[2]           7           1
    auto[3]           1           1
    auto[4]           5           1
    auto[5]           4           1
    auto[6]           2           1
    auto[7]           6           1
    =================================
```

This book gives a rough explanation of how coverage is calculated. The LRM has a very detailed explanation of coverage computation across four pages, with more details across an entire chapter. Consult it for the most accurate details.

9.5 Anatomy of a Cover Group

A cover group is similar to a class — you define it once and then instantiate it one or more times. It contains cover points, options, formal arguments, and an optional trigger. A cover group encompasses one or more data points, all of which are sampled at the same time.

You should create very clear cover group names that explicitly indicate what you are measuring and, if possible, reference to the verification plan. The name `Parity_ Errors_In_Hexaword_Cache_Fills` may seem verbose, but when you are trying to read a coverage report that has dozens of cover groups, you will appreciate the extra detail. You can also use the comment option for additional descriptive information, as shown in Section 9.9.2.

A cover group can be defined in a class or at the program or module level. It can sample any visible variable such as program/module variables, signals from an interface, or any signal in the design (using a hierarchical reference). A cover group inside a class can sample variables in that class, as well as data values from embedded objects.

Don't define the cover group in a data class, such as a transaction, as doing so can cause additional overhead when gathering coverage data. Imagine you are trying to track how many beers were consumed by patrons in a pub. Would you try to follow every bottle as it flowed from the loading dock, over the bar, and into each person? No, instead you could just have each patron check off the type and number of beers consumed, as shown in van der Schoot (2006).

In SystemVerilog, you should define cover groups at the appropriate level of abstraction. This level can be at the boundary between your testbench and the design, in the transactors that read and write data, in the environment configuration class, or wherever is needed. The sampling of any transaction must wait until it is actually received by the DUT. If you inject an error in the middle of a transaction, causing it to be aborted in transmission, you need to change how you treat it for functional coverage. You need to use a different cover point that has been created just for error handling.

A class can contain multiple cover groups. This approach allows you to have separate groups that can be enabled and disabled as needed. Additionally, each group may have a separate trigger, allowing you to gather data from many sources.

A cover group must be instantiated for it to collect data. If you forget, no error message about null handles is printed at run time, but the coverage report will not contain any mention of the cover group. This rule applies for cover groups defined either inside or outside of classes.

9.5.1   Defining a Cover Group in a Class

A cover group can be defined in a program, module, or class. In all cases, you must explicitly instantiate it to start sampling. If the cover group is defined in a class, it is known as an embedded covergroup. In this case, you do not make a separate name when you construct it; just use the original cover group name. You must construct an embedded covergroup in the class's constructor, as opposed to a non-embedded cover group that can be constructed at any time.

Sample 9.5 is very similar to the first example of this chapter except that it embeds a cover group in a transactor class, and thus does not need a separate instance name.

Sample 9.5 Functional coverage inside a class

```
class Transactor;
  Transaction tr;
  mailbox #(Transaction) mbx;
  covergroup CovDst5;
    coverpoint tr.dst;
  endgroup

  function new(input mailbox #(Transaction) mbx);
    CovDst5 = new();              // Instantiate covergroup
    this.mbx = mbx;
  endfunction

  task run();
    forever begin
      mbx.get(tr);                // Get next transaction
      @ifc.cb;
      ifc.cb.dst <= tr.dst;       // Send into DUT
      ifc.cb.data <= tr.data;
      CovDst5.sample();           // Gather coverage
    end
  endtask

endclass
```

9.6   Triggering a Cover Group

The two major parts of functional coverage are the sampled data values and the time when they are sampled. When new values are ready (such as when a transaction has completed), your testbench triggers the cover group. This can be done directly with the `sample` function, as shown in Sample 9.5, or by using a coverage event in the `covergroup` definition. The coverage event can use a @ to block on signals or events.

Use `sample` if you want to explicitly trigger coverage from procedural code, if there is no existing signal or event that tells when to sample, or if there are multiple instances of a cover group that trigger separately.

Use the coverage event in the `covergroup` declaration if you want to tap into existing events or signals to trigger coverage.

9.6.1   Sampling Using a Callback

One of the better ways to integrate functional coverage into your testbench is to use callbacks, as originally shown in Section 8.7. This technique allows you to build a flexible testbench without restricting when coverage is collected. You can decide for every point in the verification plan where and when values are sampled. And if you need an extra "hook" in the environment for a callback, you can always add one in an unobtrusive manner, as a callback only "fires" during simulations when the test registers a callback object. You can create many separate callbacks for each cover group, with little overhead. As explained in Section 8.7.4, callbacks are superior to using a mailbox to connect the testbench to the coverage objects. You might need multiple mailboxes to collect transactions from different points in your testbench. A mailbox requires a transactor to receive transactions, and multiple mailboxes cause you to juggle multiple threads. Instead of an active transactor, use a passive callback.

Sample 8.26–8.28 shows a driver class that has two callback points, before and after the transaction is transmitted. Sample 8.26 shows the base callback class, and Sample 8.28 has a test with an extended callback class that sends data to a scoreboard. Make your own extension, `Driver_cbs_coverage`, of the base callback class, `Driver_cbs`, to call the sample task for your cover group in `post_tx`. Push an instance of the coverage callback class into the driver's callback queue, and your coverage code triggers the cover group at the right time. Samples 9.6 and 9.7 define and use the callback `Driver_cbs_coverage`.

Sample 9.6 Test using functional coverage callback

```
program automatic test;
  Environment env;

  initial begin
    Driver_cbs_coverage dcc;

    env = new();
    env.gen_cfg();
    env.build();

    // Create and register the coverage callback
    dcc = new();
    env.drv.cbs.push_back(dcc); // Put into driver's Q

    env.run();
    env.wrap_up();
  end

endprogram
```

The UVM recommends gathering coverage by monitoring the DUT and sending transactions to a coverage component through an analysis port, similar to a mailbox.

Sample 9.7 Callback for functional coverage

```
class Driver_cbs_coverage extends Driver_cbs;
  covergroup CovDst7;
    ...
  endgroup

  virtual task post_tx(ref Transaction tr);
    CovDst7.sample();              // Sample coverage values
  endtask
endclass
```

9.6.2 Cover Group with a User Defined Sample Argument List

In Sample 9.5, the cover group samples a variable in transaction object that is defined inside the class. If your cover group is defined outside of a class, you can pass variables through the sample method by defining your own argument list. Now you can sample variables from anywhere in the testbench.

In Sample 9.8, the cover group is expanded to also cover the low data bit. The last statement of the run method passes the destination address and also the configuration variable for high speed mode.

Sample 9.8 Defining an argument list to the sample method

```
covergroup CovDst8 with function sample(bit [2:0] dst, bit hs);
   coverpoint dst;
   coverpoint hs;                          // High speed mode
endgroup

class Transactor;
   CovDst8 covdst;
   task run();
    forever begin
       mbx.get(tr);                        // Get next transaction
       ifc.cb.dst  <= tr.dst;              // Send into DUT
       ifc.cb.data <= tr.data;
       covdst.sample(tr.dst, high_speed);  // Gather coverage
     end
   endtask
endclass
```

9.6.3   Cover Group with an Event Trigger

In Sample 9.9, the cover group `CovDst9` is sampled when the testbench triggers the `trans_ready` event.

Sample 9.9 Cover group with a trigger

```
event trans_ready;
covergroup CovDst9 @(trans_ready);
    coverpoint ifc.cb.dst;   // Measure coverage
endgroup
```

The advantage of using an event over calling the `sample` method directly is that you may be able to use an existing event such as one triggered by an assertion, as shown in Sample 9.11.

9.6.4   Triggering on a System Verilog Assertion

If you already have an SVA that looks for useful events like a complete transaction, you can add an event trigger to wake up the cover group as shown in 9.10.

Sample 9.10 Module with SystemVerilog Assertion

```
module mem(simple_bus sb);
  bit [7:0] data, addr;
  event write_event;

  cover property
   (@(posedge sb.clk) sb.write_ena==1)
    -> write_event;
endmodule
```

Sample 9.11 Triggering a cover group with an SVA

```
program automatic test(simple_bus sb);

  covergroup Write_cg @($root.top.m1.write_event);
    coverpoint $root.top.m1.data;
    coverpoint $root.top.m1.addr;
  endgroup

  Write_cg wcg;

  initial begin
    wcg = new();
    sb.write_ena <= 1;    // Apply stimulus here
    #10000ns $finish;
  end
endprogram
```

9.7 Data Sampling

How is coverage information gathered? When you specify a variable or expression
in a cover point, SystemVerilog creates a number of "bins" to record how many
times each value has been seen. These bins are the basic units of measurement for
functional coverage. If you sample a one-bit variable, a maximum of two bins are
created. You can imagine that System Verilog drops a token in one or the other bin
every time the cover group is triggered. At the end of each simulation, a database is
created with all bins that have a token in them. You then run an analysis tool that
reads all databases and generates a report with the coverage for each part of the
design and for the total coverage.

9.7.1 Individual Bins and Total Coverage

To calculate the coverage for a point, you first have to determine the total number of
possible values, also known as the domain. There may be one value per bin or multiple

values. Coverage is the number of sampled values divided by the number of bins in the domain.

A cover point that is a 3-bit variable has the domain 0:7 and is normally divided into eight bins. If, during simulation, values belonging to seven bins are sampled, the report will show 7/8 or 87.5% coverage for this point. All these points are combined to show the coverage for the entire group, and then all the groups are combined to give a coverage percentage for all the simulation databases.

This is the status for a single simulation. You need to track coverage over time. Look for trends so you can see where to run more simulations or add new constraints or tests. Now you can better predict when verification of the design will be completed.

9.7.2   Creating Bins Automatically

As you saw in the report in Sample 9.3, System Verilog automatically creates bins for cover points. It looks at the domain of the sampled expression to determine the range of possible values. For an expression that is N bits wide, there are 2^N possible values. For the 3-bit variable dst, there are 8 possible values. The range of an enumerated type is shown in Section 9.6.8. The domain for enumerated data types is the number of named values. You can also explicitly define bins as shown in Section 9.6.5.

9.7.3   Limiting the Number of Automatic Bins Created

The cover group option auto_bin_max specifies the maximum number of bins to automatically create, with a default of 64 bins. If the domain of values in the cover point variable or expression is greater than this option, System Verilog divides the range into auto_bin_max bins. For example, a 16-bit variable has 65,536 possible values, so each of the 64 bins covers 1024 values.

In reality, you may find this approach impractical, as it is very difficult to find the needle of missing coverage in a haystack of auto-generated bins. Lower this limit to 8 or 16, or better yet, explicitly define the bins as shown in Section 9.6.5.

Sample 9.12 takes the chapter's first example and adds a cover point option that sets auto_bin_max to two bins. The sampled variable is still dst, which is three bits wide, for a domain of eight possible values. The first bin holds the lower half of the range, 0–3, and the other hold the upper values, 4–7.

Sample 9.12 Using auto_bin_max set to 2

```
covergroup CovDst12;
  coverpoint tr.dst
    { option.auto_bin_max = 2; }   // Divide into 2 bins
endgroup
```

The coverage report from VCS shows the two bins. This simulation achieved 100% coverage because the eight `dst` values were mapped to two bins. Since both bins have sampled values, your coverage is 100% as shown in Sample 9.13.

Sample 9.13 Report with `auto_bin_max` set to 2

```
Bin                   # hits      at least
====================================
auto[0:3]              15          1
auto[4:7]              17          1
```

Sample 9.12 used `auto_bin_max` as an option for the cover point only. You can also use it as an option for the entire group as shown in Sample 9.14.

Sample 9.14 Using `auto_bin_max` for all cover points

```
covergroup CovDst14;
  option.auto_bin_max = 2; // Affects dst & data
  coverpoint tr.dst;        // autobin[0:3], autobin[4:7]
  coverpoint tr.data;       // autobin[0:7], autobin[8:15]
endgroup
```

9.7.4   Sampling Expressions

You can sample expressions, but always check the coverage report to be sure you are getting the values you expect. You may have to adjust the width of the computed expression, as shown in Section 2.16. For example, sampling a 3-bit header length (0:7) plus a 4-bit payload length (0:15) creates only 2^4 or 16 bins, which may not be enough if your transactions can actually be from 0 to 22 bytes long.

Sample 9.15 Using an expression in a cover point

```
class Packet;
  rand bit [2:0] hdr_len;      // range: 0:7
  rand bit [3:0] payload_len;  // range: 0:15
  rand bit [3:0] kind;
endclass

Packet p;

covergroup CovLen15;
  len16: coverpoint (p.hdr_len + p.payload_len);
  len32: coverpoint (p.hdr_len + p.payload_len + 5'b0);
endgroup
```

Sample 9.15 has a cover group that samples the total packet length. The cover point has a label to make it easier to read the coverage report. Also, the expression has an additional dummy constant so that the transaction length is computed with 5-bit precision, for a maximum of 32 auto-generated bins.

A long run with random packets showed that the len16 had 100% coverage, but this is across only 16 bins. (The cover point only has 16 bins as the sum of a 3-bit and 4-bit value is only 4-bits in Verilog.) The cover point len32 had 72% coverage across 32 bins. (The addition of a 5-bit value to the expression for bin32 results in a 5-bit result.) Neither of these cover points are correct, as the maximum length has a domain of 0:22 (0+0:7+15). The auto-generated bins just don't work, as the maximum length is not a power of 2. You need a way to precisely define bins.

9.7.5   User-Defined Bins Find a Bug

Automatically generated bins are okay for anonymous data values, such as counter values, addresses, or values that are a power of 2. For other values, you should explicitly name the bins to improve accuracy and ease coverage report analysis. System Verilog automatically creates bin names for enumerated types, but for other variables you need to give names to the interesting states. The easiest way to specify bins is with the [] syntax, as shown in Sample 9.16.

Sample 9.16 Defining bins for transaction length

```
covergroup CovLen16;
  len: coverpoint (p.hdr_len + p.payload_len + 5'b0)
    {bins len[] = {[0:23]}; }  // Bug?? See below
endgroup
```

After sampling many random transactions, the group has 95.83% coverage. A quick look at the report in Sample 9.17 shows the problem — the length of 23 (17 hex) was never seen. The longest header is 7, and the longest payload is 15, for a total of 22, not 23! If you change to the bins declaration to use 0:22, the coverage jumps to 100%. The user-defined bins found a bug in the test.

Sample 9.17 Coverage report for transaction length

```
Bin            # hits      at least
============================
len_00         13          1
len_01         36          1
len_02         51          1
len_03         60          1
len_04         72          1
len_05         88          1
len_06         127         1
len_07         122         1
len_08         133         1
len_09         138         1
len_0a         115         1
len_0b         128         1
len_0c         125         1
len_0d         111         1
len_0e         115         1
len_0f         134         1
len_10         107         1
len_11         102         1
len_12         70          1
len_13         65          1
len_14         39          1
len_15         30          1
len_16         19          1
len_17         0           1
============================
```

9.7.6 Naming the Cover Point Bins

Sample 9.18 samples a 4-bit variable, kind, that has 16 possible values. The first bin is called zero and counts the number of times that kind is 0 when sampled. The next four values, 1–3 and 5, are all grouped into a single bin, lo. The upper eight values, 8–15, are kept in separate bins, hi_8, hi_9, hi_a, hi_b, hi_c, hi_d, hi_e, and hi_f. Note how $ in the hi bin expression is used as a shorthand notation for the largest value for the sampled variable. Lastly, misc holds all values that were not previously chosen: 4, 6, and 7.

Sample 9.18 Specifying bin names

```
covergroup CovKind18;
  coverpoint p.kind {
    bins zero = {0};           // 1 bin for kind==0
    bins lo   = {[1:3], 5};    // 1 bin for values 1:3, 5
    bins hi[] = {[8:$]};       // 8 separate bins: 8...15
    bins misc = default;       // 1 bin for rest, does not count
  }                            // No semicolon
endgroup
```

Note that the additional information about the `coverpoint` is grouped using curly braces: `{}`. This is because the bin specification is declarative code, not procedural code that would be grouped with `begin...end`. Lastly, the final curly brace is NOT followed by a semicolon, just as an `end` never is.

Now you can easily see in Sample 9.19 which bins have no hits — `hi_8` in this case.

Sample 9.19 Report showing bin names

Bin	# hits	at least
======	========	==========
hi_8	0	1
hi_9	5	1
hi_a	3	1
hi_b	4	1
hi_c	2	1
hi_d	2	1
hi_e	9	1
hi_f	4	1
lo	16	1
misc	15	1
zero	1	1

When you define the bins, you are restricting the values used for coverage to those that are interesting to you. SystemVerilog no longer automatically creates bins, and it ignores values that do not fall into a predefined bin. More importantly, only the bins you create are used to calculate functional coverage. You get 100% coverage only as long as you get a hit in every specified bin.

 Values that do not fall into any specified bin are ignored. This rule is useful if the sampled value, such as transaction length, is not a power of 2. If you are specifying bins, you can use the `default` bin specifier to catch values that you may have forgotten. However, the LRM says that `default` bins are not used in coverage calculation.

In Sample 9.18, the range for `hi` uses a dollar sign (`$`) on the right side to specify the upper value. This is a very useful shortcut - now you can let the compiler calculate the limits for a range. You can use the dollar sign on the left side of a range to specify the lower limit. In Sample 9.20, the $ in the range for bin neg represents the negative number furthest from zero: 32′h8000_0000, or -2,147,483,648, whereas the $ in bin pos represents the largest signed positive value, 32′h7FFF_FFFF, or 2,147,483,647.

Sample 9.20 Specifying ranges with $

```
int i;
covergroup range_cover;
  coverpoint i {
    bins neg  = {[$:-1]};   // Negative values
    bins zero = {0};        // Zero
    bins pos  = {[1:$]};    // Positive values
  }
endgroup
```

9.7.7   Conditional Coverage

You can use the iff keyword to add a condition to a cover point. The most common reason for doing so is to turn off coverage during reset so that stray triggers are ignored. Sample 9.21 gathers only values of dst when rst is 0, where rst is active-high.

Sample 9.21 Conditional coverage — disable during reset

```
covergroup CovDst21;
  // Don't gather coverage when rst==1
  coverpoint tr.dst iff (!bus_if.rst);
endgroup
```

Alternately, you can use the start and stop functions to control individual instances of cover groups as shown in Sample 9.22.

Sample 9.22 Using stop and start functions

```
initial begin
  CovDst22 ck = new();          // Instantiate cover group

  // Reset sequence stops collection of coverage data
  #1ns ck.stop();
  bus_if.rst <= 1;

  #100ns bus_if.rst <= 0; // End of reset
  ck.start();
  ...
end
```

9.7.8   Creating Bins for Enumerated Types

For enumerated types, SystemVerilog creates a bin for each value as you can see in Sample 9.23.

Sample 9.23 Functional coverage for an enumerated type

```
typedef enum {INIT, DECODE, IDLE} fsmstate_e;
fsmstate_e pstate, nstate;    // declare typed variables
covergroup CovFsm23;
  coverpoint pstate;
endgroup
```

Here is part of the coverage report from VCS, Sample 9.24 showing the bins for the enumerated types.

Sample 9.24 Coverage report with enumerated types

Bin	# hits	at least
auto_DECODE	11	1
auto_IDLE	11	1
auto_INIT	10	1

If you want to group multiple values into a single bin, you have to define your own bins. Any bins outside the enumerated values are ignored unless you define a bin with the default specifier. When you gather coverage on enumerated types, auto_bin_max does not apply.

9.7.9   Transition Coverage

You can specify state transitions for a cover point. In this way, you can tell not only what interesting values were seen but also the sequences. For example, you can check if dst ever went from 0 to 1, 2, or 3 as shown in Sample 9.25.

Sample 9.25 Specifying transitions for a cover point

```
covergroup CovDst25;
  coverpoint tr.dst {
    bins t1 = (0 => 1), (0 => 2), (0 => 3);
  }
endgroup
```

You can quickly specify multiple transitions using ranges. The expression (1, 2 => 3, 4) creates the four transitions (1=>3), (1=>4), (2=>3), and (2=>4).

You can specify transitions of any length. Note that you have to sample once for each state in the transition. So (0 => 1 => 2) is different from (0 => 1 => 1 => 2) or (0 => 1 => 1 => 1 => 2). If you need to repeat values, as in the last sequence, you can use the shorthand form: (0 => 1[*3] => 2). To repeat the value 1 for 3, 4, or 5 times, use 1[*3:5].

9.7.10   Wildcard States and Transitions

You use the wildcard keyword to create multiple states and transitions. Any X, Z, or ? in the expression is treated as a wildcard for 0 or 1. Sample 9.26 creates a cover point with a bin for even values and one for odd.

Sample 9.26 Wildcard bins for a cover point

```
bit [2:0] dst;
covergroup CovDst26;
  coverpoint tr.dst {
    wildcard bins even = {3'b??0};
    wildcard bins odd  = {3'b??1};
  }
endgroup
```

9.7.11   Ignoring Values

With some cover points, you never get all possible values. For instance, a 3-bit variable may be used to store just six values, 0–5. If you use automatic bin creation, you never get beyond 75% coverage. There are two ways to solve this problem. You can explicitly define the bins that you want to cover as shown in Section 9.6.5. Alternatively, you can let SystemVerilog automatically create bins, and then use ignore_bins to tell which values to exclude from functional coverage calculation like in Sample 9.27.

Sample 9.27 Cover point with ignore_bins

```
covergroup CovDst27;
  coverpoint tr.dst {
    ignore_bins hi = {6,7};    // Ignore upper 2 bins
  }
endgroup
```

The original range of low_ports_0_5, a three-bit variable is 0:7. The ignore_ bins excludes the last two bins, which reduces the range to 0:5. So total coverage for this group is the number of bins with samples, divided by the total number of bins, which is 5 in this case.

Sample 9.28   Cover point with `auto bin max` and `ignore bins`

```
covergroup CovDst28;
  coverpoint tr.dst {
    option.auto_bin_max = 4; // 0:1, 2:3, 4:5, 6:7
    ignore_bins hi = {6,7};  // Ignore upper 2 values
  }
endgroup
```

If you define bins either explicitly or by using the `auto_bin_max` option, and then ignore them, the ignored bins do not contribute to the calculation of coverage. In Sample 9.28, four bins are initially created using the `auto_bin_max` option: 0:1, 2:3, 4:5, and 6:7. However, then the uppermost bin is eliminated by `ignore_bins`, so in the end only three bins are created. This cover point can have coverage of 0%, 33%, 66%, or 100%.

9.7.12   Illegal Bins

Some sampled values not only should be ignored, but also should cause an error if they are seen. This is best done in the testbench's monitor code, but can also be done by labeling a bin with `illegal_bins` as shown in Sample 9.29. Use this to catch states that were missed by the test's error checking. This also double-checks the accuracy of your bin creation: if an illegal value is found by the cover group, it is a problem either with the testbench or with your bin definitions.

Sample 9.29   Cover point with `illegal_bins`

```
covergroup CovDst29;
  coverpoint tr.dst {
    illegal_bins hi = {6,7}; // Give error if seen
  }
endgroup
```

9.7.13   State Machine Coverage

You should have noticed that if a cover group is used on a state machine, you can use bins to list the specific states, and transitions for the arcs. However, this does not mean you should use SystemVerilog's functional coverage to measure state machine coverage. You would have to extract the states and arcs manually. Even if you did this correctly the first time, you might miss future changes to the design code. Instead, use a code coverage tool that extracts the state register, states, and arcs automatically, saving you from possible mistakes.

However, an automatic tool extracts the information exactly as coded, mistakes and all. You may want to monitor small, critical state machines manually using functional coverage.

9.8   Cross Coverage

A cover point records the observed values of a single variable or expression. You may want to know not only what bus transactions occurred but also what errors happened during those transactions, and their source and destination. For this you need cross coverage that measures what values were seen for two or more cover points at the same time. Note that when you measure cross coverage of a variable with N values, and of another with M values, SystemVerilog needs N×M cross bins to store all the combinations.

9.8.1   Basic Cross Coverage Example

Previous examples have measured coverage of the transaction kind, and destination port number, but what about the two combined? Did you try every kind of transaction into every port? The `cross` construct in SystemVerilog records the combined values of two or more cover points in a group. The `cross` statement takes only cover points or a simple variable name. If you want to use expressions, hierarchical names or variables in an object such as `handle.variable`, you must first specify the expression in a `coverpoint` with a label and then use the label in the `cross` statement.

Sample 9.30 creates cover points for `tr.kind` and `tr.dst`. Then the two points are crossed to show all combinations. SystemVerilog creates a total of 128 (8 × 16) bins. Be careful: even a simple cross can result in a very large number of bins.

Sample 9.30 Basic cross coverage

```
class Transaction;
  rand bit [3:0] kind;
  rand bit [2:0] dst;
endclass

Transaction tr;

covergroup CovDst30;
  kind: coverpoint tr.kind;     // Create cover point kind
  dst: coverpoint tr.dst;       // Create cover point dst
  cross kind, dst;              // Cross kind and dst
endgroup
```

A random testbench created 56 transactions and produced the coverage report in Sample 9.31. Note that even though all possible `kind` and `dst` values were generated, only 1/3 of the cross combinations were seen. This is a very typical result. Also note that the total coverage for the group is the cross coverage plus the coverage for `kind` and `dst`.

Sample 9.31   Coverage summary report for basic cross coverage

```
Cumulative report for Transaction::CovDst30
Summary:
    Coverage: 78.91
    Goal: 100


Coverpoint              Coverage      Goal          Weight
==========================================================
kind                    100.00        100           1
dst                     100.00        100           1
==========================================================
Cross                   Coverage      Goal          Weight
==========================================================
Transaction::CovDst30   78.91         100           1

Cross Coverage report
CoverageGroup: Transaction::CovDst30
    Cross: Transaction::CovDst30
Summary
    Coverage: 36.72
    Goal: 100
    Coverpoints Crossed: kind dst
    Number of Expected Cross Bins: 128
    Number of User Defined Cross Bins: 0
    Number of Automatically Generated Cross Bins: 47


    Automatically Generated Cross Bins

    kind        dst             # hits        at least
    ==========================================================
    auto[0]     auto[0]         1             1
    auto[0]     auto[1]         2             1
    auto[0]     auto[2]         1             1
    auto[0]     auto[5]         1             1
...
```

9.8.2   Labeling Cross Coverage Bins

If you want more readable cross coverage bin names, you can label the individual cover point bins as demonstrated in Sample 9.32, and SystemVerilog will use these names when creating the cross bins.

Sample 9.32 Specifying cross coverage bin names

```
covergroup CovDstKind32;
  dst: coverpoint tr.dst
  {bins dst[] = {[0:$]};
  }
  kind: coverpoint tr.kind
  {bins zero = {0};          // 1 bin for kind==0
   bins lo   = {[1:3]};      // 1 bin for values 1:3
   bins hi[] = {[8:$]};      // 8 separate bins
   bins misc = default;      // 1 bin for rest, does not count
  }
  cross kind, dst;
endgroup
```

If you define bins that contain multiple values, the coverage statistics change. In the report below, the number of bins has dropped from 128 to 80. This is because kind has 10 bins: zero, lo, hi_8, hi_9, 9hi_a, hi_b, hi_c, hi_d, hi_e, and hi_f. Remember that the misc bin, which defined its values with default, does not add to the coverage total. The percentage of coverage jumped from 87.5% to 90.91% as shown in Sample 9.33 because any single value in the lo bin, such as 2, allows that bin to be marked as covered, even if the other values, 1 or 3, are not seen.

Sample 9.33 Cross coverage report with labeled bins

```
Summary
     Coverage: 90.91
     Number of Coverpoints Crossed: 2
     Coverpoints Crossed: kind dst
     Number of Expected Cross Bins: 88
     Number of Automatically Generated Cross Bins: 80
     Automatically Generated Cross Bins
```

dst	kind	# hits	at least
dst_0	hi_8	3	1
dst_0	hi_a	1	1
dst_0	hi_b	4	1
dst_0	hi_c	4	1
dst_0	hi_d	4	1
dst_0	hi_e	1	1
dst_0	lo	7	1
dst_0	zero	1	1
dst_1	hi_8	3	1

. . .

9.8.3   Excluding Cross Coverage Bins

To reduce the number of bins, use `ignore_bins`. With cross coverage, you specify the cover point with `binsof` and the set of values with `intersect` so that a single `ignore_bins` construct can sweep out many individual bins.

Sample 9.34   Excluding bins from cross coverage

```
covergroup CovDst34;
  dst: coverpoint tr.dst
  {bins dst[] = {[0:$]};
  }
  kind: coverpoint tr.kind {
    bins zero = {0};        // 1 bin for kind==0
    bins lo   = {[1:3]};    // 1 bin for values 1:3
    bins hi[] = {[8:$]};    // 8 separate bins
    bins misc = default;    // 1 bin for rest, does not count
  }
  cross kind, dst {
    ignore_bins hi = binsof(dst) intersect {7};
    ignore_bins md = binsof(dst) intersect {0} &&
                     binsof(kind) intersect {[9:11]};
    ignore_bins lo = binsof(kind.lo);
  }
endgroup
```

The first `ignore_bins` in Sample 9.34 just excludes bins where `dst` is 7 and any value of `kind`. Since `kind` is a 4-bit value, this statement excludes 12 bins, as `misc`'s values of 4–7 don't count because of the `default`. The second `ignore_bins` is more selective, ignoring bins where `dst` is 0 and `kind` is 9, 10, or 11, for a total of 3 bins.

The `ignore_bins` can use the bins defined in the individual cover points. The `ignore_bins lo` uses bin names to exclude `kind.lo` that is 1, 2, or 3. The bins must be names defined at compile time, such as `zero` and `lo`. The bins `hi_8`, `hi_9`, `hi_a`,... `hi_f`, and any automatically generated bins do not have names that can be used at compile time in other statements such as `ignore_bins`; these names are created at run time or during the report generation.

Note that `binsof` uses parentheses `()` while `intersect` specifies a range and therefore uses curly braces `{}`.

9.8.4   Excluding Cover Points from the Total Coverage Metric

The total coverage for a group is based on all simple cover points and cross coverage. If you are only sampling a variable or expression in a `coverpoint` to be used in a `cross` statement, you should set its weight to 0 so that it does not contribute to the total coverage.

Sample 9.35 Specifying cross coverage weight

```
covergroup CovDst35;
  kind: coverpoint tr.kind
    {bins zero = {0};
     bins lo   = {[1:3]};
     bins hi[] = {[8:$]};
     type_option.weight = 5;      // Count in total
  }
  dst: coverpoint tr.dst
    {bins dst[] = {[0:$]};
     type_option.weight = 0;      // Don't count towards total
  }
  cross kind, dst
    {type_option.weight = 10;}  // Give cross extra weight
endgroup
```

There are two types of options: those that are specific to an instance of a cover-group and those that specify an option for the covergroup type as a whole. The instance specific options are like local variables and are specified with the `option` keyword, as in `option.auto_bin_max=2` from Sample 9.12. The alternatives are specified with the `type_option` keyword and are tied to the cover group, like static variables in a class. In Sample 9.35, `type_option.weight` applies to all instances of this group. The LRM has a detailed explanation of the difference, and this book shows the most common options and their usage.

9.8.5 Merging Data from Multiple Domains

One problem with cross coverage is that you may need to sample values from dif-ferent timing domains. You might want to know if your processor ever received an interrupt in the middle of a cache fill. The interrupt hardware is separate from and may use different clocks than the cache hardware, making it difficult to know when to trigger the cover group. On the other hand, you want to make sure you have tested this case, as a previous design had a bug of this very sort.

The solution is to create a timing domain separate from the cache or interrupt hardware. Make copies of the signals into temporary variables and then sample them in a new coverage group that measures the cross coverage.

9.8.6 Cross Coverage Alternatives

As your cross coverage definition becomes more elaborate, you may spend consid-erable time specifying which bins should be used and which should be ignored. You may have two random bits, a and b with three interesting states, {a==0, b==0}, {a==1, b==0}, and {b==1}.

Sample 9.36 shows how you can name bins in the cover points and then gather cross coverage using those bins.

Sample 9.36   Cross coverage with bin names

```
class Sample;
  rand bit a, b;
endclass

Sample sam;

covergroup CrossBinNames;
  a: coverpoint sam.a
     { bins a0 = {0};
       bins a1 = {1};
       option.weight=0;}    // Don't count this coverpoint
  b: coverpoint sam.b
     { bins b0 = {0};
       bins b1 = {1};
       option.weight=0;}    // Don't count this coverpoint
  ab: cross a, b
     { bins a0b0 = binsof(a.a0) && binsof(b.b0);
       bins a1b0 = binsof(a.a1) && binsof(b.b0);
       bins b1   = binsof(b.b1); }
endgroup
```

Sample 9.37 gathers the same cross coverage, but now uses binsof to specify the cross coverage values.

Sample 9.37   Cross coverage with binsof

```
covergroup CrossBinsofIntersect;
  a: coverpoint sam.a
     { option.weight=0; }   // Don't count this coverpoint
  b: coverpoint sam.b
     { option.weight=0; }   // Don't count this coverpoint
  ab: cross a, b
       { bins a0b0 = binsof(a) intersect {0} &&
                     binsof(b) intersect {0};
         bins a1b0 = binsof(a) intersect {1} &&
                     binsof(b) intersect {0};
         bins b1   = binsof(b) intersect {1}; }
endgroup
```

Alternatively, you can make a cover point that samples a concatenation of values. Then you only have to define bins using the less complex cover point syntax.

Sample 9.38 Mimicking cross coverage with concatenation

```
covergroup CrossManual;
  ab: coverpoint {sam.a, sam.b}
    { bins a0b0 = {2'b00};
      bins a1b0 = {2'b10};
      wildcard bins b1 = {2'b?1};
    }
endgroup
```

Use the style in Sample 9.36 if you already have bins defined for the individual cover points and want to use them to build the cross coverage bins. Use Sample 9.37 if you need to build cross coverage bins but have no pre-defined cover point bins. Use Sample 9.38 if you want the tersest format.

9.9   Generic Cover Groups

As you start writing cover groups, you will find that some are very similar to one another. SystemVerilog allows you to create a generic cover group so that you can specify a few unique details when you instantiate it.

9.9.1   Pass Cover Group Arguments by Value

Sample 9.39 shows a cover group that uses an argument to split the range into two halves. Just pass the midpoint value to the cover groups' new function.

Sample 9.39 Covergroup with simple argument

```
class Transaction;
  bit [2:0] dst;        // Values: 0:7
endclass
Transaction tr;

covergroup CovDst39 (int mid);
  coverpoint tr.dst
    {bins lo = {[0:mid-1]};
     bins hi = {[mid:$]};
  }
endgroup

CovDst39 cp;
initial
  cp = new(5);                    // lo=0:4, hi=5:7
```

9.9.2   Pass Cover Group Arguments by Reference

You can specify a variable to be sampled with pass-by-reference. Here you want the cover group to sample the value during the entire simulation, not just to use the value when the constructor is called.

Sample 9.40 Pass-by-reference

```
bit [2:0] dst_a, dst_b;

covergroup CovDst40 (ref bit [2:0] dst, input int mid);
  coverpoint dst {
    bins lo = {[0:mid-1]};
    bins hi = {[mid:$]};
  }
endgroup

CovDst40 cpa, cpb;
initial
  begin
    cpa = new(dst_a, 4);  // dst_a, lo=0:3, hi=4:7
    cpb = new(dst_b, 2);  // dst_b, lo=0:1, hi=2:7
  end
```

Like a task or function, the arguments to a cover group have a sticky direction. In Sample 9.40, if you forgot the `input` direction, the `mid` argument will have the direction `ref`. The example would not compile because you cannot pass a constant (4 or 2) into a `ref` argument.

9.10   Coverage Options

You can specify additional information in the cover group using options. There are two flavors of options: instance options that apply to a specific cover group instance and type options that apply to all instances of the cover group, and are analogous to static data members of classes. Options can be placed in the cover group so that they apply to all cover points in the group, or they can be put inside a single cover point for finer control. You have already seen the `auto_bin_max` and `weight` options. Here are several more.

9.10.1   Per-Instance Coverage

If your testbench instantiates a coverage group multiple times, by default System-Verilog groups together all the coverage data from all the instances. However, if you have several generators, each creating very different streams of transactions, you will need to see separate reports. For example, one generator may be creating long transactions while another makes short ones. The cover group in Sample 9.41 can be instantiated in each separate generator. It keeps track of coverage for each instance, and has a unique comment string with the hierarchical path to the cover group instance.

Sample 9.41 Specifying per-instance coverage

```
covergroup CoverLength(ref bit [2:0] len);
  coverpoint len;
  option.per_instance = 1;
endgroup
```

The per-instance option can only be given in the cover group, not in the cover point or cross point.

9.10.2   Cover Group Comment

You can add a comment into coverage reports to make them easier to analyze. A comment could be as simple as the section number from the verification plan to tags used by a report parser to automatically extract relevant information from the sea of data. If you have a cover group that is only instantiated once, use the `type_option` as shown in Sample 9.42.

Sample 9.42 Specifying comments for a cover group

```
covergroup CovDst42;
  type_option.comment = "Section 3.2.14 Dst port numbers";
  coverpoint tr.dst;
endgroup
```

However, if you have multiple instances, you can give each a separate comment, as long as you also use the `per-instance` option as shown in Sample 9.43.

Sample 9.43 Specifying comments for a cover group instance

```
covergroup CovDst43(int lo,hi, string comment);
  option.comment = comment;
  option.per_instance = 1;
  coverpoint tr.dst
    {bins range = {[lo:hi]};
    }
endgroup
...
CovDst43 cd_lo = new(0,3, "Low dst numbers");
CovDst43 cd_hi = new(4,7, "High dst numbers");
```

9.10.3   Coverage Threshold

You may not have sufficient visibility into the design to gather robust coverage information. Suppose you are verifying that a DMA state machine can handle bus errors. You don't have access to its current state, but you know the range of cycles that are needed for a transfer. So if you repeatedly cause errors during that range, you have probably covered all the states. So you could set option.at_least to 8 or more to specify that after 8 hits on a bin, you are confident that you have exercised that combination.

If you define option.at_least at the cover group level, it applies to all cover points. If you define it inside a point, it only applies to that single point.

However, as Sample 9.2 showed, even after 32 attempts, the random kind variable still did not hit all possible values. So only use at_least if there is no direct way to measure coverage, like when the testbench can not probe the DUT details.

9.10.4   Printing the Empty Bins

By default, the coverage report shows only the bins with samples. Your job is to verify all that is listed in the verification plan, so you are actually more interested in the bins without samples. Use the option cross_num_print_missing to tell the simulation and report tools to show you all bins, especially the ones with no hits. Set it to a large value, as shown in Sample 9.44, but no larger than you are willing to read.

Sample 9.44 Report all bins including empty ones

```
covergroup CovDst44;
  kind: coverpoint tr.kind;
  dst: coverpoint tr.dst;
  cross kind, dst;
  option.cross_num_print_missing = 1_000;
endgroup
```

9.10.5   Coverage Goal

The goal for a cover group or point is the level at which the group or point is considered fully covered. The default is 100% coverage. If you set this level below 100% like in Sample 9.45, you are requesting less than complete coverage, which is probably not desirable. This option affects only the coverage report.

Sample 9.45 Specifying the coverage goal

```
covergroup CovDst45;
  coverpoint tr.dst;
  option.goal = 90;   // Settle for partial coverage
endgroup
```

9.11   Analyzing Coverage Data

In general, assume you need more seeds and fewer constraints. After all, it is easier
to run more tests than to construct new constraints. If you are not careful, new constraints can easily restrict the search space.

If your cover point has only zero or one sample, your constraints are probably
not targeting these areas at all. You need to add constraints that "pull" the solver
into new areas. In Sample 9.16, the transaction length had an uneven distribution.
Sample 9.46 shows the full class. This situation is similar to the distribution seen
when you roll two dice and look at the total value.

Sample 9.46 Original class for packet length

```
class Packet;
  rand bit [2:0] hdr_len;
  rand bit [3:0] payload_len;
  rand bit [4:0] len;
  constraint length {len == hdr_len + payload_len; }
endclass
```

The problem with this class is that `len` is not evenly weighted. Look in the coverage report and note how the low and high values are rarely hit. Figure 9.5 is a
graph of the values from the report.

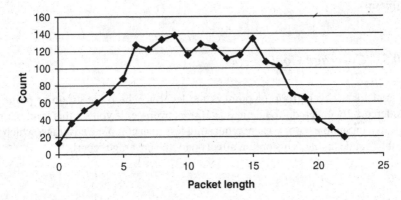

Fig. 9.5 Uneven probability for packet length

If you want to make the total length be evenly distributed, use a `solve...before` constraint as shown in Sample 9.47 and plotted in Fig. 9.6.

Sample 9.47 `solve...before` constraint for packet length

```
constraint length
  {len == hdr_len + payload_len;
   solve len before hdr_len, payload_len; }
```

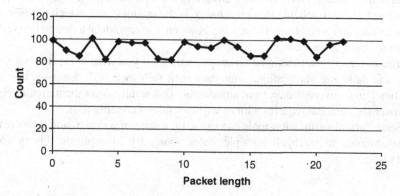

Fig. 9.6 Even probability for packet length with `solve...before`

The normal alternative to `solve...before` is the `dist` constraint. However, this does not work, as `len` is also being constrained by the sum of the two lengths.

9.12 Measuring Coverage Statistics During Simulation

You can query the level of functional coverage on the fly during simulation. This allows you to check whether you have reached your coverage goals, and possibly to control a random test.

At the global level, you can get the total coverage of all cover groups with `$get_coverage`, which returns a real number between 0. and 100. This system task looks across all cover groups.

You can narrow down your measurements with the `get_coverage()` and `get_inst_coverage()` methods. The first function works with both cover group names and instances to give coverage across all instances of a cover group, for example `CoverGroup::get_coverage()` or `cgInst.get_coverage()`. The second function returns coverage for a specific cover group instance, for example `cgInst.get_inst_coverage()`. You need to specify `option.per_instance=1` if you want to gather per-instance coverage.

The most practical use for these functions is to monitor coverage over a long test. If the coverage level does not advance after a given number of transactions or cycles, the test should stop. Hopefully, another seed or test will increase the coverage.

While it would be nice to have a test that can perform some sophisticated actions based on functional coverage results, it is very hard to write this sort of test. Each test + random seed pair may uncover new functionality, but it may take many runs to reach a goal. If a test finds that it has not reached 100% coverage, what should it do? Run for more cycles? How many more? Should it change the stimulus being generated? How can you correlate a change in the input with the level of functional coverage? The one reliable thing to change is the random seed, which you should only do once per simulation. Otherwise, how can you reproduce a design bug if the stimulus depends on multiple random seeds?

You can query the functional coverage statistics if you want to create your own coverage database. Verification teams have built their own SQL databases that are fed functional coverage data from simulation. This setup allows them greater control over the data, but requires a lot of work outside of creating tests.

Some formal verification tools can extract the state of a design and then create input stimulus to reach all possible states. Don't try to duplicate this in your testbench!

9.13   Conclusion

When you switch from writing directed tests, hand-crafting every bit of stimulus, to constrained-random testing, you might worry that the tests are no longer under your command. By measuring coverage, especially functional coverage, you regain control by knowing what features have been tested.

Using functional coverage requires a detailed verification plan and much time creating the cover groups, analyzing the results, and modifying tests to create the proper stimulus. This may seem like a lot of work, but is less effort than would be required to write the equivalent directed tests. Additionally, the time spent in gathering coverage helps you better track your progress in verifying your design.

9.14   Exercises

1. For the class below, write a covergroup to collect coverage on the test plan requirement, "All ALU opcodes must be tested." Assume the opcodes are valid on the positive edge of signal clk.

```
typedef enum {ADD, SUB, MULT, DIV} opcode_e;

class Transaction;
  rand opcode_e opcode;
  rand byte operand1;
  rand byte operand2;
endclass

Transaction tr;
```

2. Expand the solution to Exercise 1 to cover the test plan requirement, "Operand1 shall take on the values maximum negative (−128), zero, and maximum positive (127)." Define a coverage bin for each of these values as well as a default bin. Label the coverpoint `operand1_cp`.

3. Expand the solution to Exercise 2 to cover the following test plan requirements:

   a. "The opcode shall take on the values ADD or SUB" (hint: this is 1 coverage bin).
   b. "The opcode shall take on the values ADD followed by SUB" (hint: this is a second coverage bin).
      Label the coverpoint `opcode_cp`.

4. Expand the solution to Exercise 3 to cover the test plan requirement, "Opcode must not equal DIV" (hint: report an error using `illegal_bins`).

5. Expand the solution to Exercise 4 to collect coverage on the test plan requirement, "The opcode shall take on the values ADD or SUB when operand1 is maximum negative or maximum positive value." Weight the cross coverage by 5.

6. Assuming that your covergroup is called `Covcode` and the instantiation name of the covergroup is `ck`, expand Exercise 4 to:

   a. Display the coverage of coverpoint `operand1_cp` referenced by the instantiation name.
   b. Display the coverage of coverpoint `opcode_cp` referenced by the covergroup name.

Chapter 10
Advanced Interfaces

In Chapter 4 you learned how to connect the design and testbench with interfaces. These physical interfaces represent real signals, similar to the wires that connected ports in Verilog-1995. A testbench uses these interfaces by statically connecting to them through ports. However, for many designs, the testbench needs to connect dynamically to the design.

For example, in a network switch, a single driver class may connect to many interfaces, one for each input channel of the DUT. You wouldn't want to write a unique driver for each channel — instead you want to write a generic driver, instantiate it N times, and have it connect to each of the N physical interfaces. You can do this in SystemVerilog by using a virtual interface, which is merely a handle or pointer to a physical interface. A better name for a virtual interface would be a "ref interface."

You may need to write a testbench that attaches to several different configurations of your design. In another example, a chip may have multiple configurations. In one, the pins might drive a USB bus, whereas in another the same pins may drive an I2C serial bus. Once again, you can use a virtual interface so you can decide at run time which drivers to run in your testbench.

A SystemVerilog interface is more than just signals — you can put executable code inside. This might include routines to read and write to the interface, initial and always blocks that run code inside the interface, and assertions to constantly check the status of the signals. However, do not put testbench code in an interface. Program blocks have been created expressly for building a testbench, including scheduling their execution in the Reactive region, as described in the SystemVerilog LRM.

C. Spear and G. Tumbush, *SystemVerilog for Verification: A Guide to Learning the Testbench Language Features*, DOI 10.1007/978-1-4614-0715-7_10,
© Springer Science+Business Media, LLC 2012

10.1   Virtual Interfaces with the ATM Router

The most common use for a virtual interface is to allow objects in a testbench to refer to items in a replicated interface using a generic handle rather than the actual name. Virtual interfaces are the only mechanism that can bridge the dynamic world of objects with the static world of modules and interfaces.

10.1.1   The Testbench with Just Physical Interfaces

Chapter 4 showed how to build an interface to connect a 4x4 ATM router to a testbench. Sample 10.1 and 10.2 show the ATM interfaces for the receive and transmit directions.

Sample 10.1 Rx interface with clocking block

```
// Rx interface with modports and clocking block
interface Rx_if (input logic clk);
  logic [7:0] data;
  logic soc, en, clav, rclk;

  clocking cb @(posedge clk);
    output data, soc, clav;  // Directions are relative
    input  en;               // to the testbench
  endclocking : cb

  modport TB (clocking cb);

  modport DUT (output en, rclk,
               input  data, soc, clav);
endinterface : Rx_if
```

Sample 10.2 Tx interface with clocking block

```
// Tx interface with modports and clocking block
interface Tx_if (input logic clk);
  logic [7:0] data;
  logic soc, en, clav, tclk;

  clocking cb @(posedge clk);
      input  data, soc, en;
      output clav;
  endclocking : cb
  modport TB (clocking cb);

  modport DUT (output data, soc, en, tclk,
               input  clav);
endinterface : Tx_if
```

These interfaces can be used in a program block shown in Sample 10.3. This procedural code is hard coded with interface names such as Rx0 and Tx0. Note that in these examples, the top module does not pass a clock to the testbench; instead the tests synchronize with clocking blocks in the interfaces, thus allowing you to work at a higher level of abstraction.

Sample 10.3 Testbench using physical interfaces

```
program automatic test(Rx_if.TB Rx0, Rx1, Rx2, Rx3,
                       Tx_if.TB Tx0, Tx1, Tx2, Tx3,
                       output logic rst);

  bit [7:0] bytes[`ATM_SIZE];

  initial begin
    // Reset the device
    rst <= 1;
    Rx0.cb.data <= '0;
    ...
    receive_cell0;
    ...
  end

  task receive_cell0();
    @(Tx0.cb);
    Tx0.cb.clav <= 1;          // Assert ready to receive
    wait (Tx0.cb.soc == 1);    // Wait for Start of Cell

    for (int i=0; i<`ATM_SIZE; i++) begin
      wait (Tx0.cb.en == 0);   // Wait for enable
        @(Tx0.cb);

      bytes[i] = Tx0.cb.data;
      @(Tx0.cb);
      Tx0.cb.clav <= 0;        // Deassert flow control
    end
  endtask

endprogram
```

Figure 10.1 shows the testbench communicating with the design through virtual interfaces.

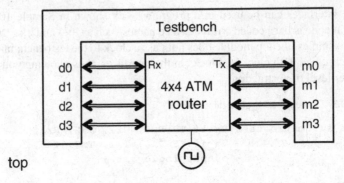

Fig. 10.1 Router and testbench with interfaces

The top level module must connect an array of interfaces to work with the testbench in Sample 10.6. The module in Sample 10.4 instantiates an array of interfaces, and passes this array to the testbench. Since the DUT was written with four RX and four TX interfaces, you need to pass the individual interface array elements into the DUT instance.

Sample 10.4 Top level module with array of interfaces

```
module top;
  logic clk, rst;

  Rx_if Rx[4] (clk);
  Tx_if Tx[4] (clk);

  test        t1 (Rx, Tx, rst);  // See testbench in Sample 10-6
  atm_router a1 (Rx[0], Rx[1], Rx[2], Rx[3],
                 Tx[0], Tx[1], Tx[2], Tx[3],
                 clk, rst);

  initial begin
    clk = 0;
    forever #20 clk = !clk;
    end
endmodule : top
```

10.1.2   Testbench with Virtual Interfaces

A good OOP technique is to create a class that uses a handle to reference an object, rather than a hard-coded object name. In this case, you can make a single Driver class and a single Monitor class, have them operate on a handle to the data, and then pass in the handle at run time.

The program block in Sample 10.5 is still passed the 4 Rx and 4 Tx interfaces as ports, as in Sample 10.3, but it creates an array of virtual interfaces, vRx and vTx. These can now be passed into the constructors for the drivers and monitors.

Sample 10.5 Testbench using virtual interfaces

```
program automatic test(Rx_if.TB Rx0, Rx1, Rx2, Rx3,
                       Tx_if.TB Tx0, Tx1, Tx2, Tx3,
                       output logic rst);

  Driver drv[4];
  Monitor mon[4];
  Scoreboard scb[4];

  virtual Rx_if.TB vRx[4] = '{Rx0, Rx1, Rx2, Rx3};
  virtual Tx_if.TB vTx[4] = '{Tx0, Tx1, Tx2, Tx3};

  initial begin
    foreach (scb[i]) begin
      scb[i] = new(i);
      drv[i] = new(scb[i].exp_mbx, i, vRx[i]);
      mon[i] = new(scb[i].rcv_mbx, i, vTx[i]);
    end
    ...
  end
endprogram
```

You can also skip the virtual interface array variables, and make an array in the port list. These interfaces are passed to the constructors as shown in Sample 10.6.

Sample 10.6 Testbench using virtual interfaces

```
program automatic test(Rx_if.TB Rx[4], Tx_if.TB Tx[4],
                       output logic rst);
...
  initial begin
    foreach (scb[i]) begin
      scb[i] = new(i);
      drv[i] = new(scb[i].exp_mbx, i, Rx[i]);
      mon[i] = new(scb[i].rcv_mbx, i, Tx[i]);
    end
    ...
  end
endprogram
```

The task monitor::receive_cell in Sample 10.7 is similar to the task **receive_cell0** in Sample 10.3, except it uses the virtual interface name Tx instead of the physical interface Tx0.

Sample 10.7 Monitor class using virtual interfaces

```
typedef virtual Tx_if vTx_t;

class Monitor;
    int      stream_id;
    mailbox rcv_mbx;
    vTx_t    Tx;

function new(input mailbox rcv_mbx,
             input int      stream_id,
             input vTx_t    Tx);
    this.rcv_mbx = rcv_mbx;
    this.stream_id = stream_id;
    this.Tx = Tx;
endfunction // new


task run();
    ATM_Cell ac;

    fork begin

      // Initialize output signals
      Tx.cb.clav <= 0;                    // Not ready to receive
      @Tx.cb;

      $display("@%0d: Monitor::run[%0d] starting",
               $time, stream_id);
      forever begin
        receive_cell(ac);
      end
    end
    join_none

  endtask : run


  task receive_cell(inout ATM_Cell ac);
    bit [7:0] bytes[];

    bytes = new[ATM_CELL_SIZE];
    ac = new();                           // Initialize the cell

    @Tx.cb;
    Tx.cb.clav <= 1;                      // Assert ready to receive
    while (Tx.cb.soc !== 1'b1)            // Wait for Start of Cell
      @Tx.cb;
```

```
    foreach (bytes[i]) begin
      while (Tx.cb.en != 0)              // Wait if enable goes away
        @Tx.cb;

      bytes[i] = Tx.cb.data;
      @Tx.cb;
      Tx.cb.clav <= 0;                   // Deassert flow control
    end

    ac.byte_unpack(bytes);
    $display("@%0d: Monitor::run(%0d) received cell vci=%h",
             $time, stream_id, ac.vci);

    // Send cell to scoreboard
    rcv_mbx.put(ac);
  endtask : receive_cell

endclass : Monitor
```

A common mistake when creating a testbench is to leave off the modport name from a virtual interface declaration. The program in Sample 10.5 declares Tx_if.TB Tx0 in the port list, so it can only assign Tx0 to a virtual interface declared with the TB modport. See the declaration of the virtual interface Tx in Sample 10.7.

10.1.3   Connecting the Testbench to an Interface in Port List

This book shows tests that connect to the DUT with interfaces in the port list. This style is comfortable to Verilog users who have always connected modules using signals in ports. Sample 10.8 is the top level module, also known as a test harness, which connects the DUT and test using an interface in the port list.

Sample 10.8 Test harness using an interface in the port list

```
module top;
  bus_ifc bus();     // Instantiate the interface
  test t1(bus);      // Pass to test through port list
  dut  d1(bus);      // Pass to DUT through port list
  ...
endmodule
```

Sample 10.9 shows the program block with an interface in the port list.

Sample 10.9   Test with an interface in the port list

```
program automatic test(bus_ifc bus);
  initial $display(bus.data);  // Use an interface signal
endprogram
```

What happens if you add a new interface to your design? The test harness in Sample 10.10 declares the new bus and puts it in the port lists.

Sample 10.10   Top module with a second interface in the test's port list

```
module top;
  bus_ifc bus();       // Instantiate the interface
  new_ifc newb();      // and a new one
  test t1(bus, newb);  // Test with two interfaces
  dut  d1(bus, newb);  // DUT with two interfaces
  ...
endmodule
```

Now you have to change the test in Sample 10.9 to include another interface in the port list, giving the test in Sample 10.11.

Sample 10.11   Test with two interfaces in the port list

```
program automatic test(bus_ifc bus, new_ifc newb);
  initial $display(bus.data);  // Use an interface signal
endprogram
```

Adding a new interface to your design means you need to edit all existing tests so they can plug into the test harness. How can you avoid this extra work? Avoid port connections!

10.1.4   Connecting the Test to an Interface with an XMR

Your test needs to connect to the physical interface in the harness, so use a cross module reference (XMR) and a virtual interface in the program block as shown in Sample 10.12. You must use a virtual interface so you can assign it the physical interface in the top level module.

Sample 10.12   Test with virtual interface and XMR

```
program automatic test();
  virtual bus_ifc bus = top.bus; // Cross module reference
  initial $display(bus.data);     // Use an interface signal
endprogram
```

The program connects to the test harness shown in Sample 10.13.

Sample 10.13   Test harness without interfaces in the port list

```
module top;
  bus_ifc bus();    // Instantiate the interface
  test t1();        // Don't use port list for test
  dut  d1(bus);     // Still use port list for DUT
  ...
endmodule
```

This approach is recommended by methodologies such as the VMM to make your test code more reusable. If you add a new interface to your design, as shown in Sample 10.14, the test harness changes, but existing tests don't have to change.

Sample 10.14   Test harness with a second interface

```
module top;
  bus_ifc bus();    // Instantiate the interface
  new_ifc newb();   // and a new one
  test t1();        // Instantiaton remains the same
  dut  d1(bus, newb);
  ...
endmodule
```

The harness in Sample 10.14 works with the test in Sample 10.12 that does not know about the new interface, as well as the test in Sample 10.15 that does.

Sample 10.15   Test with two virtual interfaces and XMRs

```
program automatic test();
  virtual bus_ifc bus = top.bus;
  virtual new_ifc newb = top.newb

  initial begin
    $display(bus.data);  // Use existing interface
    $display(newb.addr); // and new one
  end
endprogram
```

Some methodologies have a rule that makes the connection between tests and harnesses slightly more complicated than with traditional ports, but means you won't have to modify existing tests, even if the design changes. The examples in this book use the simple style of interfaces in the port lists, but you should decide if test reuse is important enough to change your coding style.

10.2   Connecting to Multiple Design Configurations

A common challenge to verifying a design is that it may have several configurations. You could make a separate testbench for each configuration, but this could lead to a combinatorial explosion as you explore every alternative. Instead, you can use virtual interfaces to dynamically connect to the optional interfaces.

10.2.1   A Mesh Design

Sample 10.16 is built of a simple replicated component, an 8-bit counter. This resembles a DUT that has a device such as a network chip or processor that is instantiated repeatedly in a mesh configuration. The key idea is that the top-level module creates an array of interfaces and counters. Now the testbench can connect its array of virtual interfaces to the physical ones.

Sample 10.16 shows the code for the counter's interface, X_if. If the code printed the signal values with a $monitor, they would display when any signal changed. Instead, the always block waits until the clocking block changes, then prints the values of the signals at the end of the time slot with $strobe. The result is you are now working at a higher level of abstraction, seeing the values cycle by cycle instead of the individual events.

Sample 10.16   Interface for 8-bit counter

```
interface X_if (input logic clk);
  logic [7:0] din, dout;
  logic reset_l, load;

  clocking cb @(posedge clk);
    output din, load;
    input dout;
  endclocking

  always @cb
    $strobe("@%0t: %m: out=%0d, in=%0d, ld=%0d, r=%0d",
            $time, dout, din, load, reset_l);

  modport DUT (input clk, din, reset_l, load,
               output dout);

  modport TB (clocking cb, output reset_l);
endinterface
```

The simple counter is shown in Sample 10.17.

Sample 10.17 Counter model using X_if interface

```
// Simple 8-bit counter with load and active-low reset
module counter(X_if.DUT xi);
  logic [7:0] count;
  assign xi.dout = count;

  always @(posedge xi.clk or negedge xi.reset_l)
    begin
      if (!xi.reset_l)  count <= '0;
      else if (xi.load) count <= xi.din;
      else              count <= count+1;
    end
endmodule
```

The top-level module in Sample 10.18 uses a generate statement to instantiate NUM_XI interfaces and counters, but only one testbench.

Sample 10.18 Top-level module with an array of virtual interfaces

```
module top;
  parameter NUM_XI = 2;  // Number of design instances

  // Clock generator
  bit clk;
  initial begin
    clk <= '0;
    forever #20 clk = ~clk;
  end

  // Instantiate NUM_XI interfaces
  X_if xi[NUM_XI] (clk);

  // Instantiate the testbench, passing the number of interfaces
  test #(.NUM_XI(NUM_XI)) tb();

  // Generate NUM_XI counter instances
  generate
  for (genvar i=0; i<NUM_XI; i++)
    begin : count_blk
      counter c (xi[i]);
    end
  endgenerate

endmodule : top
```

In Sample 10.19, the key line in the testbench is where the local virtual interface array, vxi, is assigned to point to the array of physical interfaces in the top module, top.xi. (Note that this example takes some shortcuts compared to the recommendations in Chapter 8. To simplify Sample 10.18, the environment class has been merged with the test, whereas the generator, agent, and driver layers have been compressed into the driver.)

The testbench assumes there is at least one counter and thus at least one X interface. If your design could have zero counters, you would have to use a dynamic array to hold the virtual interfaces, as a fixed-size array cannot have a size of zero. The actual number of interfaces is passed as a parameter from the top-level module.

Sample 10.19   Counter testbench using virtual interfaces

```
program automatic test #(NUM_XI=2);

  virtual X_if.TB vxi[NUM_XI]; // Virtual interface array
  Driver driver[];

  initial begin
    // Connect local virtual interface to top
    vxi = top.xi;

    // Create N drivers
    driver = new[NUM_XI];
    foreach (driver[i])
      driver[i] = new(vxi[i], i);

    foreach (driver[i]) begin
      automatic int j = i;
      fork
        begin
          driver[j].reset();
          driver[j].load_op();
        end
      join_none
    end

    repeat (10) @(vxi[0].cb);
  end

endprogram
```

Of course in this simple example, you could just pass the interface directly into the Driver's constructor, rather than make a separate variable.

In Sample 10.20, the Driver class uses a single virtual interface to drive and sample signals from the counter.

Sample 10.20 Driver class using virtual interfaces

```
class Driver;
  virtual X_if.TB xi;
  int id;

  function new(input virtual X_if.TB xi, input int id);
    this.xi = xi;
    this.id = id;
  endfunction

  task reset();
    $display("@%0t: Driver[%0d]: Start reset", $time, id);
    // Reset the device
    xi.reset_l <= 1;
    xi.cb.load <= 0;
    xi.cb.din <= '0;
    @(xi.cb) xi.reset_l <= 0;
    @(xi.cb) xi.reset_l <= 1;
    $display("@%0t:Driver[%0d]: End reset", $time, id);
  endtask : reset

  task load_op();
    $display("@%0t: Driver[%0d]: Start load", $time, id);
    ##1 xi.cb.load <= 1;
    xi.cb.din <= id + 10;

    ##1 xi.cb.load <= 0;
    repeat (5) @(xi.cb);
    $display("@%0t: Driver[%0d]: End load", $time, id);
  endtask : load_op

endclass : Driver
```

10.2.2   Using *Typedefs* with *Virtual Interfaces*

You can reduce the amount of typing, and ensure you always use the correct modport by replacing "virtual X_if.TB" with a typedef, as shown in Sample 10.21
through 10.23, of the interface, testbench, and driver.

Sample 10.21 Interface with a typedef

```
interface X_if (input logic clk);
  //...
endinterface
typedef virtual X_if.TB vx_if;
```

Sample 10.22 Testbench using a typedef for virtual interfaces

```
program automatic test #(NUM_XI=2);
  vx_if vxi[NUM_XI];          // Virtual interface array
  Driver driver[];
  // ...
endprogram
```

Sample 10.23 Driver using a typedef for virtual interfaces

```
class Driver;
  vx_if xi;
  int id;

  function new(input vx_if xi, input int id);
    this.xi = xi;
    this.id = id;
  endfunction
  // ...
endclass : Driver
```

10.2.3   Passing Virtual Interface Array Using a Port

The previous examples passed the array of virtual interfaces using a cross module reference (XMR). An alternative is to pass the array in a port. Since the array in the top module is static and so only needs to be referenced once, the XMR style makes more sense than using a port that normally is used to pass changing values.

Sample 10.24 uses a global parameter to define the number of X interfaces. Here is a snippet of the top module.

Sample 10.24 Testbench using an array of virtual interfaces

```
parameter NUM_XI = 2;  // Number of instances

module top;
  // Instantiate N interfaces
  X_if xi [NUM_XI] (clk);

  ...
  // Instantiate the testbench
  test tb(xi);

endmodule : top
```

The testbench that uses the virtual interfaces is shown in Sample 10.25. It creates an array of virtual interfaces so that it can pass them into the constructor for the driver class, or just pass the interface directly from the port.

Sample 10.25 Testbench passing virtual interfaces with a port

```
program automatic test(X_if xi[NUM_XI]);

  vx_if vxi[NUM_XI];
  Driver driver[];

  initial begin
    // Build phase
    // Connect the local virtual interfaces to the top
    vxi = xi;                       // Assign the interface array
    driver = new[NUM_XI];

    foreach (vxi[i])                // Create NUM_XI drivers
      driver[i] = new(vxi[i], i);

    // Reset phase
    foreach (vxi[i])
      fork
        begin
          driver[i].reset();
          driver[i].load_op();
        end
      join
    //...
  end

endprogram
```

10.3 Parameterized Interfaces and Virtual Interfaces

The example in Section 10.2 shows an 8-bit counter and matching busses. What if you want to vary the counter's width? Verilog-1995 allows you to parameterize modules, and System Verilog extends this concept with parameterized interfaces and virtual interfaces.

First, update the counter, originally shown in Sample 10.17 with parameters. This only requires changing the first few lines. Sample 10.26 now passes the number of interfaces in as a parameter too.

Sample 10.26 Parameterized counter model using X_if interface

```
// Simple N-bit counter with load and active-low reset
module counter #(BIT_WIDTH = 8) (X_if.DUT xi);
  logic [BIT_WIDTH-1:0] count;
...
```

Next, Sample 10.27 adds the bit width parameter to the interface in Sample 10.16.

Sample 10.27 Parameterized interface for 8-bit counter

```
interface X_if #(BIT_WIDTH=8) (input logic clk);
  logic [BIT_WIDTH-1:0] din, dout;
...
```

Sample 10.28 shows the parameter being passed into the testbench.

Sample 10.28 Parameterized top-level module with an array of virtual interfaces

```
module top;
  parameter NUM_XI = 2;     // Number of design instances
  parameter BIT_WIDTH = 4; // Width of counter and bus

  // Clock generator
  bit clk;
  initial begin
    clk <= '0;
    forever #20 clk = ~clk;
  end

  // Instantiate N interfaces
  X_if #(.BIT_WIDTH(BIT_WIDTH)) xi[NUM_XI] (clk);

  // Testbench with the number of interfaces and bit width
  test #(.NUM_XI(NUM_XI), .BIT_WIDTH(BIT_WIDTH)) tb();

  // Generate N counter instances
  generate
  for (genvar i=0; i<NUM_XI; i++)
    begin : count_blk
      counter #(.BIT_WIDTH(BIT_WIDTH)) c (xi[i]);
    end
  endgenerate

endmodule : top
```

Lastly are the testbench module and Driver class are shown in Samples 10.29 and 10.30. These have virtual interfaces that must be parameterized. The syntax for this is a little tricky, especially when you have a modport. First, the testbench, updated from Sample 10.19. Notice how the parameter goes between the type name and the modport.

Sample 10.29 Parameterized counter testbench using virtual interfaces

```
program automatic test #(NUM_XI=2, BIT_WIDTH=8);
  virtual X_if #(.BIT_WIDTH(BIT_WIDTH)).TB vxi[NUM_XI];

...
```

Sample 10.30 Driver class using virtual interfaces

```
class Driver;
  virtual X_if  #(.BIT_WIDTH(BIT_WIDTH)) xi;

  //...

endclass
```

10.4   Procedural Code in an Interface

Just as a class contains both variables and routines, an interface can contain code such as routines, assertions, and initial and always blocks. Recall that an interface includes the signals and functionality of the communication between two blocks. So the interface block for a bus can contain the signals and also routines to perform commands such as a read or write. The inner workings of these routines are hidden from the external blocks, allowing you to defer the actual implementation. Access to these routines is controlled using the modport statement, just as with signals. A task or function is imported into a modport so that it is then visible to any block that uses the modport.

These routines can be used by both the design and the testbench. This approach ensures that both are using the same protocol, eliminating a common source of testbench bugs. However, not all synthesis tools can handle routines in an interface.

A problem with sharing code between the design and testbench is that the independence between the design and verification teams is lost. If only one person implements the interface protocol for both parts, who checks it?

You can verify a protocol with assertions in an interface. An assertion can check for illegal combinations, such as protocol violations and unknown values. These can display state information and stop simulation immediately so that you can easily debug the problem. An assertion can also fire when good transactions occur. Functional coverage code uses this type of assertion to trigger the gathering of coverage information.

10.4.1   Interface with Parallel Protocol

When creating your system, you may not know whether to choose a parallel or serial protocol. The interface in Sample 10.31 has two tasks, initiatorSend and targetRcv, that send a transaction between two blocks using the interface signals. It sends the address and data in parallel across two 8-bit buses.

Sample 10.31   Interface with tasks for parallel protocol

```
interface simple_if(input logic clk);
  logic [7:0] addr;
  logic [7:0] data;
  bus_cmd_e cmd;
  modport TARGET
    (input   addr, cmd, data,
      import task targetRcv (output bus_cmd_e c,
                              logic [7:0] a, d));
   modport INITIATOR
     (output addr, cmd, data,
       import task initiatorSend(input bus_cmd_e c,
                                  logic [7:0] a, d)
     );

  // Parallel send
  task initiatorSend(input bus_cmd_e c,
                      logic [7:0] a, d);
    @(posedge clk);
    cmd <= c;
    addr <= a;
    data <= d;
  endtask

  // Parallel receive
  task targetRcv(output bus_cmd_e c, logic [7:0] a, d);
    @(posedge clk);
    a = addr;            // Use non-blocking assignments to
    d = data;            // immediately sample the bus values
    c = cmd;             // and avoid race conditions
  endtask
endinterface: simple_if
```

10.4.2   Interface with Serial Protocol

The interface in Sample 10.32 implements a serial interface for sending and receiving the address and data values. It has the same interface and routine names as Sample 10.31, so you can swap between the two without having to change any design or testbench code.

Sample 10.32 Interface with tasks for serial protocol

```
interface simple_if(input logic clk);
  logic addr;
  logic data;
  logic start = 0;
  bus_cmd_e cmd;

  modport TARGET(input  addr, cmd, data,
                 import task targetRcv (output bus_cmd_e c,
                                        logic [7:0] a, d));
  modport INITIATOR(output addr, cmd, data,
                    import task initiatorSend(input bus_cmd_e c,
                                              logic [7:0] a, d));

  // Serial send
  task initiatorSend(input bus_cmd_e c, logic [7:0] a, d);
    @(posedge clk);
    start <= 1;
    cmd <= c;
    foreach (a[i]) begin
      addr <= a[i];
      data <= d[i];
      @(posedge clk);
      start <= 0;
    end
    cmd <= IDLE;
  endtask

  // Serial receive
  task targetRcv(output bus_cmd_e c, logic [7:0] a, d);
    @(posedge start);
    c = cmd;
    foreach (a[i]) begin
      @(posedge clk);
      a[i] = addr;
      d[i] = data;
    end
  endtask

endinterface: simple_if
```

10.4.3 Limitations of Interface Code

Tasks in interfaces are fine for RTL, where the functionality is strictly defined. However, these tasks are a poor choice for any type of verification IP. Interfaces and their code cannot be extended, overloaded, or dynamically instantiated based on configuration. An interface cannot have private data. Any code for verification needs maximum flexibility and configurability, and so should go in classes that run in a program block.

10.5   Conclusion

The interface construct in SystemVerilog provides a powerful technique to group together the connectivity, timing, and functionality for the communication between blocks. In this chapter you saw how you can create a single testbench that connects to many different design configurations containing multiple interfaces. Your signal layer code can connect to a variable number of physical interfaces at run time with virtual interfaces. Additionally, an interface can have routines that drive the signals and assertions to check the protocol, but put the test in a program block, not an interface.

In many ways, an interface can resemble a class with pointers, encapsulation, and abstraction. This lets you create an interface to model your system at a higher level than Verilog's traditional ports and wires. Just remember to keep the testbench in the program block.

10.6   Exercises

1. Complete the following code, as indicated by the comments.

```
class Driver;
  ...
  // Declare a virtual interface for the DUT
  function new(input inst_mbox #(Instruction) agt2drv,
              /* complete the argument list */);
    this.agt2drv = agt2drv;
    // Save the virtual interface argument in
    // the class-level variable
  endfunction
endclass

class Environment;
  Driver drv;
  .....
  drv = new(agt2drv, /* complete the argument list */);
  .....
endclass
```

2. Using the solution to Exercise 1, complete the following code as indicated by the comments.

```
program automatic test(risc_spm_if risc_bus);
    import my_package::*;
    Environment env;
    initial begin
        // Create object referenced by env handle
    end
endprogram
```

3. Modify the following program declaration to use cross module references (XMR). Assume the top module that contains the interface is named top.

```
program automatic test(risc_spm_if risc_bus);
    ...
endprogram
```

Modify the following instantiation of program test to use cross module references (XMR).

```
`include "risc_spm_if.sv"
module top;
    ....
    test t1(risc_bus);
    ....
endmodule
```

4. Expand the solution to Exercise 3 to create NUM_RISC_BUS environments and create NUM_RISC_BUS interfaces.

5. Expand the solution to Exercise 3 to use a typedef for the virtual interface.

6. Modify the following interface to use a parameter, ADDRESS_WIDTH. By default the addressing space supports 256 words.

```
interface risc_spm_if (input bit clk);

  bit rst;
  bit     [7:0] data_out;

  logic [7:0] address;
  logic [7:0] data_in;
  logic       write;
  modport DUT (input clk, data_out,
               output address, data_in, write);
endinterface
```

Chapter 11
A Complete SystemVerilog Testbench

This chapter applies the many concepts you have learned about SystemVerilog features to verify a design. The testbench creates constrained random stimulus, and gathers functional coverage. It is structured according to the guidelines from Chapter 8 so you can inject new behavior without modifying the lower-level blocks.

The design is an ATM switch that was shown in Sutherland [2006], who based his SystemVerilog description on an example from Janick Bergeron's Verification Guild. Sutherland took the original Verilog design and used SystemVerilog design features to create a switch that can be configured from 4×4 to 16×16. The testbench in the original example creates ATM cells using $urandom, overwrites certain fields with ID values, sends them through the device, then checks that the same values were received.

The entire example, with the testbench and ATM switch, is available for download at http://chris.spear.net/systemverilog. This chapter shows just the testbench code.

11.1 Design Blocks

The overall connection between the design and testbench, shown in Fig. 11.1, follows the pattern shown in Chapter 4.

C. Spear and G. Tumbush, *SystemVerilog for Verification: A Guide to Learning the Testbench Language Features*, DOI 10.1007/978-1-4614-0715-7_11,
© Springer Science+Business Media, LLC 2012

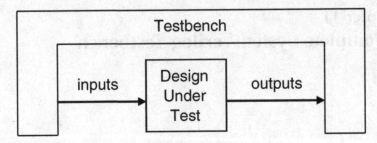

Fig. 11.1 The testbench — design environment

The top level of the design is called squat, as shown in Fig. 11.2. The module has 1..N Utopia Rx interfaces that are sending UNI formatted cells. Inside the DUT, cells are stored, converted to NNI format, and forwarded to the Tx interfaces. The forwarding is done according to a lookup table that is addressed with the VPI field of the incoming cell. The table is programmed through the management interface.

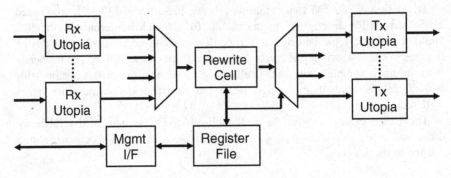

Fig. 11.2 Block diagram for the squat design

The top level module in Sample 11.1 defines arrays of interfaces for the Rx and Tx ports.

Sample 11.1 Top level module

```
`timescale 1ns/1ns
`define TxPorts 4  // set number of transmit ports
`define RxPorts 4  // set number of receive ports

module top;
  parameter int NumRx = `RxPorts;
  parameter int NumTx = `TxPorts;

  logic rst, clk;
  // System Clock and Reset
  initial begin
    rst = 0; clk = 0;
    #5ns rst = 1;
    #5ns clk = 1;
    #5ns rst = 0; clk = 0;
    forever
      #5ns clk = ~clk;
  end

  Utopia Rx[0:NumRx-1] ();// NumRx x Level 1 Utopia Rx Interface
  Utopia Tx[0:NumTx-1] ();// NumTx x Level 1 Utopia Tx Interface
  cpu_ifc mif();             // Utopia management interface
  squat #(NumRx, NumTx) squat(Rx, Tx, mif, rst, clk);  // DUT
  test  #(NumRx, NumTx) t1(Rx, Tx, mif, rst);          // Test
endmodule : top
```

The testbench program in Sample 11.2 passes the interfaces and signals through the port list. See Section 10.1.4 for a discussion on ports vs. cross module references. The actual testbench code is in the Environment class. The program steps through the phases of the environment. In order to work at a higher level of abstraction, the testbench only uses clocking blocks in the interfaces to synchronize with the DUT, not low level clocks.

Sample 11.2 Testbench program

```
program automatic test
  #(parameter int NumRx = 4, parameter int NumTx = 4)
    (Utopia.TB_Rx Rx[0:NumRx-1],
     Utopia.TB_Tx Tx[0:NumTx-1],
     cpu_ifc.Test mif,
     input logic rst);

`include "environment.sv"
  Environment env;

  initial begin
    env = new(Rx, Tx, NumRx, NumTx, mif);
    env.gen_cfg();
    env.build();
    env.run();
    env.wrap_up();
  end
endprogram // test
```

The testbench loads control information into the ATM switch through the Management interface, also known as the CPU interface, shown in Sample 11.3. In this chapter's examples, the interface is only used to load the lookup table that maps VPI to forwarding masks.

Sample 11.3 CPU Management Interface

```
interface cpu_ifc;
  logic        BusMode, Sel, Rd_DS, Wr_RW, Rdy_Dtack;
  logic [11:0] Addr;
  CellCfgType  DataIn, DataOut;     // Defined in Sample 11-11

  modport Peripheral
          (input  BusMode, Addr, Sel, DataIn, Rd_DS, Wr_RW,
           output DataOut, Rdy_Dtack);

  modport Test
          (output BusMode, Addr, Sel, DataIn, Rd_DS, Wr_RW,
           input  DataOut, Rdy_Dtack);

endinterface : cpu_ifc

typedef virtual cpu_ifc.Test vCPU_T;
```

Sample 11.4 shows the Utopia interface, which is used by the testbench to communicate with the squat design by transmitting and receiving ATM cells. The interface has clocking blocks for the transmit and receive paths, and modports for the design and testbench connections to the interface.

Sample 11.4 Utopia interface

```
interface Utopia #(IfWidth = 8);

  logic [IfWidth-1:0] data;
  bit clk_in, clk_out;
  bit soc, en, clav, valid, ready, reset, selected;

  ATMCellType ATMcell;   // union of structures for ATM cells

  modport TopReceive (
    input  data, soc, clav,
    output clk_in, reset, ready, clk_out, en, ATMcell, valid );

  modport TopTransmit (
    input  clav,
    inout  selected,
    output clk_in, clk_out, ATMcell, data, soc, en, valid,
           reset, ready );

  modport CoreReceive (
    input  clk_in, data, soc, clav, ready, reset,
    output clk_out, en, ATMcell, valid );

  modport CoreTransmit (
    input  clk_in, clav, ATMcell, valid, reset,
    output clk_out, data, soc, en, ready );

  clocking cbr @(negedge clk_out);
    input clk_in, clk_out, ATMcell, valid, reset, en, ready;
    output data, soc, clav;
  endclocking : cbr
  modport TB_Rx (clocking cbr);

  clocking cbt @(negedge clk_out);
    input clk_out, clk_in, ATMcell, soc, en, valid,
          reset, data, ready;
    output clav;
  endclocking : cbt
  modport TB_Tx (clocking cbt);

endinterface

typedef virtual Utopia vUtopia;
typedef virtual Utopia.TB_Rx vUtopiaRx;
typedef virtual Utopia.TB_Tx vUtopiaTx;
```

11.2   Testbench Blocks

The environment class, as shown in Section 8.2.1, is the scaffolding that supports
the testbench structure. Inside this class lies the blocks of your layered testbench,
such as generators, drivers, monitors, and scoreboard. The environment also controls the sequencing of the four testbench steps: generate a random configuration,
build the testbench environment, run the test and wait for it to complete, and a wrap-
up phase to shut down the system and generate reports. Sample 11.5 shows the ATM
environment class. It uses the virtual interface vCPU_T defined in Sample 11.3.

Sample 11.5 Environment class header

```
class Environment;
  UNI_generator gen[];
  mailbox gen2drv[];
  event    drv2gen[];
  Driver drv[];
  Monitor mon[];
  Config cfg;
  Scoreboard scb;
  Coverage cov;
  virtual Utopia.TB_Rx Rx[];
  virtual Utopia.TB_Tx Tx[];
  int numRx, numTx;
  vCPU_T mif;
  CPU_driver cpu;

  extern function new(input vUtopiaRx Rx[],
                      input vUtopiaTx Tx[],
                      input int numRx, numTx,
                      input vCPU_T mif);
  extern virtual function void gen_cfg();
  extern virtual function void build();
  extern virtual task run();
  extern virtual function void wrap_up();

endclass : Environment
```

With the $test$plusargs() system task, the Environment class constructor
in Sample 11.6 looks for the VCS switch +ntb_random_seed, which sets the
random seed for the simulation. The system task $value$plusargs() extracts
the value from the switch. Your simulator may have a different way to set the seed.
It is important to print the seed in the log file so that if the test fails, you can run it
again with the same value.

Sample 11.6 Environment class methods

```
//-------------------------------------------------------------------
// Construct an environment instance
function Environment::new(input vUtopiaRx Rx[],
                          input vUtopiaTx Tx[],
                          input int numRx, numTx,
                          input vCPU_T mif);
  this.Rx = new[Rx.size()];
  foreach (Rx[i]) this.Rx[i] = Rx[i];
  this.Tx = new[Tx.size()];
  foreach (Tx[i]) this.Tx[i] = Tx[i];
  this.numRx = numRx;
  this.numTx = numTx;
  this.mif = mif;
  cfg = new(numRx,numTx);

  if ($test$plusargs("ntb_random_seed")) begin
    int seed;
    $value$plusargs("ntb_random_seed=%d", seed);
    $display("Simulation run with random seed=%0d", seed);
  end
  else
    $display("Simulation run with default random seed");
endfunction : new

//-------------------------------------------------------------------
// Randomize the configuration descriptor
function void Environment::gen_cfg();
  `SV_RAND_CHECK(cfg.randomize());
  cfg.display();
endfunction : gen_cfg

//-------------------------------------------------------------------
// Build the environment objects for this test
// Note that objects are built for every channel,
// even if they are not used.  This reduces null handle bugs.
function void Environment::build();
  cpu = new(mif, cfg);
  gen = new[numRx];
  drv = new[numRx];
  gen2drv = new[numRx];
  drv2gen = new[numRx];
  scb = new(cfg);
  cov = new();

  // Build generators
  foreach(gen[i]) begin
    gen2drv[i] = new();
```

```systemverilog
      gen[i] = new(gen2drv[i], drv2gen[i],
                  cfg.cells_per_chan[i], i);
      drv[i] = new(gen2drv[i], drv2gen[i], Rx[i], i);
    end

    // Build monitors
    mon = new[numTx];
    foreach (mon[i])
      mon[i] = new(Tx[i], i);

    // Connect scoreboard to drivers & monitors with callbacks
    begin
      Scb_Driver_cbs sdc = new(scb);
      Scb_Monitor_cbs smc = new(scb);
      foreach (drv[i]) drv[i].cbsq.push_back(sdc);
      foreach (mon[i]) mon[i].cbsq.push_back(smc);
    end

    // Connect coverage to monitor with callbacks
    begin
      Cov_Monitor_cbs smc = new(cov);
     foreach (mon[i])
        mon[i].cbsq.push_back(smc);
    end
endfunction : build

//----------------------------------------------------------------
// Start the transactors: generators, drivers, monitors
// Channels that are not in use don't get started
task Environment::run();
  int num_gen_running;

  // The CPU interface initializes before anyone else
  cpu.run();

  num_gen_running = numRx;

  // For each input RX channel, start generator and driver
  foreach(gen[i]) begin
    int j=i;       // Automatic var holds index in spawned threads
    fork
      begin
        if (cfg.in_use_Rx[j])
          gen[j].run();            // Wait for generator to finish
        num_gen_running--;// Decrement driver count
      end
      if (cfg.in_use_Rx[j]) drv[j].run();
    join_none
  end
```

```
// For each output TX channel, start monitor
foreach(mon[i]) begin
  int j=i;       // Automatic var holds index in spawned threads
  fork
    mon[j].run();
  join_none
end

// Wait for all generators to finish, or time-out
fork : timeout_block
  wait (num_gen_running == 0);
  begin
    repeat (1_000_000) @(Rx[0].cbr);
    $display("@%0t: %m ERROR: Generator timeout ", $time);
    cfg.nErrors++;
  end
join_any
disable timeout_block;

// Wait for the data to flow through switch, into monitors,
// and scoreboards
repeat (1_000) @(Rx[0].cbr);
endtask : run

//-----------------------------------------------------------------
// Post-run cleanup / reporting
function void Environment::wrap_up();
  $display("@%0t: End of sim, %0d errors, %0d warnings",
           $time, cfg.nErrors, cfg.nWarnings);
  scb.wrap_up();
endfunction : wrap_up
```

The method `Environment::build` in Sample 11.6 connects the scoreboard to the driver and monitor with the callback class, which is shown in Sample 11.7, `Scb_Driver_cbs`. This class sends the expected values to the scoreboard. The base driver callback class, `Driver_cbs`, is shown in Sample 11.20.

Sample 11.7 Callback class connects driver and scoreboard

```
class Scb_Driver_cbs extends Driver_cbs;
  Scoreboard scb;

  function new(input Scoreboard scb);
    this.scb = scb;
  endfunction : new

  // Send received cell to scoreboard
  virtual task post_tx(input Driver drv,
    input UNI_cell c);
    scb.save_expected(c);
  endtask : post_tx
endclass : Scb_Driver_cbs
```

The callback class in Sample 11.8, Scb_Monitor_cbs, connects the monitor with the scoreboard. The base monitor callback class, Monitor_cbs, is shown in Sample 11.21.

Sample 11.8 Callback class connects monitor and scoreboard

```
class Scb_Monitor_cbs extends Monitor_cbs;
  Scoreboard scb;

  function new(input Scoreboard scb);
    this.scb = scb;
  endfunction : new

  // Send received cell to scoreboard
  virtual task post_rx(input Monitor mon,
                       input NNI_cell c);
    scb.check_actual(c, mon.PortID);
  endtask : post_rx
endclass : Scb_Monitor_cbs
```

The environment connects the monitor to the coverage class with the final callback class, Cov_Monitor_cbs, shown in Sample 11.9.

Sample 11.9 Callback class connects the monitor and coverage

```
class Cov_Monitor_cbs extends Monitor_cbs;
  Coverage cov;

  function new(input Coverage cov);
    this.cov = cov;
  endfunction : new

  // Send received cell to coverage
  virtual task post_rx(input Monitor mon,
                       input NNI_cell c);
    CellCfgType CellCfg = top.squat.lut.read(c.VPI);
    cov.sample(mon.PortID, CellCfg.FWD);
  endtask : post_rx
endclass : Cov_Monitor_cbs
```

The random configuration class header is shown in Sample 11.10. It starts with nCells, a random value for the total number of cells that flow through the system. The constraint c_nCells_valid ensures that the number of cells is valid by being greater than zero, whereas c_nCells_reasonable limits the test to a reasonable size, 1000 cells. You can disable or override this if you want longer tests.

Next is a dynamic bit array, in_use_Rx, to specify which Rx channels into the switch are active. This is used in Sample 11.6 in the run method so that only active channels run.

The array cells_per_chan is used to randomly divide the total number of cells across the active channels. The constraint zero_unused_channels sets the number of cells to zero for inactive channels. To help the solver, the active channel mask is solved before dividing up the cells between channels. Otherwise, a channel would be inactive only if the number of cells assigned to it was zero, which is very unlikely.

Sample 11.10 Environment configuration class

```
class Config;
  int nErrors, nWarnings;  // Number of errors, warnings
  bit [31:0] numRx, numTx; // Copy of parameters

  rand bit [31:0] nCells;  // Total cells
  constraint c_nCells_valid
    {nCells > 0; }
  constraint c_nCells_reasonable
    {nCells < 1000; }

  rand bit in_use_Rx[];       // Input / output channel enabled
  constraint c_in_use_valid
    {in_use_Rx.sum() > 0; }    // At least one RX is enabled

  rand bit [31:0] cells_per_chan[];
  constraint c_sum_ncells_sum  // Split cells over all channels
    {cells_per_chan.sum() == nCells;}   // Total number of cells

  // Set the cell count to zero for any channel not in use
  constraint zero_unused_channels
    {foreach (cells_per_chan[i])
      {
        // Needed for even dist of in_use
        solve in_use_Rx[i] before cells_per_chan[i];
        if (in_use_Rx[i])
          cells_per_chan[i] inside {[1:nCells]};
        else cells_per_chan[i] == 0;
      }
    }

  extern function new(input bit [31:0] numRx, numTx);
  extern virtual function void display(input string prefix="");
endclass : Config
```

The cell rewriting and forwarding configuration type is shown in Sample 11.11.

Sample 11.11 Cell configuration type

```
typedef struct packed {
  bit [`TxPorts-1:0] FWD;
  bit [11:0] VPI;
} CellCfgType;
```

The methods for the configuration class are shown in Sample 11.12

Sample 11.12 Configuration class methods

```
function Config::new(input bit [31:0] numRx, numTx);
  this.numRx = numRx;
  in_use_Rx = new[numRx];
  this.numTx = numTx;
  cells_per_chan = new[numRx];
endfunction : new

function void Config::display(input string prefix);
  $write("%sConfig: numRx=%0d, numTx=%0d, nCells=%0d (",
          prefix, numRx, numTx, nCells);
  foreach (cells_per_chan[i])
    $write("%0d ", cells_per_chan[i]);
  $write("), enabled RX: ", prefix);
  foreach (in_use_Rx[i]) if (in_use_Rx[i]) $write("%0d ", i);
  $display;
endfunction : display
```

The ATM switch accepts UNI formatted cells and sends out NNI formatted cells. These cells are sent through both an OOP testbench and a structural design, so they are defined using `typedef`. The major difference between the two formats is that the UNI's GFC and VPI field are merged into the NNI's VPI. The definitions in Sample 11.13 through 11.15 are from Sutherland [2006].

Sample 11.13 UNI cell format

```
typedef struct packed {
  bit         [3:0]   GFC;
  bit         [7:0]   VPI;
  bit         [15:0]  VCI;
  bit                 CLP;
  bit         [2:0]   PT;
  bit         [7:0]   HEC;
  bit [0:47]  [7:0]   Payload;
} uniType;
```

Sample 11.14 NNI cell format

```
typedef struct packed {
  bit         [11:0]  VPI;
  bit         [15:0]  VCI;
  bit                 CLP;
  bit         [2:0]   PT;
  bit         [7:0]   HEC;
  bit [0:47]  [7:0]   Payload;
} nniType;
```

The UNI and NNI cells are merged with a byte memory to form a universal type, shown in Sample 11.15.

Sample 11.15 ATMCellType

```
typedef union packed {
  uniType uni;
  nniType nni;
  bit [0:52] [7:0] Mem;
} ATMCellType;
```

The testbench generates constrained random ATM cells, shown in Sample 11.16, that are extended from the BaseTr class, defined in Sample 8.24.

Sample 11.16 UNI_cell definition

```
class UNI_cell extends BaseTr;
  // Physical fields
  rand bit        [3:0]   GFC;
  rand bit        [7:0]   VPI;
  rand bit        [15:0]  VCI;
  rand bit                CLP;
  rand bit        [2:0]   PT;
       bit        [7:0]   HEC;
  rand bit [0:47] [7:0]   Payload;

  // Meta-data fields
  static bit [7:0] syndrome[0:255];
  static bit syndrome_not_generated = 1;

  extern function new();
  extern function void post_randomize();
  extern virtual function bit compare(input BaseTr to);
  extern virtual function void display(input string prefix="");
  extern virtual function BaseTr copy(input BaseTr to=null);
  extern virtual function void pack(output ATMCellType to);
  extern virtual function void unpack(input ATMCellType from);
  extern function NNI_cell to_NNI();
  extern function void generate_syndrome();
  extern function bit [7:0] hec (bit [31:0] hdr);
endclass : UNI_cell
```

Sample 11.17 shows the methods for the UNI cell.

Sample 11.17 UNI_cell methods

```
function UNI_cell::new();
  if (syndrome_not_generated)
    generate_syndrome();
endfunction : new


// Compute the HEC value after all other data has been chosen
function void UNI_cell::post_randomize();
  HEC = hec({GFC, VPI, VCI, CLP, PT});
endfunction : post_randomize


// Compare this cell with another
// This could be improved by telling what field mismatched
function bit UNI_cell::compare(input BaseTr to);
  UNI_cell c;
  $cast(c, to);
  if (this.GFC != c.GFC)         return 0;
  if (this.VPI != c.VPI)         return 0;
  if (this.VCI != c.VCI)         return 0;
  if (this.CLP != c.CLP)         return 0;
  if (this.PT  != c.PT)          return 0;
  if (this.HEC != c.HEC)         return 0;
  if (this.Payload != c.Payload) return 0;
  return 1;
endfunction : compare


// Print a "pretty" version of this object
function void UNI_cell::display(input string prefix);
  ATMCellType p;

  $display("%sUNI id:%0d GFC=%x, VPI=%x, VCI=%x, CLP=%b, PT=%x,
HEC=%x, Payload[0]=%x",
           prefix, id, GFC, VPI, VCI, CLP, PT, HEC, Payload[0]);
  this.pack(p);
  $write("%s", prefix);
  foreach (p.Mem[i]) $write("%x ", p.Mem[i]);
  $display;
endfunction : display


// Make a copy of this object
function BaseTr UNI_cell::copy(input BaseTr to);
  if (to == null) copy = new();
  else            $cast(copy, to);
```

```
  copy.GFC       = this.GFC;
  copy.VPI       = this.VPI;
  copy.VCI       = this.VCI;
  copy.CLP       = this.CLP;
  copy.PT        = this.PT;
  copy.HEC       = this.HEC;
  return copy;
endfunction : copy


// Pack this object's properties into a byte array
function void UNI_cell::pack(output ATMCellType to);
  to.uni.GFC     = this.GFC;
  to.uni.VPI     = this.VPI;
  to.uni.VCI     = this.VCI;
  to.uni.CLP     = this.CLP;
  to.uni.PT      = this.PT;
  to.uni.HEC     = this.HEC;
  to.uni.Payload = this.Payload;
endfunction : pack


// Unpack a byte array into this object
function void UNI_cell::unpack(input ATMCellType from);
  this.GFC      = from.uni.GFC;
  this.VPI      = from.uni.VPI;
  this.VCI      = from.uni.VCI;
  this.CLP      = from.uni.CLP;
  this.PT       = from.uni.PT;
  this.HEC      = from.uni.HEC;
  this.Payload  = from.uni.Payload;
endfunction : unpack


// Generate a NNI cell from an UNI cell - used in scoreboard
function NNI_cell UNI_cell::to_NNI();
  NNI_cell copy;
  copy = new();
  copy.VPI      = this.VPI;     // NNI has wider VPI
  copy.VCI      = this.VCI;
  copy.CLP      = this.CLP;
  copy.PT       = this.PT;
  copy.HEC      = this.HEC;
  copy.Payload  = this.Payload;
  return copy;
endfunction : to_NNI
```

```
// Generate the syndrome array, which is used to compute HEC
function void UNI_cell::generate_syndrome();
   bit [7:0] sndrm;
   for (int i = 0; i < 256; i = i + 1 ) begin
      sndrm = i;
      repeat (8) begin
         if (sndrm[7] === 1'b1)
            sndrm = (sndrm << 1) ^ 8'h07;
         else
            sndrm = sndrm << 1;
      end
      syndrome[i] = sndrm;
   end
   syndrome_not_generated = 0;
endfunction : generate_syndrome


// Compute the HEC value for this object
function bit [7:0] UNI_cell::hec (input bit [31:0] hdr);
   hec = 8'h00;
   repeat (4) begin
      hec = syndrome[hec ^ hdr[31:24]];
      hdr = hdr << 8;
   end
   hec = hec ^ 8'h55;
endfunction : hec
```

The NNI_cell class is almost identical to UNI_cell, except that it does not have a GFC field, or a method to convert to a UNI_cell.

Sample 11.18 shows the UNI cells random atomic generator, as originally shown in Section 8.2. The generator randomizes the blueprint instance of the UNI cell, and then sends out a copy of the cell to the driver.

Sample 11.18 UNI_generator class

```
class UNI_generator;
  UNI_cell blueprint; // Blueprint for generator
  mailbox  gen2drv;   // Mailbox to driver for cells
  event    drv2gen;   // Event from driver when done with cell
  int      nCells;    // Num cells for this generator to create
  int      PortID;    // Which Rx port are we generating?

  function new(input mailbox gen2drv,
              input event drv2gen,
              input int nCells, PortID);
    this.gen2drv = gen2drv;
    this.drv2gen = drv2gen;
    this.nCells  = nCells;
    this.PortID  = PortID;
    blueprint = new();
  endfunction : new

  task run();
    UNI_cell c;
    repeat (nCells) begin
      `SV_RAND_CHECK(blueprint.randomize());
      $cast(c, blueprint.copy());
      c.display($sformatf("@%0t: Gen%0d: ", $time, PortID));
      gen2drv.put(c);
      @drv2gen;// Wait for driver to finish with it
    end
  endtask : run

endclass : UNI_generator
```

Sample 11.19 shows the Driver class that sends UNI cells into the ATM switch. This class uses the driver callbacks in Sample 11.20. Note that there is a circular relationship here. The Driver class has a queue of Driver_cbs objects, and the pre_tx() and post_tx() methods in Driver_cbs are passed Driver objects. When you compile the two classes, you may need either typedef class Driver; before the Driver_cbs class definition, or typedef class Driver_cbs; before the Driver class definition.

Sample 11.19 driver class

```
typedef class Driver_cbs;

class Driver;

  mailbox gen2drv;    // For cells sent from generator
  event   drv2gen;    // Tell generator when I am done with cell
  vUtopiaRx Rx;       // Virtual ifc for transmitting cells
```

```
   Driver_cbs cbsq[$];   // Queue of callback objects
   int PortID;

   extern function new(input mailbox gen2drv,
                       input event drv2gen,
                       input vUtopiaRx Rx,
                       input int PortID);
   extern task run();
   extern task send (input UNI_cell c);

endclass : Driver


// new(): Construct a driver object
function Driver::new(input mailbox gen2drv,
                    input event drv2gen,
                    input vUtopiaRx Rx,
                    input int PortID);
  this.gen2drv = gen2drv;
  this.drv2gen = drv2gen;
  this.Rx      = Rx;
  this.PortID  = PortID;
endfunction : new


// run(): Run the driver.
// Get transaction from generator, send into DUT
task Driver::run();
  UNI_cell c;
  bit drop = 0;

  // Initialize ports
  Rx.cbr.data  <= 0;
  Rx.cbr.soc   <= 0;
  Rx.cbr.clav  <= 0;

  forever begin
    // Read the cell at the front of the mailbox
    gen2drv.peek(c);
    begin: Tx
      // Pre-transmit callbacks
      foreach (cbsq[i]) begin
        cbsq[i].pre_tx(this, c, drop);
        if (drop) disable Tx; // Don't transmit this cell
      end

      c.display($sformatf("@%0t: Drv%0d: ", $time, PortID));
      send(c);
```

```
    // Post-transmit callbacks
    foreach (cbsq[i])
      cbsq[i].post_tx(this, c);
    end : Tx

    gen2drv.get(c);   // Remove cell from the mailbox
    ->drv2gen;   // Tell the generator we are done with this cell
  end
endtask : run


// send(): Send a cell into the DUT
task Driver::send(input UNI_cell c);
  ATMCellType Pkt;

  c.pack(Pkt);
  $write("Sending cell: ");
  foreach (Pkt.Mem[i])
  $write("%x ", Pkt.Mem[i]); $display;

  // Iterate thru bytes of cell
  @(Rx.cbr);
  Rx.cbr.clav <= 1;
  for (int i=0; i<=52; i++) begin
    // If not enabled, loop
    while (Rx.cbr.en === 1'b1) @(Rx.cbr);

    // Assert Start Of Cell, assert enable, send byte 0 (i==0)
    Rx.cbr.soc  <= (i == 0);
    Rx.cbr.data <= Pkt.Mem[i];
    @(Rx.cbr);
  end
  Rx.cbr.soc <= 'z;
  Rx.cbr.data <= 8'bx;
  Rx.cbr.clav <= 0;
endtask
```

Sample 11.20 shows the driver callback class which has simple callbacks that are called before & after a cell is transmitted. This class has empty tasks, which are used by default. A test case can extend this class to inject new behavior in the driver without having to change any code in the driver

Sample 11.20 Driver callback class

```
typedef class Driver;

class Driver_cbs;
  virtual task pre_tx(input Driver drv,
                      input UNI_cell c,
                      inout bit drop);
  endtask : pre_tx

  virtual task post_tx(input Driver drv,
                       input UNI_cell c);
  endtask : post_tx
endclass : Driver_cbs
```

The Monitor class in Sample 11.21 has a very simple callback, with just one task that is called after a cell is received.

Sample 11.21 Monitor callback class

```
typedef class Monitor;

class Monitor_cbs;
  virtual task post_rx(input Monitor mon,
                       input NNI_cell c);
  endtask : post_rx
endclass : Monitor_cbs
```

Sample 11.22 shows the Monitor class. Like the Driver class, this uses a typedef to break the circular compile dependency with Monitor_cbs.

Sample 11.22 The Monitor class

```
typedef class Monitor_cbs;

class Monitor;

  vUtopiaTx Tx;              // Virtual interface with output of DUT
  Monitor_cbs cbsq[$];       // Queue of callback objects
  bit [1:0] PortID;

  extern function new(input vUtopiaTx Tx, input int PortID);
  extern task run();
  extern task receive (output NNI_cell c);
endclass : Monitor


// new(): construct an object
function Monitor::new(input vUtopiaTx Tx, input int PortID);
  this.Tx     = Tx;
```

```
   this.PortID = PortID;
endfunction : new


// run(): Run the monitor
task Monitor::run();
  NNI_cell c;

  forever begin
    receive(c);
    foreach (cbsq[i])
      cbsq[i].post_rx(this, c);   // Post-receive callback
  end
endtask : run


// receive(): Read cell from the DUT, pack into a NNI cell
task Monitor::receive(output NNI_cell cell);
  ATMCellType Pkt;

  Tx.cbt.clav <= 1;
  while (Tx.cbt.soc !== 1'b1 && Tx.cbt.en !== 1'b0)
    @(Tx.cbt);
  for (int i=0; i<=52; i++) begin
    // If not enabled, loop
    while (Tx.cbt.en !== 1'b0) @(Tx.cbt);

    Pkt.Mem[i] = Tx.cbt.data;
    @(Tx.cbt);
  end

  Tx.cbt.clav <= 0;

  c = new();
  c.unpack(Pkt);
  c.display($sformatf("@%0t: Mon%0d: ", $time, PortID));
endtask : receive
```

The scoreboard in Sample 11.23 gets expected cells from the driver through the function save_expected, and the cells actually received by the monitor with the function check_actual. The function save_expected() is called from the callback Scb_Driver_cbs::post_tx(), shown in Sample 11.7. The function check_actual() is called from Scb_Monitor_cbs::post_rx() in Sample 11.8.

Sample 11.23 The Scoreboard class

```
class Expect_cells;
  NNI_cell q[$];
  int iexpect, iactual;
endclass : Expect_cells


class Scoreboard;
  Config cfg;
  Expect_cells expect_cells[];
  NNI_cell cellq[$];
  int iexpect, iactual;

  extern function new(Config cfg);
  extern virtual function void wrap_up();
  extern function void save_expected(UNI_cell ucell);
  extern function void check_actual(input NNI_cell c,
                                    input int portn);
  extern function void display(string prefix="");
endclass : Scoreboard


function Scoreboard::new(input Config cfg);
  this.cfg = cfg;
  expect_cells = new[NumTx];
  foreach (expect_cells[i])
    expect_cells[i] = new();
endfunction : Scoreboard


function void Scoreboard::save_expected(input UNI_cell ucell);
  NNI_cell ncell = ucell.to_NNI;
  CellCfgType CellCfg = top.squat.lut.read(ncell.VPI);

  $display("@%0t: Scb save: VPI=%x, Forward=%b",
           $time, ncell.VPI, CellCfg.FWD);
  ncell.display($sformatf("@%0t: Scb save: ", $time));

  // Find all Tx ports where this cell will be forwarded
  for (int i=0; i<NumTx; i++)
    if (CellCfg.FWD[i]) begin
      expect_cells[i].q.push_back(ncell); // Save cell in this q
      expect_cells[i].iexpect++;
      iexpect++;
    end
endfunction : save_expected
```

```systemverilog
function void Scoreboard::check_actual(input NNI_cell c,
                                       input int portn);
  NNI_cell match;
  int match_idx;

  c.display($sformatf("@%0t: Scb check: ", $time));

  if (expect_cells[portn].q.size() == 0) begin
    $display("@%0t: ERROR: %m cell not found, SCB TX%0d empty",
             $time, portn);
    c.display("Not Found: ");
    cfg.nErrors++;
    return;
  end

  expect_cells[portn].iactual++;
  iactual++;

  foreach (expect_cells[portn].q[i]) begin
    if (expect_cells[portn].q[i].compare(c)) begin
      $display("@%0t: Match found for cell", $time);
      expect_cells[portn].q.delete(i);
      return;
    end
  end

  $display("@%0t: ERROR: %m cell not found", $time);
  c.display("Not Found: ");
  cfg.nErrors++;
endfunction : check_actual


// Print end of simulation report
function void Scoreboard::wrap_up();
  $display("@%0t: %m %0d expected cells, %0d actual cells rcvd",
           $time, iexpect, iactual);

  // Look for leftover cells
  foreach (expect_cells[i]) begin
    if (expect_cells[i].q.size()) begin
      $display("@%0t: %m cells in SCB Tx[%0d] at end of test",
               $time, i);
      this.display("Unclaimed: ");
      cfg.nErrors++;
    end
  end
endfunction : wrap_up
```

```
// Print the contents of the scoreboard, mainly for debugging
function void Scoreboard::display(input string prefix);
  $display("@%0t: %m so far %0d expected cells, %0d actual
rcvd", $time, iexpect, iactual);
  foreach (expect_cells[i]) begin
    $display("Tx[%0d]: exp=%0d, act=%0d",
           i, expect_cells[i].iexpect, expect_cells[i].iactual);
  foreach (expect_cells[i].q[j])
    expect_cells[i].q[j].display(
                 $sformatf("%sScoreboard: Tx%0d: ", prefix, i));
  end
endfunction : display
```

Sample 11.24 shows the class used to gather functional coverage. Since the coverage only looks at data in a single class, the cover group is defined and instantiated inside the Coverage class. The data values are read by the class's sample() method, then the cover group's sample() method is called to record the values.

Sample 11.24 Functional coverage class

```
class Coverage;
  bit [1:0] src;
  bit [NumTx-1:0] fwd;

  covergroup CG_Forward;
    coverpoint src
      {bins src[] = {[0:3]};
       option.weight = 0;}
    coverpoint fwd
      {bins fwd[] = {[1:15]}; // Ignore fwd==0
       option.weight = 0;}
    cross src, fwd;
  endgroup : CG_Forward

  function new();
    CG_Forward = new;      // Instantiate the covergroup
  endfunction : new

  // Sample input data
  function void sample(input bit [1:0] src,
                       input bit [NumTx-1:0] fwd);
    $display("@%0t: Coverage: src=%d. FWD=%b", $time, src, fwd);
    this.src = src;
    this.fwd = fwd;
    CG_Forward.sample();
  endfunction : sample
endclass : Coverage
```

Sample 11.25 shows the `CPU_driver` class that contains the methods to drive the CPU interface.

Sample 11.25 The CPU_driver class

```
class CPU_driver;
  vCPU_T mif;
  CellCfgType lookup [255:0]; // copy of look-up table
  Config cfg;
  bit [NumTx-1:0] fwd;

  extern function new(vCPU_T mif, Config cfg);
  extern task Initialize_Host ();
  extern task HostWrite (int a, CellCfgType d); // configure
  extern task HostRead (int a, output CellCfgType d);
  extern task run();
endclass : CPU_driver


function CPU_driver::new(input vCPU_T mif, Config cfg);
  this.mif = mif;
  this.cfg = cfg;
endfunction : new


task CPU_driver::Initialize_Host();
  mif.BusMode <= 1;
  mif.Addr <= 0;
  mif.DataIn <= 0;
  mif.Sel <= 1;
  mif.Rd_DS <= 1;
  mif.Wr_RW <= 1;
endtask : Initialize_Host


task CPU_driver::HostWrite (int a, CellCfgType d); // configure
  #10 mif.Addr <= a; mif.DataIn <= d; mif.Sel <= 0;
  #10 mif.Wr_RW <= 0;
  while (mif.Rdy_Dtack!==0) #10;
  #10 mif.Wr_RW <= 1; mif.Sel <= 1;
  while (mif.Rdy_Dtack==0) #10;
endtask : HostWrite


task CPU_driver::HostRead (input int a, output CellCfgType d);
  #10 mif.Addr <= a; mif.Sel <= 0;
  #10 mif.Rd_DS <= 0;
  while (mif.Rdy_Dtack!==0) #10;
  #10 d = mif.DataOut; mif.Rd_DS <= 1; mif.Sel <= 1;
```

```
   while (mif.Rdy_Dtack==0) #10;
endtask : HostRead


task CPU_driver::run();
  CellCfgType CellFwd;
  Initialize_Host();

  // Configure through Host interface
  repeat (10) @(negedge clk);
  $write("Memory: Loading ... ");
  for (int i=0; i<=255; i++) begin
    CellFwd.FWD = $urandom();
`ifdef FWDALL
    CellFwd.FWD = '1
`endif
    CellFwd.VPI = i;
    HostWrite(i, CellFwd);
    lookup[i] = CellFwd;
  end

  // Verify memory
  $write("Verifying ...");
  for (int i=0; i<=255; i++) begin
    HostRead(i, CellFwd);
    if (lookup[i] != CellFwd) begin
    $display("FATAL, Mem Loc 0x%x contains 0x%x, expected 0x%x",
                i, CellFwd, lookup[i]);
      $finish;
    end
  end
  $display("Verified");

endtask : run
```

11.3   Alternate Tests

The simplest test program is shown in Sample 11.2 and runs with very few constraints. During verification, you will be creating many tests, depending on the
major functionality to be tested. Each test can then be run with different seeds.

11.3.1   Your First Test - Just One Cell

The first test you run should probably have just one cell, such as the test in Sample
11.26. You can add a new constraint to the Config class by extending it, and then

injecting a new object into the environment before randomization. Once this test works, you can try two cells, then rewrite the constraint on the number of cells to run longer sequences.

Sample 11.26 Test with one cell

```
program automatic test
  #(parameter int NumRx = 4, parameter int NumTx = 4)
   (Utopia.TB_Rx Rx[0:NumRx-1],
    Utopia.TB_Tx Tx[0:NumTx-1],
    cpu_ifc.Test mif,
    input logic rst, clk);

`include "environment.sv"
  Environment env;

class Config_1_cell extends Config;
  constraint one_cells {nCells == 1; }

  function new(input int NumRx,NumTx);
    super.new(NumRx,NumTx);
  endfunction : new
endclass : Config_1_cells


  initial begin
    env = new(Rx, Tx, NumRx, NumTx, mif);

    begin // Just simulate for 1 cell
      Config_1_cells c1 = new(NumRx,NumTx);
      env.cfg = c1;
    end

    env.gen_cfg();      // Config will have just 1 cell
    env.build();
    env.run();
    env.wrap_up();
  end

endprogram // test
```

11.3.2 Randomly Drop Cells

The next test you may run creates errors by occasionally dropping cells, as shown in Sample 11.27. You need to make a callback for the driver that sets the drop bit. Then, in the test, inject this new functionality after the driver class has been constructed during the build phase.

Sample 11.27   Test that drops cells using driver callback

```
program automatic test
  #(parameter int NumRx = 4, parameter int NumTx = 4)
    (Utopia.TB_Rx Rx[0:NumRx-1],
     Utopia.TB_Tx Tx[0:NumTx-1],
     cpu_ifc.Test mif,
     input logic rst, clk);

`include "environment.sv"
  Environment env;

class Driver_cbs_drop extends Driver_cbs;
  virtual task pre_tx(input ATM_cell cell, ref bit drop);
    // Randomly drop 1 out of every 100 transactions
    drop = ($urandom_range(0,99) == 0);
  endtask
endclass


  initial begin
    env = new(Rx, Tx, NumRx, NumTx, mif);
    env.gen_cfg();
    env.build();

    begin                 // Create error injection callback
      Driver_cbs_drop dcd = new();
      env.drv.cbs.push_back(dcd); // Put into driver's Q
    end

    env.run();
    env.wrap_up();
  end

endprogram // test
```

11.4   Conclusion

This chapter shows how you can build a layered testbench, following the guidelines in this book. You can then create new tests by just modifying a single file and injecting new behavior, utilizing the hooks such as callbacks and multiple environment phases.

The testbench was able to get to 100% functional coverage of the ATM switch, at least for the basic cover group. You can use this example to explore more about SystemVerilog testbenches.

11.5   Exercises

1. In Sample 11.2, why is `clk` not passed into the port list of program `test`?

2. In Sample 11.6, could `numRx` be substituted for `Rx.size()` ? Why or why not?

3. For the following code snippet from Sample 11.6, explain what is being created for each statement.

```
function void Environment::build();
  cpu = new(mif, cfg);
  gen = new[numRx];
  drv = new[numRx];
  gen2drv = new[numRx];
  drv2gen = new[numRx];
  scb = new(cfg);
  cov = new();
  foreach(gen[i]) begin
    gen2drv[i] = new();
    gen[i] = new(gen2drv[i], drv2gen[i],
                 cfg.cells_per_chan[i], i);
    drv[i] = new(gen2drv[i], drv2gen[i], Rx[i], i);
  end
  ...
endfunction : build
```

4. In Sample 11.9, what coverage object does the handle `cov` point to?

5. In Sample 11.17, the function `UNI_cell::copy` assumes that the handle to the object `UNI_cell` points to an object of class `UNI_cell` as depicted in the following drawing. Draw what object the handle `dst` points to for the following function calls.

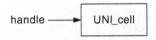

   a. `copy();`

   b. `copy(handle);`

6. In Sample 11.18, why are the `$cast()` required?

7. In Sample 11.19 and 11.20, why are the `typedef` declaration needed?

8. In Sample 11.19, why is `peek()` used first and then later a `get()`?

9. In Sample 11.23, is the error message "...cell not found..." in the function `check_actual` printed every time it are called? Why or why not?

10. Why do classes `Environment`, `Scoreboard`, and `CPU_driver` all define a handle to class `Config`? Are 3 objects of class `Config` created?

Chapter 12
Interfacing with C/C++

In Verilog, you can communicate with C routines using the Programming Language Interface. With the three generations of the PLI: TF (Task / Function), ACC (Access), and VPI (Verification Procedural Interface), you can create delay calculators, connect and synchronize multiple simulators, and add debug tools such as waveform displays. However the PLI's greatest strength is also its greatest weakness. If you just want to connect a simple C routine using the PLI, you need to write dozens of lines of code, and understand many different concepts such as synchronizing with multiple simulation phases, call frames, and instance pointers. Additionally, the PLI adds overhead to your simulation as it copies data between the Verilog and C domains, in order to protect Verilog data structures from corruption.

SystemVerilog introduces the Direct Programming Interface (DPI), an easier way to interface with C, C++, or any other foreign language. Once you declare or "import" the C routine with the `import` statement, you can call it as if it were any SystemVerilog routine. Additionally, your C code can call SystemVerilog routines. With the DPI you can connect C code that reads stimulus, contains a reference model, or just extends SystemVerilog with new functionality. Currently SystemVerilog only supports an interface to the C language. C++ code has to be wrapped to look like C.

If you have a SystemC model that does not consume time, and that you want to connect to SystemVerilog, you can use the DPI. SystemC models with time-consuming methods are best connected with the utilities built into your favorite simulator.

The first half of this chapter is data-centric and shows how you can pass different data types between SystemVerilog and C. The second half is control centric, showing how you can pass control back and forth between SystemVerilog and C. While the actual C code is trivial, with the factorial function, the Fibonacci series, and counters, they are easy to understand so you can quickly substitute your own code.

C. Spear and G. Tumbush, *SystemVerilog for Verification: A Guide to Learning the Testbench Language Features*, DOI 10.1007/978-1-4614-0715-7_12,
© Springer Science+Business Media, LLC 2012

12.1   Passing Simple Values

The first few examples in this chapter show you how to pass integral values between SystemVerilog and C, and the mechanics of how to declare routines and their arguments on both sides. Later sections show how to pass arrays and structures.

12.1.1   Passing Integer and Real Values

The most basic data type that you can pass between SystemVerilog and C is an `int`, the 2-state, 32-bit type. Sample 12.1 shows the SystemVerilog code that calls a C factorial routine, shown in Sample 12.2.

Sample 12.1 SystemVerilog code calling C factorial routine

```
import "DPI-C" function int factorial(input int i);

program automatic test;
  initial begin
    for (int i=1; i<=10; i++)
      $display("%0d! = %0d", i, factorial(i));
  end
endprogram
```

The `import` statement declares that a SystemVerilog routine `factorial` is implemented in a foreign language such as C. The modifier `"DPI-C"` specifies that this is a Direct Programing Interface routine, and the rest of the statement describes the routine arguments.

Sample 12.1 passes 32-bit signed values using the SystemVerilog `int` data type that maps directly to the C `int` type. The SystemVerilog int is always 32 bits, whereas the width of an int in C is operating system dependent. The C function in Sample 12.2 takes an integer as an input and so the DPI passes the argument by value.

Sample 12.2 C factorial function

```
int factorial(int i) {
  if (i<=1) return 1;
  else       return i*factorial(i-1);
}
```

12.1.2   The Import Declaration

The `import` declaration defines the prototype of the C task or function, but using SystemVerilog types. A C function with a return value is mapped to a SystemVerilog function. A void C function can be mapped to a SystemVerilog task or void function. If the name of the imported C function conflicts with a SystemVerilog name, you can import the function with a new name. In Sample 12.3, the C function `expect` is mapped to the SystemVerilog name `fexpect`, as the name `expect` is a reserved keyword in SystemVerilog. The name `expect` becomes a global symbol, used to link with the C code, whereas `fexpect` is a local SystemVerilog symbol. In the second half of the example, the C function `stat` is given a new name in SystemVerilog, `file_exists`. SystemVerilog does not support overloading a routine, for example by importing `expect` once with a `real` argument and once with an `int`.

Sample 12.3 Changing the name of an imported function

```
program automatic test;

  // C function has same name as reserved keyword, change it
  import "DPI-C" \expect = function int fexpect();
  ...
    if (actual != fexpect()) $display("ERROR");
  ...

  // Change name of C function "stat" to "file_exists"
  import "DPI-C" stat = function int file_exists
                 (input string fname, output int buff[1000]);
  initial begin
    int buff[1000];
    $display("file_exists(\"none.such\") = %0d",
             file_exists("none.such", buff));
  end
endprogram
```

You can import routines anywhere in your SystemVerilog code where you can declare a routine including inside programs, modules, interfaces, packages, and `$unit`, the compilation-unit space. The imported routine will be local to the declaration space in which it is declared. If you need to call an imported routine in several locations in your code, put the `import` statement in a package which you import where it is needed. Any changes to the `import` statements are localized to the package.

12.1.3 Argument Directions

Imported C routines can have zero or more arguments. By default the argument direction is `input` (data goes from SystemVerilog to C), but can also be `output` and `inout`. The direction `ref` is not supported. A function can return a simple value such as an integer or real number, or have no return value if you make it `void`. Sample 12.4 shows how to specify argument directions.

Sample 12.4 Argument directions

```
import "DPI-C" function int addmul (input int a, b,
                                    output int sum);
import "DPI-C" function void stop_model();
```

You can reduce the chances of bugs in your C code by declaring any input arguments as const as shown in Sample 12.5 so the C compiler will give an error for any write to an input.

Sample 12.5 C factorial routine with const argument

```
int factorial(const int i) {
  if (i<=1) return 1;
  else      return i*factorial(i-1);
}
```

12.1.4   Argument Types

Each variable that is passed through the DPI has two matching definitions: one for the SystemVerilog side, and one for the C side. It is your responsibility to use compatible types. The SystemVerilog simulator cannot compare the types as it is unable to read the C code. (The VCS compiler produces vc_hdrs.h and Questa creates incl.h with the C header for any routine that you have imported. You can use this file as a guide to matching the types.)

Table 12.1 shows the data type mapping between SystemVerilog and the inputs and outputs of C routines. The C structures are defined in the include file svdpi.h. Arrays mapping is discussed in Section 12.4 and 12.5, and structures are discussed in Section 12.6.

Table 12.1 Data types mapping between SystemVerilog and C

SystemVerilog	C (input)	C (output)
byte	char	char*
shortint	short int	short int*
int	int	int*
longint	long long int	long int*
shortreal	float	float*
real	double	double*
string	const char*	char**
string [N]	const char**	char**
bit	svBit or unsigned char	svBit* or unsigned char*
logic, reg	svLogic or unsigned char	svLogic* or unsigned char*
bit[N:0]	const svBitVecVal*	svBitVecVal*
reg[N:0] logic[N:0]	const svLogicVecVal*	svLogicVecVal*
unsized array[]	const svOpenArrayHandle	svOpenArrayHandle
chandle	const void*	void*

Note that some mappings are not exact. For example, a bit in SystemVerilog maps to svBit in C, which ultimately maps to unsigned char in the svdpi.h include file. As a result, you could write illegal values into the upper bits.

The LRM limits imported function results "small values", which include: void, byte, shortint, int, longint, real, shortreal, chandle, and string, plus single bit values of type bit and logic. A function cannot return a vector such as bit [6:0] as this would require returning a pointer to a svBitVecVal structure.

12.1.5   Importing a Math Library Routine

Sample 12.6 shows how you can call many functions in the C math library directly, without a C wrapper, thereby reducing the amount of code that you need to write. The Verilog real type maps to a C double.

Sample 12.6   Importing a C math function

```
import "DPI-C" function real fabs(input real r);
...
initial $display("fabs(0)=%f", fabs(-1.0));
```

12.2   Connecting to a Simple C Routine

Your C code might contain a simulation model, such as a processor, that is instantiated side by side with Verilog models. Or your code could be a reference model that is compared to a Verilog model at the transaction or cycle level. Many examples in this chapter show an 7-bit counter written in C or C++. Though very simple, the counter has the same parts as a complex model, with inputs, outputs, storage of internal values between calls, and the need to support multiple instances. The counter is 7 bits to show what happens when a hardware type does not match a C type.

12.2.1   A Counter with Static Storage

Sample 12.7 is the C code for an 7-bit counter. The count is stored in a static variable, as you might do if you wrote the model before thinking about simulation.

Sample 12.7 Counter routine using a static variable

```
#include <svdpi.h>

void counter7(svBitVecVal *o,
              const svBitVecVal *i,
              const svBit reset,
              const svBit load) {
  static unsigned char count = 0; // Static count storage

  if (reset)      count = 0;    // Reset
  else if (load)  count = *i;   // Load value
  else            count++;      // Count
  count &= 0x7f;                // Mask off upper bit

  *o = count;
}
```

The reset and load signals are 2-state single bit signals, and so they are passed as svBit which reduces to unsigned char. Your code could declare the value either way, but play it safe by using the SystemVerilog DPI types. The input i is 2-state, and 7 bits wide, and is passed as svBitVecVal. Notice that it is passed as a const pointer, which means the underlying value can change, but you cannot change the value of the pointer, such as making it point to another value. Likewise, the reset and load inputs are also marked as const. In this example, the 7-bit counter value is stored in a char, so you have to mask off the upper bit.

The file svdpi.h contains the definitions for SystemVerilog DPI structures and methods. The C code examples in the rest of this chapter leave off the #include statements, unless they are important to the discussion.

Sample 12.8 shows a SystemVerilog program that imports and calls the C function for the 7-bit counter.

Sample 12.8   Testbench for an 7-bit counter with static storage

```
import "DPI-C" function void counter7(output bit [6:0] out,
                                       input bit [6:0] in,
                                       input bit reset, load);
program automatic counter;
  bit [6:0] out, in;
  bit       reset, load;

  initial begin
    $monitor("SV: out=%3d, in=%3d, reset=%0d, load=%0d\n",
             out, in, reset, load);
    reset = 0;                          // Default values
    load = 0;
    in = 126;
    counter7(out, in, reset, load);     // Apply default values

    #10 reset = 1;
    counter7(out, in, reset, load);     // Apply reset

    #10 reset = 0;
    load = 1;
    counter7(out, in, reset, load);     // Load in=126

    #10 load = 0;
    counter7(out, in, reset, load);     // Count
  end
endprogram
```

12.2.2   The Chandle Data Type

The chandle data type allows you to store a C or C++ pointer in your SystemVerilog code. A chandle variable is wide enough to hold a pointer on the machine where the code was compiled, i.e. 32- or 64-bits. The counter in Sample 12.7 works well as long as it is the only one in the design. You could wrap the counter7 calls from Sample 12.8 in a module, and instantiate multiple copies in a design. However, since the counter value is stored in a C static, every instance shares a single value. If you need more than one instance of a module that calls C code, the C code needs to store its variables somewhere other than in static variables. A better way is to allocate storage, and pass a handle to it, along with the input and output signal values. Sample 12.9 shows a counter that stores the 7-bit count in the structure c7. This is overkill for a simple counter, but if you are creating a model for a larger device, you can build from this example.

Sample 12.9 Counter routine using instance storage

```
#include <svdpi.h>
#include <malloc.h>
#include <veriuser.h>

typedef struct {  // Structure to hold counter value
  unsigned char cnt;
} c7;

// Construct a counter structure
void* counter7_new() {
  c7* c = (c7*) malloc(sizeof(c7)); // Cast malloc value to c7
  c->cnt = 0;
  return c;
}

// Process the counter inputs
void counter7(c7 *inst,
              svBitVecVal* count,
              const svBitVecVal* i,
              const svBit reset,
              const svBit load) {

  if (reset)      inst->cnt = 0;   // Reset
  else if (load)  inst->cnt = *i;  // Load value
  else            inst->cnt++;     // Count
  inst->cnt &= 0x7f;               // Mask upper bit

  *count = inst->cnt;              // Write to output
  io_printf("C: count=%d, i=%d, reset=%d, load=%d\n",
            *count, *i, reset, load);
}
```

The routine counter7_new constructs the counter instance. This returns a chandle that must be passed into future calls to counter7. The counter value is stored in a struct of type c7. The function counter7_new calls malloc to allocate the struct, and casts the result into a local pointer c.

The C code uses the PLI task io_printf to display debug messages. The routine is helpful when you are debugging C and SystemVerilog code side-by-side as it writes to the same outputs, including log files, as $display, including the simulator's log file. The routine is defined in veriuser.h.

The testbench for this counter in Sample 12.10 differs from the static one in several ways. First, the counter must be constructed before it can be used. Next, the counter is called on a clock edge, rather than calling it in-line with the stimulus. For simplicity, the counter is invoked when the clock goes high, and stimulus is applied when the clock goes low, to avoid any race conditions.

Sample 12.10 Testbench for an 7-bit counter with per-instance storage

```
import "DPI-C" function chandle counter7_new();
import "DPI-C" function void counter7
      (input chandle inst,
        output bit [6:0] out,
        input  bit [6:0] in,
        input  bit  reset, load);

// Test two instances of the counter
program automatic test;

  bit [6:0] o1, o2, i1, i2;
  bit        reset, load, clk;
  chandle    inst1, inst2;      // Points to storage in C

  initial begin
    inst1 = counter7_new();
    inst2 = counter7_new();
    fork
      forever #10 clk = ~clk;
      forever @(posedge clk) begin
        counter7(inst1, o1, i1, reset, load);
        counter7(inst2, o2, i2, reset, load);
      end
    join_none

    reset = 0;                       // Initialize signals
    load = 0;
    i1 = 120;
    i2 = 10;

    @(negedge clk) load = 1;    // Load inputs
    @(negedge clk) load = 0;    // Count
    @(negedge clk) $finish;
  end
endprogram
```

12.2.3 Representation of Packed Values

The string "DPI-C"[1] specifies that you are using the canonical representation of packed values. This representation stores a SystemVerilog variable as a C array of one or more elements. A 2-state variable is stored using the type svBitVecVal. A 2-state array is stored with multiple elements of this type.

---

[1] Early versions of the LRM used "DPI" but this is now obsolete and should not be used.

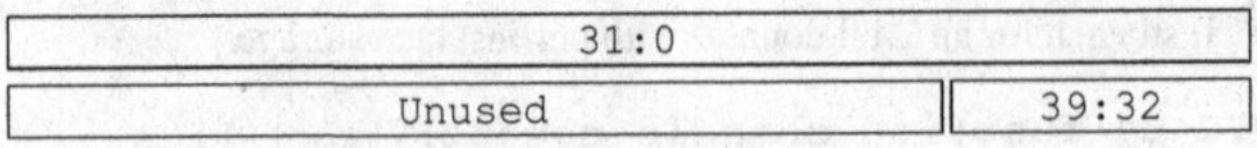

31:0	
Unused	39:32

Fig. 12.1 Storage of a 40-bit 2-state variable

For performance reasons. the SystemVerilog simulator may not mask the upper bits after calling a DPI routine, and so the SystemVerilog variable could be corrupted. Make sure your C code treats these values properly.

If you need to convert between bits and words, use the macro SV_PACKED_ DATA_NELEMS. For example, to convert 40 bits to two 32-bit words (as seen in Fig. 12.1), use SV_PACKED_DATA_NELEMS(40).

12.2.4 4-State Values

Each 4-state bit in SystemVerilog is stored in the simulator using two bits known as aval and bval, as shown in Table 12.2

Table 12.2 4-state bit encoding

4-state value	bval	aval
0	0	0
1	0	1
Z	1	0
X	1	1

A single bit 4-state variable, such as logic f, is stored in an unsigned byte, with the aval bit stored in the least significant bit, and the bval in the next higher bit. So the value 1'b0 is seen as 0x0 in C, 1'b1 is 0x1, 1'bz is 0x2, and 1'bx is 0x3.

A 4-state vector such as logic [31:0] lword is stored using pairs of 32 bits, svLogicVecVal, which contains the aval and bval bits as shown in Figure 12.2. The 32-bit variable lword is stored in a single svLogicVecVal. Variables wider than 32-bits are stored in multiple svLogicVecVal elements, with the first element contains the 32 least significant bits, the next element contains the next 32 bits, up to the most significant bits. A 40-bit logic variable is stored as one svLogicVecVal for the least significant 32 bits, and a second for the upper 8 bits (Fig. 12.2). The unused 24-bits in this upper value are undetermined, and you are responsible for masking or extending the sign bit, as needed. The svLogicVecVal type is equivalent to s_vpi_vecval, which is used to represent 4-state types such as logic in the VPI.

aval 31:0	
bval 31:0	
Unused	aval 39:32
Unused	bval 39:32

Fig. 12.2 Storage of a 40-bit 4-state variable

 Beware of arguments declared without bit subscripts or those declared with a single bit. An argument declared as `input logic a` is stored in an `unsigned char`. The argument `input logic [0:0] b` is `svLogicVecVal`, even though it contains only a single bit.

Sample 12.11 shows the import statements for a 4-state counter. The only difference from Sample 12.10 is that the `bit` types are now `logic`.

Sample 12.11 Testbench for counter that checks for Z or X values

```
import "DPI-C" function chandle counter7_new();
import "DPI-C" function void counter7
      (input chandle inst,
       output logic [6:0] out,
       input  logic [6:0] in,
       input  logic  reset, load);
```

The counter previously shown in Sample 12.9 assumes all the inputs are 2-state. Sample 12.12 extends this code to check for Z and X values on `reset`, `load`, and `i`. The actual count is still kept as a 2-state value.

Sample 12.12 Counter routine that checks for Z and X values

```
// 4-state replacement for counter7 from Sample 12-9
void counter7(c7 *inst,
              svLogicVecVal* count,
              const svLogicVecVal* i,
              const svLogic reset,
              const svLogic load) {

  if (reset & 0x2) {   // Check just the bval bit of scalar
    io_printf("Error: Z or X detected on reset\n\n");
    return;
  }
  if (load & 0x2) {   // Check just the bval bit of scalar
    io_printf("Error: Z or X detected on load\n\n");
    return;
  }
  if (i->bval) {      // Check just the bval bits of 7-bit vector
    io_printf("Error: Z or X detected on i\n\n");
    return;
  }

  if (reset)      inst->cnt = 0;        // Reset
  else if (load) inst->cnt = i->aval;  // Load value
  else           inst->cnt++;          // Count
  inst->cnt &= 0x7f;                    // Mask upper bit

  count->aval = inst->cnt;             // Write to output
  count->bval = 0;
}
```

If you want to force the simulation to terminate cleanly because of a condition found in an imported routine, you can call the VPI routine vpi_control(vpiFinish, 0). This routine and constant are defined in the include file vpi_user.h. The value vpiFinish tells the simulator to execute a $finish after your imported routine returns.

12.2.5   Converting from 2-State to 4-State

If you have a DPI application that works with 2-state types and you want to convert it to work with 4-state types, follow the following guidelines.

On the SystemVerilog side, change the import declaration from using 2-state types such as bit and int to 4-state types such as logic and integer. Make sure you are using 4-state variables in the function call.

On the C side, switch the argument declarations from svBitVecVal to svLogicVecVal. Any reference to the arguments will have to use the .aval suffix to correctly access the data. When you read from a 4-state variable, check the `bval` bits to see if there are any Z or X values. When you write to a 4-state variable, clear the `bval` bits unless you need to write Z or X values.

12.3 Connecting to C++

You can use the DPI to connect routines written in C or C++ to SystemVerilog. There are several ways your C++ code can communicate using the DPI, depending on your model's level of abstraction.

12.3.1 The Counter in C++

Sample 12.13 shows a C++ class for the 7-bit counter, with 2-state inputs. It connects to the SystemVerilog testbench in Sample 12.10 and the C++ wrapper code in Sample 12.14.

Sample 12.13 Counter class

```
class Counter7 {
public:
  Counter7();
  void counter7_signal(svBitVecVal* count,
                       const svBitVecVal* i,
                       const svBit reset,
                       const svBit load);
private:
  unsigned char cnt;
};

Counter7::Counter7() {
  cnt = 0;                      // Initialize counter
}

void Counter7::counter7_signal(svBitVecVal* count,
                               const svBitVecVal* i,
                               const svBit reset,
                               const svBit load) {
  if (reset)       cnt = 0;    // Reset
  else if (load)   cnt = *i;   // Load
  else             cnt++;      // Count
  cnt &= 0x7F;                 // Mask upper bit
  *count = cnt;
}
```

12.3.2   Static Methods

The DPI can only call a C or C++ function that is known at link time. As a result, your SystemVerilog code cannot call a C++ routine in an object as the object does not exist when the linker runs.

So what if you need to call a method in a C++ class? The solution, as shown in Sample 12.14, is to create a function with a fixed address, that then can communicate with the C++ dynamic objects and methods. The first routine, `counter7_new`, constructs an object for the counter and returns a handle to the object. The second static routine, `counter7`, calls the C++ method that performs the counter logic, using the object handle.

Sample 12.14   Static methods and linkage

```
extern "C" void* counter7_new()
{
  return new Counter7;
}

// Call a counter instance, passing the signal values
extern "C" void counter7(void* inst,
                         svBitVecVal* count,
                         const svBitVecVal* i,
                         const svBit reset,
                         const svBit load)
{
  Counter7 *c7 = (Counter7 *) inst;
  c7->counter7_signal(count, i, reset, load);
}
```

The `extern` "C" code tells the C++ compiler that the external information sent to the linker should use C calling conventions and not perform name mangling. You can put this before each routine that is called by SystemVerilog, or put `extern` "C" { ... } around a set of methods.

From the testbench point of view, the C++ counter looks the same as the counter that stored the value in per-instance storage, shown in Sample 12.9, so you can use the same testbench, Sample 12.10, for both.

12.3.3   Communicating with a Transaction Level C++ Model

The previous C / C++ code examples were low-level models that communicated with the SystemVerilog at the signal level. This is not efficient; for example the counter is called every clock cycle, even if the data or control inputs have not changed. When you create models for complex devices such as processors and networking devices, communicate with them at the transaction level for faster simulations.

The C++ counter model in Sample 12.15 has a transaction-level interface, communicating with methods instead of signals and a clock.

Sample 12.15 C++ counter communicating with methods

```
class Counter7 {
public:
  Counter7();
  void count();
  void load(const svBitVecVal* i);
  void reset();
  int get();
private:
  unsigned char cnt;
};

Counter7::Counter7() {            // Initialize counter
  cnt = 0;
}

void Counter7::count() {          // Increment counter
  cnt = cnt + 1;
  cnt &= 0x7F;                    // Mask upper bit
}

void Counter7::load(const svBitVecVal* i) {
  cnt = *i;
  cnt &= 0x7F;                    // Mask upper bit
}

void Counter7::reset() {
  cnt = 0;
}

// Get the counter value in a pointer to a svBitVecVal
int Counter7::get() {
  return cnt;
}
```

The dynamic C++ methods such as reset, load, and count are wrapped in static methods that use the object handle, passed from SystemVerilog, as shown in Sample 12.16.

Sample 12.16 Static wrapper for C++ transaction level counter

```
#ifdef __cplusplus
extern "C" {
#endif

void* counter7_new() {
  return new Counter7;
}

void counter7_count(void* inst){
  Counter7 *c7 = (Counter7 *) inst;
  c7->count();
}

void counter7_load(void* inst, const svBitVecVal* i) {
  Counter7 *c7 = (Counter7 *) inst;
  c7->load(i);
}

void counter7_reset(void* inst) {
  Counter7 *c7 = (Counter7 *) inst;
  c7->reset();
}

int counter7_get(void* inst) {
  Counter7 *c7 = (Counter7 *) inst;
  return c7->get();
}

#ifdef __cplusplus
}
#endif
```

The OOP interface for the transaction level counter is carried up to the testbench.
Sample 12.17 has the SystemVerilog import statements and a class to wrap the C++
object. This allows you to hide the C++ handle inside the class.

Note that the `counter7_get()` function returns an `int` (32-bit, signed) rather
than `bit [6:0]`, as the latter would require returning a pointer to a `svBitVecVal`, as
shown in Table 12.1. An imported function can not return a pointer. It can only return
a value of type `void`, `byte`, `shortint`, `int`, `longint`, `real`, `shortreal`,
`chandle`, and `string`, plus single bit values of type `bit` and `logic`.

Sample 12.17 Testbench for C++ model using methods

```
import "DPI-C" function chandle counter7_new();
import "DPI-C" function void counter7_count(input chandle inst);
import "DPI-C" function void counter7_load(input chandle inst,
                                           input bit [6:0] i);
import "DPI-C" function void counter7_reset(input chandle inst);
import "DPI-C" function int counter7_get(input chandle inst);

// Wrap the static C static wrapper functions with a
// SystemVerilog class to hide the C++ instance handle
class Counter7;
   chandle inst;

   function new();
      inst = counter7_new();
   endfunction

   function void count();
      counter7_count(inst);
   endfunction

   function void load(input bit [6:0] val);
      counter7_load(inst, val);
   endfunction

   function void reset();
      counter7_reset(inst);
   endfunction

   function bit [6:0] get();
      return counter7_get(inst);
   endfunction
endclass : Counter7

program automatic test;
   Counter7 c1;

   initial begin
      c1 = new;

      c1.reset();
      $display("SV: Post reset: counter1=%0d", c1.get());

      c1.load(126);
      if (c1.get() == 126)
        $display("Successful load");
      else
        $display("Error: load, expect 126, got %0d", c1.get());
```

```
    c1.count();   // count = 127
    if (c1.get() == 127)
      $display("Successful count");
    else
      $display("Error: load, expect 127, got %0d", c1.get());

    c1.count();   // count = 0
    if (c1.get() == 0)
      $display("Successful rollover");
    else
      $display("Error: rollover, exp 127, got %0d", c1.get());
    end

endprogram
```

12.4   Simple Array Sharing

So far you have seen examples of passing scalar and vectors between SystemVerilog and C. A typical C model might read an array of values, perform some computation, and return another array with the results.

12.4.1   *Single Dimension Arrays - 2-State*

Sample 12.18 shows a routine that computes the first 20 values in the Fibonacci series. It is called by the SystemVerilog code in Sample 12.19.

Sample 12.18 C routine to compute Fibonacci series

```
void fib(svBitVecVal data[20]) {
  int i;
  data[0] = 1;
  data[1] = 1;
  for (i=2; i<20; i++)
    data[i] = data[i-1] + data[i-2];
}
```

Note that in C, you could have alternatively declared the argument as a pointer, *data or an array, data[20]. In this example, they are interchangeable.

Sample 12.19 Testbench for Fibonacci routine

```
import "DPI-C" function void fib(output bit [31:0] data[20]);

program automatic test;
  bit [31:0] data[20];

  initial begin
    fib(data);
    foreach (data[i]) $display(i,,data[i]);
  end
endprogram
```

Notice that the array of Fibonacci values is allocated and stored in SystemVerilog, even though it is calculated in C. There is no way to allocate an array in C and reference it in SystemVerilog.

12.4.2 Single Dimension Arrays - 4-State

Sample 12.20 shows the Fibonacci C routine for a 4-state array with the testbench in Sample 12.21.

Sample 12.20 C routine to compute Fibonacci series with 4-state array

```
void fib(svLogicVecVal data[20]) {
  int i;
  data[0].aval = 1;    // Write to both aval
  data[0].bval = 0;    //     and bval
  data[1].aval = 1;
  data[1].bval = 0;
  for (i=2; i<20; i++) {
    data[i].aval = data[i-1].aval + data[i-2].aval;
    data[i].bval = 0; // Don't forget to clear bval
  }
}
```

Sample 12.21 Testbench for Fibonacci routine with 4-state array

```
import "DPI-C" function void fib(output logic [31:0] data[20]);

program automatic test;
  logic [31:0] data[20];

  initial begin
    fib(data);
    foreach (data[i]) $display(i,,data[i]);
  end
endprogram
```

Section 12.2.5 describes how to convert a 2-state application to 4-state.

12.5   Open arrays

When sharing arrays between SystemVerilog and C, you have two options. For the fastest simulations, you can reverse-engineer the layout of the elements in System-Verilog, and write your C code to use this mapping. This approach is fragile, meaning that you will have to rewrite and debug your C code if any of the array sizes change. A more robust approach is to use "open arrays", and their associated SystemVerilog routines to manipulate them. These allow you to write generic C routines that can operate on any size array.

12.5.1   Basic Open Array

Sample 12.22 and 12.23 show how to pass a simple array between SystemVerilog and C with open arrays. Use the empty square brackets [] in the SystemVerilog import statement to specify that you are passing an open array.

Sample 12.22 Testbench code calling a C routine with an open array

```
import "DPI-C" function void fib_oa(output bit [31:0] data[]);

program automatic test;
  localparam SIZE = 20;
  bit [31:0] data[SIZE], r;

  initial begin
    fib_oa(data, SIZE);
    foreach (data[i])
      $display(i,,data[i]);
  end
endprogram
```

Your C code references the open array with a handle of type svOpenArray-Handle. This points to a structure with information about the array such as the declared word range. You can locate the actual array elements with calls such as svGetArrayPtr. Note that svSize() is an open array query method, as described in the next section.

Sample 12.23 C code using a basic open array

```
void  fib_oa(const svOpenArrayHandle data_oa) {
  int i, *data;
  data = (int *) svGetArrayPtr(data_oa);
  data[0] = 1;
  data[1] = 1;
  for (i=2; i<=svSize(data_oa, 1); i++)
    data[i] = data[i-1] + data[i-2];
}
```

12.5.2   Open Array Methods

There are many DPI methods to access their contents and ranges, as defined in svdpi.h. These only work with open array handles declared as svOpenArray-Handle, not with pointers such as svBitVecVal or svLogicVecVal. The methods in Table 12.3 give you information about the size of an open array.

Table 12.3 Open array query functions

function	Description
int svLeft(h, d)	Left bound for dimension d
int svRight(h, d)	Right bound for dimension d
int svLow(h, d)	Low bound for dimension d
int svHigh(h, d)	High bound for dimension d
int svIncrement(h, d)	If left >= right, 1, else −1
int svSize(h, d)	Number of elements in dimension d: svHigh−svLow+1
int svDimensions(h)	Number of dimensions in open array
int svSizeOfArray(h)	Total size of array in bytes

In Table 12.3, the variable h is a svOpenArrayHandle and d is an int. The dimensions are numbered starting with d=1.

The functions in Table 12.4 return the locations of the C storage for the entire array or a single element.

Table 12.4 Open array locator functions

Function	Returns pointer to:
void *svGetArrayPtr(h)	storage for the entire array
void *svGetArrElemPtr(h, i1, ...)	an element in the array
void *svGetArrElemPtr1(h, i1)	an element in a 1-D array
void *svGetArrElemPtr2(h, i1, i2)	an element in a 2-D array
void *svGetArrElemPtr3(h, i1, i2, i3)	an element in a 3-D array

12.5.3   Passing Unsized Open Arrays

Sample 12.24 calls C code with a 2-dimensional array. The C code uses the svLow and svHigh methods to find the array ranges, which, in this example, don't follow the usual 0..size-1.

Sample 12.24 Testbench calling C code with multi-dimensional open array

```
import "DPI-C" function void mydisplay(inout int h[][]);

program automatic test;
  int a[6:1][8:3];        // Note word ranges are high:low
  initial begin
    foreach (a[i,j]) a[i][j] = i+j;
    mydisplay(a);
    foreach (a[i,j])
      $display("V: a[%0d][%0d] = %0d", i, j, a[i][j]);
  end
endprogram
```

This calls the C code in Sample 12.25 that reads the array using the open array methods. The routine `svLow(handle, dimension)` returns the lowest index number for the specified dimension. So `svLow(h,1)` returns 1 for the array declared with the range [6:1]. Likewise, `svHigh(h, 1)` returns 6. You should use `svLow` and `svHigh` with C `for` loops.

The methods `svLeft` and `svRight` return the left and right index from the array declaration, 6 and 1 respectively for the range [6:1]. At the center of Sample 12.25, the call `svGetArrElemPtr2` returns a pointer to an element in a two dimensional array.

Sample 12.25 C code with multi-dimensional open array

```
void mydisplay(const svOpenArrayHandle h) {
  int i, j;
  int lo1 = svLow(h, 1);
  int hi1 = svHigh(h, 1);
  int lo2 = svLow(h, 2);
  int hi2 = svHigh(h, 2);
  for (i=lo1; i<=hi1; i++) {
    for (j=lo2; j<=hi2; j++) {
      int *a = (int*) svGetArrElemPtr2(h, i, j);
      io_printf("C: a[%d][%d] = %d\n", i, j, *a);
      *a = i * j;
    }
  }
}
```

12.5.4 Packed Open Arrays in DPI

An open array in the DPI is treated as having a single packed dimension and one or more unpacked dimensions. You can pass an array with multiple packed dimensions, as long as they pack into an element that is the same size as a single element of the formal argument. For example, if you have the formal argument bit[63:0] b64[] in the import statement, you could pass in the actual argument bit [1:0] [0:3][6:-1] bpack [9:1]. Sample 12.26 shows the SystemVerilog code with packed open arrays.

Sample 12.26 Testbench for packed open arrays

```
import "DPI-C" function void view_pack(input bit [63:0] b64[]);

program automatic test;
  bit [1:0][0:3][6:-1] bpack[9:1];

  initial begin
    foreach(bpack[i]) bpack[i] = i;
    bpack[2] = 64'h12345678_90abcdef;

    $display("SV: bpack[2]=%h", bpack[2]);   // 64 bits
    $display("SV: bpack[2][0]=%h", bpack[2][0]); // 32 bits
    $display("SV: bpack[2][0][0]=%h", bpack[2][0][0]); // 8 bits

    view_pack(bpack);
  end
endprogram : test
```

Sample 12.27 C code using packed open arrays

```
void view_pack(const svOpenArrayHandle h) {
  int i;

  for (i=svLow(h,1); i<svHigh(h,1); i++)
    io_printf("C: b64[%d]=%llx\n",
              i, *(long long int *)svGetArrElemPtr1(h, i));
}
```

Notice that the C code in Sample 12.27 prints a 64-bit value using %11x, and casts the result from svGetArrayElemPtr1 to long long int.

12.6   Sharing Composite Types

By this point you may wonder how to pass objects between SystemVerilog and C. The layout of class properties does not match between the two languages, so you cannot share objects directly. Instead, you must create similar structures on each side, plus pack and unpack methods to convert between the two formats. Once you have all this in place, you can share composite types.

12.6.1 Passing Structures Between SystemVerilog and C

The following example shares a simple structure for a pixel made of three bytes packed into a word. Sample 12.28 shows the C structure. Notice that C treats a char as signed variable, which can give you unexpected results, so the structure marks the char as unsigned. The bytes are in reverse order from the SystemVerilog because this code was written for a Intel x86 processor that is little-endian, which means that the least significant byte is stored at a lower address than the most significant. A Sun SPARC is big endian, so the bytes are stored in the same order as in SystemVerilog: r, g, b.

Sample 12.28 C code to share a structure

```
typedef struct {
  unsigned char b, g, r;  // x86 litle-endian
//unsigned char r, g, b;  // SPARC format
} *p_rgb;

void invert(p_rgb rgb) {
  rgb->r = ~rgb->r;       // Invert the color values
  rgb->g = ~rgb->g;
  rgb->b = ~rgb->b;
  io_printf("C:  Invert rgb=%02x,%02x,%02x\n",
            rgb->r, rgb->g, rgb->b);
}
```

The SystemVerilog testbench in Sample 12.29 has a packed struct that holds a single pixel, and class to encapsulate the pixel operations. The RGB_T struct is packed so SystemVerilog will store the bytes in consecutive locations. Without the packed modifier, each 8-bit value would be stored in a separate word.

Sample 12.29 Testbench for sharing structure

```
typedef struct packed { bit [ 7:0] r, g, b; } RGB_T;
import "DPI-C" function void invert(inout RGB_T pstruct);

program automatic test;

class RGB;
  rand bit [ 7:0] r, g, b;
  function void display(input string prefix="");
    $display("%sRGB=%x,%x,%x", prefix, r, g, b);
  endfunction : display

  // Pack the class properties into a struct
  function RGB_T pack();
    pack.r = r; pack.g = g; pack.b = b;
  endfunction : pack

  // Unpack a struct into the class properties
  function void unpack(input RGB_T pstruct);
    r = pstruct.r; g = pstruct.g; b = pstruct.b;
  endfunction : unpack
endclass : RGB

  initial begin
    RGB pixel;
    RGB_T pstruct;

    pixel = new;
    repeat (5) begin
      `SV_RAND_CHECK(pixel.randomize()); // Create random pixel
      pixel.display("\nSV: before "); // Print it
      pstruct = pixel.pack();          // Convert to a struct
      invert(pstruct);                 // Call C to invert bits
      pixel.unpack(pstruct);           // Unpack struct to class
      pixel.display("SV: after  ");    // Print it
    end
  end
endprogram
```

12.6.2 Passing Strings Between SystemVerilog and C

Using the DPI, you can pass strings from C back to SystemVerilog. You might need to pass a string for the symbolic value of a structure, or get a string representing the internal state of your C code for debug.

The easiest way to pass a string from C to SystemVerilog is for your C function to return a pointer to a static string as shown in Sample 12.30. The string must be declared as `static` in C, and not as a local string. Non-static variables are stored on the stack and are reclaimed when the function returns.

Sample 12.30 Returning a string from C

```c
char *print(p_rgb rgb) {
  static char s[12];
  sprintf(s, "%02x,%02x,%02x", rgb->r, rgb->g, rgb->b);
  return s;
}
```

A danger with static storage is that multiple concurrent calls could end up sharing storage. For example, a SystemVerilog $display statement that is printing several pixels might call the above `print` routine multiple times. Depending on how the SystemVerilog compiler orders these calls, later calls to `print()` could overwrite results from earlier calls, unless the SystemVerilog compiler makes a copy of the string. Note that a call to an imported routine can never be interrupted by the SystemVerilog scheduler. Sample 12.31 stores the strings in a heap to support concurrent calls.

Sample 12.31 Returning a string from a heap in C

```c
#define PRINT_SIZE 12
#define MAX_CALLS 16
#define HEAP_SIZE PRINT_SIZE * MAX_CALLS

char *print(p_rgb rgb) {
  static char print_heap[HEAP_SIZE + PRINT_SIZE];
  char *s;
  static int heap_idx = 0;
  int nchars;

  s = &print_heap[heap_idx];
  nchars = sprintf(s, "%02x,%02x,%02x",
                      rgb->r, rgb->g, rgb->b);
  heap_idx += nchars + 1;              // Don't forget null!
  if (heap_idx > HEAP_SIZE)
    heap_idx = 0;
  return s;
}
```

12.7   Pure and Context Imported Methods

Imported methods are classified as `pure`, `context`, or generic. A `pure` function calculates its output strictly based on its inputs, with no outside interactions. Specifically, a `pure` function does not access any global or static variables, perform any file operations, or interact with anything outside the function such as the operating system, processes, shared memory, sockets, etc. The SystemVerilog compiler may optimize away calls to a `pure` function if the result is not needed, or replace the call with the results from a previous call with the same arguments. The `factorial` function in Sample 12.5, and the `sin` function in 12.6 are both `pure` functions as their result is only based on their inputs. Sample 12.32 shows how to import a pure function.

Sample 12.32 Importing a pure function

```
import "DPI-C" pure function int factorial(input int i);
import "DPI-C" pure function real sin(input real in);
```

An imported routine may need to know the context of where it is called so it can call a PLI TF, ACC, or VPI methods, or a SystemVerilog task that has been exported. Use the `context` attribute for these methods as shown in Sample 12.33.

Sample 12.33 Imported context tasks

```
import "DPI-C" context task call_sv(bit [31:0] data);
```

An imported routine may use global storage, so it is not `pure`, but might not have any PLI references, so it does not need the overhead of a `context` routine. Sutherland (2004) uses the term "generic" for these methods as the SystemVerilog LRM does not have a specific name. By default, an imported routine is generic, as are many of the examples in this chapter.

There is overhead invoking a `context` imported routine as the simulator needs to record the calling context, so only declare a routine as `context` if needed. On the other hand, if a generic imported routine calls an exported task or a PLI routine that accesses SystemVerilog data objects, the simulator could crash.

A context-aware PLI routine is one that needs to know where it was called from so that it can access information relative to that location.

12.8 Communicating from C to SystemVerilog

The examples so far have shown you how to call C code from your SystemVerilog models. The DPI also allows you to call SystemVerilog routines from C code. The SystemVerilog routine can be a simple task to record the result from an operation in C, or a time-consuming task representing part of a hardware model.

12.8.1 A simple Exported Function

Sample 12.34 shows a module that imports a context function, and exports a System-Verilog function.

Sample 12.34 Exporting a SystemVerilog function

```
module block;
  import "DPI-C" context function void c_display();
  export "DPI-C" function sv_display;  // No type or args

  initial c_display();

  function void sv_display();
    $display("SV: in sv_display");
  endfunction
endmodule : block
```

The export declaration in Sample 12.34 looks naked because the LRM forbids putting a return value declaration or any arguments. You can't even give the usual empty parentheses. This information in the export declaration would duplicate the information in the function declaration at the end of the module and could thus become out of sync if you ever changed the function.

Sample 12.35 shows the C code that calls the exported function.

Sample 12.35 Calling an exported SystemVerilog function from C

```
extern void sv_display();

void c_display() {
  io_printf("C:  in c_display\n");
  sv_display();
}
```

This example prints the line from the C code, followed by the $display output from the SystemVerilog, as shown in Sample 12.36.

Sample 12.36 Output from simple export

```
C:  in c_display
SV: in sv_display
```

12.8.2   C function Calling SystemVerilog Function

While the majority of your testbench should be in SystemVerilog, you may have legacy testbenches in C or other languages, or applications that you want to reuse. This section creates a SystemVerilog memory model that is stimulated by C code that reads transactions from an external file.

The first version of the memory model, shown in Sample 12.38 and 12.37, is coded with just functions, so everything runs in zero time. The C code in

Sample 12.37 opens the file, reads a command, and calls the exported function. Error checking has been removed for compactness.

Sample 12.37   C code to read simple command file and call exported function

```c
#include <svdpi.h>
#include <stdio.h>
extern void mem_build(int);

void read_file(char *fname){
  char cmd;
  FILE *file;

  file = fopen(fname, "r");
  while (!feof(file)) {
    cmd = fgetc(file);
    switch (cmd)
      {
      case 'M': {
        int hi;
        fscanf(file, "%d", &hi);
        mem_build(hi);
        break;
      }
    }
  }
  fclose(file);
}
```

The SystemVerilog code calls the C task read_file which opens a file. The only command in the file sets the memory size, so the C code calls an exported function.

Sample 12.38   SystemVerilog module for simple memory model

```verilog
module memory;
  import "DPI-C" context function void read_file(string fname);
  export "DPI-C" function mem_build;  // No type or args

  initial
    read_file("mem.dat");

  int mem[];

  function void mem_build(input int size);
    mem = new[size];      // Allocate dynamic memory elements
  endfunction

endmodule : memory
```

Notice that in Sample 12.38, the `export` statement does not have any arguments as this information is already in the function declaration.

The command file is trivial, with one command to construct a memory with 100 elements as shown in Sample 12.39.

Sample 12.39 Command file for simple memory model

```
M 100
```

12.8.3   C Task Calling SystemVerilog Task

A real memory model has operations such as read and write that consume time, and thus must be modeled with tasks.

Sample 12.40 shows the SystemVerilog code for the second version of the memory model. It has several improvements compared to Sample 12.38. There are two new tasks, `mem_read` and `mem_write`, which respectively take 20ns and 10ns to complete. The imported routine `read_file` is now a SystemVerilog task as it is calling other tasks that consume time. The `import` statement now specifies that `read_file` is a context task, as the simulator needs to create a separate stack when it is called.

Sample 12.40 SystemVerilog module for memory model with exported tasks

```
module memory;
  import "DPI-C" context task read_file(string fname);
  export "DPI-C" task mem_read;
  export "DPI-C" task mem_write;
  export "DPI-C" function mem_build;

  initial read_file("mem.dat");

  int mem[];

  function void mem_build(input int size);
    mem = new[size];
  endfunction

  task mem_read(input int addr, output int data);
    #20 data = mem[addr];
  endtask

  task mem_write(input int addr, input int data);
    #10 mem[addr] = data;
  endtask
endmodule : memory
```

The C code in Sample 12.41 primarily expands the case statement that decodes commands and calls the exported tasks, which are declared as `extern int` according to the LRM.[2]

Sample 12.41 C code to read command file and call exported function

```c
extern int mem_read(int, int*);
extern int mem_write(int, int);
extern void mem_build(int);

void read_file(const char *fname) {
  char cmd;
  FILE *file;

  file = fopen(fname, "r");
  while (!feof(file)) {
    cmd = fgetc(file);
    switch (cmd) {
      case 'M': {
        int hi;
        fscanf(file, "%d ", &hi);
        mem_build(hi);
        break;
      }

      case 'R': {
        int addr, data, exp;
        fscanf(file, "%d %d ", &addr, &exp);
        mem_read(addr, &data);
        if (data != exp)
          io_printf("C: Data=%d, exp=%d\n", data, exp);
        break;
      }

      case 'W': {
        int addr, data;
        fscanf(file, "%d %d ", &addr, &data);
        mem_write(addr, data);
        break;
      }
    }
  }
  fclose(file);
}
```

The command file in Sample 12.42 has new commands that write two locations, and then reads back one of them, and includes the expected value.

---

[2] VCS declared exported tasks as void functions in C.

Sample 12.42 Command file for simple memory model

```
M 100
W 12 34
W 99 8
R 12 34
```

12.8.4   Calling Methods in Objects

You can export SystemVerilog methods, except for those defined inside a class. This restriction is similar to the restriction of importing static C methods, as shown in Section 12.3.2 as objects do not exist when SystemVerilog elaborates your code. The solution is to pass a reference to the object between the SystemVerilog and C code. However, unlike a C pointer, a SystemVerilog handle cannot be passed through the DPI. You can instead have an array of handles, and pass the array index between the two languages.

The following examples build on the previous versions of the memory. The SystemVerilog code in Sample 12.44 has a class that encapsulates the memory. Now you can have multiple memories, each in a separate object. The command file in Sample 12.43 creates two memories, M0, and M1. Then it performs several writes to initialized locations in both memories, and lastly tries to read back the values. Notice that location 12 is used for both memories.

Sample 12.43 Command file for exported methods with OOP memories

```
M0 1000
M1 2000
W0 12 34
W1 12 88
W0 99 18
R1 22 44
R0 12 34
R1 12 88
```

The SystemVerilog code in Sample 12.44 constructs a new object for every M command in the file. The exported function mem_build calls the Memory constructor. It then stores the handle to the Memory object in a SystemVerilog queue, and returns the queue index, idx, to the C code as shown in Sample 12.45. The handles are stored in a queue so you can dynamically add new memories. The exported tasks mem_read and mem_write now have an additional argument, the index of the memory handle in the queue.

Sample 12.44 SystemVerilog module with memory model class

```systemverilog
module memory;
  import "DPI-C" context task read_file(string fname);
  export "DPI-C" task mem_read;
  export "DPI-C" task mem_write;
  export "DPI-C" function mem_build;

  initial read_file("mem.dat");  // Call C code to read file


  class Memory;
    int mem[];

    function new(input int size);
      mem = new[size];
    endfunction

    task mem_read(input int addr, output int data);
      #20 data = mem[addr];
    endtask

    task mem_write(input int addr, input int data);
      #10 mem[addr] = data;
    endtask : mem_write
  endclass : Memory

  Memory memq[$];              // Queue of Memory objects

  // Construct a new memory instance & push on the queue
  function void mem_build(input int size);
    Memory m;
    m = new(size);
    memq.push_back(m);
  endfunction

  // idx is the index of the memory handle in memq
  task mem_read(input int idx, addr, output int data);
    memq[idx].mem_read(addr, data);
  endtask

  task mem_write(input int idx, addr, input int data);
    memq[idx].mem_write(addr, data);
  endtask

endmodule : memory
```

Sample 12.45 C code to call exported tasks with OOP memory

```c
extern int mem_read(int, int, int*);
extern int mem_write(int, int, int);
extern void mem_build(int);

void read_file(char *fname) {
  char cmd;
  int idx;
  FILE *file;

  file = fopen(fname, "r");
  while (!feof(file)) {
    cmd = fgetc(file);
    fscanf(file, "%d ", &idx);
    switch (cmd)
      {
      case 'M': {
        int hi;
        fscanf(file, "%d ", &hi);
        mem_build(hi);
        break;
      }

      case 'R': {
        int addr, data, exp;
        fscanf(file, "%d %d ", &addr, &exp);
        mem_read(idx, addr, &data);
        if (data != exp)
          io_printf("C: Error Data=%d, exp=%d\n", data, exp);
        break;
      }

      case 'W': {
        int addr, data;
        fscanf(file, "%d %d ", &addr, &data);
        mem_write(idx, addr, data);
        break;
      }
    }
  }
  fclose(file);
}
```

12.8.5   The Meaning of Context

The context of an imported routine is the location where it was defined, such as $unit, module, program, or package scope, just like a normal SystemVerilog routine. If you import a routine in two different scopes, the corresponding C code executes in the context of where the import statement occurred. This is similar to defining a SystemVerilog run() task in each of two separate modules. Each task accesses variables in its own module, with no ambiguity.

Sample 12.46 shows that if you add a second module to Sample 12.34 that imports the same C code and exports its own function, the C routine will call different SystemVerilog methods, depending on the context of the import and export statements.

Sample 12.46   Second module for simple export example

```
module top;
  import "DPI-C" context function void c_display();
  export "DPI-C" function sv_display;

  block b1();
  initial c_display();

  function void sv_display();
    $display("SV: In %m");
  endfunction
endmodule : top


module block;
  import "DPI-C" context function void c_display();
  export "DPI-C" function sv_display;

  initial c_display();

  function void sv_display();
    $display("SV: In %m");
  endfunction
endmodule : block
```

The output in Sample 12.47 shows that one C routine calls two separate SystemVerilog methods, depending on where the C routine was called.

Sample 12.47   Output from simple example with two modules

```
C:  in c_display
SV: In top.b1.sv_display
C:  in c_display
SV: In top.sv_display
```

12.8.6   Setting the Scope for an Imported Routine

Just as your SystemVerilog code can call a routine in the local scope, an imported C routine can call a routine outside its default context. Use the routine svGetScope to get a handle to the current scope, and then use that handle in a call to svGetScope to make the C code think it is inside another context. Sample 12.48 shows the C code for two methods. The first, save_my_scope( ), saves the scope of where it was called from the SystemVerilog side. The second routine, c_display(), sets its scope to the saved one, prints a message, then calls your function, sv_display().

Sample 12.48 C code getting and setting context

```
extern void sv_display();
svScope my_scope;

void save_my_scope() {
  my_scope = svGetScope();
}

void c_display() {
  // Print the current scope
  io_printf("\nC: c_display called from scope %s\n",
            svGetNameFromScope(svGetScope()));

  // Set a new scope
  svSetScope(my_scope);
  io_printf("C: calling %s.sv_display\n",
            svGetNameFromScope(svGetScope()));
  sv_display();
}
```

The C code calls svGetNameFromScope() that returns a string of the current scope. The scope is printed twice, once with the scope where the C code was first called from, and again with the scope that was previously saved. The routine svGetScopeFromName() takes a string with a SystemVerilog scope and returns a pointer to a svScope handle that can be used with svSetScope().

In the SystemVerilog code in Sample 12.49, the first module, block, calls a C routine that saves the context. When the module top calls c_display(), the routine sets scope back to block, and so it calls the sv_display() routine in the block module, not the top module.

Sample 12.49 Modules calling methods that get and set context

```
module block;
  import "DPI-C" context function void c_display();
  import "DPI-C" context function void save_my_scope();
  export "DPI-C" function sv_display;

  function void sv_display();
    $display("SV: In %m");
  endfunction : sv_display

  initial begin
    save_my_scope();
    c_display();
  end

endmodule : block


module top;
  import "DPI-C" context function void c_display();
  export "DPI-C" function sv_display;

  function void sv_display();
    $display("SV: In %m");
  endfunction : sv_display

  block b1();

  initial #1 c_display();

endmodule : top
```

This produces the output shown in Sample 12.50.

Sample 12.50 Output from svSetScope code

```
C: c_display called from top.b1
C: Calling top.b1.sv_display
SV: In top.b1.sv_display

C: c_display called from top
C: Calling top.b1.sv_display
SV: In top.b1.sv_display
```

You could use this concept of scope to allow a C model to know where it was instantiated from, and differentiate each instance. For example, a memory model may be instantiated several times, and needs to allocate unique storage for every instance.

12.9   Connecting Other Languages

This chapter has shown the DPI working with C and C++. With a little work, you can connect other languages. The easiest way is to call the Verilog $system() task. If you need the return value from the command, use the Unix system() function and the WEXITSTATUS macro. The SystemVerilog code in Sample 12.51 calls a C wrapper for system().

Sample 12.51 SystemVerilog code calling C wrapper for Perl

```
import "DPI-C" function int call_perl(string s);

program automatic perl_test;
  int ret_val;
  string script;

  initial begin
    $value$plusargs("script=%s", script);
    $display("SV: Running '%0s'", script);
    ret_val = call_perl(script);
    $display("SV: Perl script returned %0d", ret_val );
  end
endprogram : perl_test
```

Sample 12.52 is the C wrapper that calls system() and translates the return value.

Sample 12.52 C wrapper for Perl script

```
#include <svdpi.h>
#include <stdlib.h>
#include <wait.h>

int call_perl(const char* command) {
  int result = system(command);
  return WEXITSTATUS(result);
}
```

Sample 12.53 is a Perl script that prints a message and returns a value.

Sample 12.53 Perl script called from C and SystemVerilog

```
#!/usr/local/bin/perl
print "Perl: Hello world!\n" ;
exit (3)
```

Now you can run the Unix command in Sample 12.54 to run the simulation and call the hello.pl script.

Sample 12.54 VCS command line to run Perl script

```
> simv +script="perl hellp.pl"
```

12.10   Conclusion

The Direct Programing Interface allows you to call C routines as if they are just another SystemVerilog routine, passing SystemVerilog types directly into C. This has less overhead than the PLI, which builds argument lists, and always has to keep track of the calling context, not to mention the complexity of having up to four C routines for every system task.

Additionally, with the DPI, your C code can call SystemVerilog routines, allowing external applications to control simulation. With the PLI you would need trigger variables and more argument lists, and you have to worry about subtle bugs from multiple calls to time-consuming tasks.

The most difficult part of the DPI is mapping SystemVerilog types to C, especially if you have structures and classes that are shared between the two languages. If you can master this problem, you can connect almost any application to SystemVerilog.

12.11   Exercises

1. Create a C function, shift_c, that has two input arguments: a 32-bit unsigned input value i and an integer for the shift amount n. The input i is shifted n places. When n is positive, values are shifted left, when n is negative, shifted right, and when n is 0, no shift is performed. The function returns the shifted value. Create a SystemVerilog module that calls the C function and tests each feature. Provide the output.

2. Expand Exercise 1 to add a third argument to shift_c, a load flag ld. When ld is true, i is shifted by n places and then loaded into an internal 32-bit register. When ld is false, the register is shifted n places. The function returns the value of the register after these operations. Create a SystemVerilog module that calls the C function and tests each feature. Provide the output.

3. Expand Exercise 2 to create multiple instances of the shift_c function. Each instance in C needs a unique identifier, so use the address where the internal register is stored. Print this address along with the arguments when the function shift_c is called. Instantiate the function twice, and call each instance twice. Provide the output.

4. Expand the C code from Exercise 3 to display the total number of times the `shift_c` function has been called, even if the function is instantiated more than once.

5. Expand Exercise 4 to provide the ability to initialize the stored value at instantiation.

6. Expand Exercise 5 to encapsulate the `shift_c` function in a class.

7. For the code in Sample 12.24 and 12.25, what is returned by the following open array methods?

```
svLeft(h, 1);
svLeft(h, 2);
svRight(h, 1);
svRight(h, 2);
svSize(h, 1);
svSize(h, 2);
svDimensions(h);
svSizeOfArray(h);
```

8. Modify Exercise 1 so that instead of shifting the value in C, the function calls an exported SystemVerilog void function named `shift_sv` that does the shifting.

9. Expand Exercise 8 to call the SystemVerilog function `shift_sv` for two different SystemVerilog objects as demonstrated in Section 12.8.4 of the text. Assume the SystemVerilog function `shift_build` has been exported to the C code.

10. Expand Exercise 8 to:

   a. Create a SystemVerilog class `Shift` containing the function `shift_sv` that stores the result in a class-level variable, and a `shift_print` function that displays the stored result.
   b. Define and export SystemVerilog function `shift_build`.
   c. Support the creation of multiple `Shift` objects with the handles to these objects stored in a queue.
   d. Create a testbench that constructs multiple `Shift` objects. Demonstrate that each object holds a separate result after performing calculations.

References

Bergeron, Janick. *Writing Testbenches Using SystemVerilog.* Norwell, MA: Springer, 2006

Bergeron, Janick, Cerny, Eduard, Hunter, Alan, and Nightingale, Andrew. *Verification Methodology Manual for SystemVerilog.* Norwell, MA: Springer, 2006

Cohen, Ben, Venkataramanan, Srinivasan, and Kumari, Ajeetha. *SystemVerilog Assertions Handbook for Formal and Dynamic Verification*: VhdlCohen Publishing 2005

Cummings, Cliff. *Nonblocking Assignments in Verilog Synthesis, Coding Styles That Kill!* Synopsys User Group, San Jose, CA, 2000

Cummings, Cliff, Salz, Arturo. *SystemVerilog Event Regions, Race Avoidance & Guidelines,* Synopsys User Group, Boston, CA, 2006

Denning, Peter. *The Locality Principle,* Communications of the ACM, 48(7), July 2005, pp. 19–24

Haque, Faisal, Michelson, Jonathan. *The Art of Verification with SystemVerilog Assertions.* Verification Central 2006

IEEE *IEEE Standard for SystemVerilog — Unified Hardware Design, Specification, and Verification Language.* New York: IEEE 2009 (a.k.a. SystemVerilog Language Reference Manual, or LRM.)

IEEE *IEEE Standard Verilog Hardware Design, Description Language.* New York: IEEE 2001

Rich, Dave *Are SystemVerilog Program Blocks Needed?* http://blogs.men-tor.com/verificationhorizons/blog/2009/05/07/programblocks/ 2009

Sutherland, Stuart. *Integrating SystemC Models with Verilog and SystemVerilog Using the SystemVerilog Direct Programing Interface.* Synopsys User Group Europe, 2004

Sutherland, Stuart, Davidmann, Simon, Flake, Peter, and Moorby, Phil. *System-Verilog for Design: A Guide to Using SystemVerilog for Hardware Design and Modeling.* Norwell, MA: Springer, 2006

Sutherland, Stuart, Mills, Don. *Verilog and SystemVerilog Gotchas.* Norwell, MA: Springer, 2007

Synopsys, Inc., *Hybrid RTL Formal Verification Ensures Early Detection of Corner-Case Bugs,* http://synopsys.com/products/magellan/magellan_wp.html, 2003

van der Schoot, Hans, and Bergeron, Janick *Transaction-Level Functional Coverage in SystemVerilog.* San Jose, CA: DVCon, February 2006

Vijayaraghavan, Srikanth, and Ramanathan, Meyyappan. *A Practical Guide for SystemVerilog Assertions.* Norwell, MA: Springer, 2005

Wachowski Andy, and Wachowski Larry. *The Matrix.* Hollywood, CA: Warner Brothers Studios, 1999

C. Spear and G. Tumbush, *SystemVerilog for Verification: A Guide to Learning the Testbench Language Features,* DOI 10.1007/978-1-4614-0715-7, © Springer Science+Business Media, LLC 2012

Index

Printed in the United States
By Bookmasters

Printed in the United States
By Bookmasters